THE COMPLETE METRIC SYSTEM
with
THE INTERNATIONAL SYSTEM OF UNITS

Compiled and Edited by

A. L. Le MARAIC

Foreign Marketing Specialist

and

JOHN P. CIARAMELLA*

Foreign Trade Consultant

*Formerly Manager Foreign Bond Dept.–Crown Bond Company
Manager Foreign Exchange Arbitrage Dept.–J.V. Reddy & Co.
Manager Foreign Trade Dept.–L. W. Dumont & Co.
President–Crude Rubber Inspection Bureau, Inc.
President–Pan-American Rotenone Corporation
Purchasing Agent United States Government Board of Economic Warfare, War Production Board, Commodity Credit Corporation.
Manager Foreign Trade Division–Horner Iron & Metals Corporation.
Publisher & Editor of the International Trader.

PUBLISHED BY ABBEY BOOKS

SOMERS, N. Y. 10589

Revised and Expanded Edition

Library of Congress Catolog Card Number: 72-97799

ISBN Number: 0-913768-00-6

Manufactured in the U.S.A.

PREFACE

The International System of Units (SI) is a new concept updating the traditional metric system of the past. Its newness, however, does not represent a detachment from the old system, rather, a continuity establishing simplified, consistent, and coherent standards of metric weights and measures to be used by all countries of the world henceforth.

The United States Government has already taken legislative action to legalize the change-over to the new International System of Units (SI) adopted by the 11th General Conference of Weights and Measures (CGPM) in 1960 by its Resolution 12 thus concurring in the establishment of a universal specification for units of measurement. A joint resolution proposed by the United States Department of Commerce was introduced in the 92nd Congress as Senate Resolution 219, and House Joint Resolution 1092, 1132, and 1169.

This compilation is intended as a guide to the use of the new International Metric System (SI) in the United States and to disseminate its basic principles so as to enable educators and others to put it to use as soon as possible in keeping with the aims and efforts of the United States Government and the international community of nations.

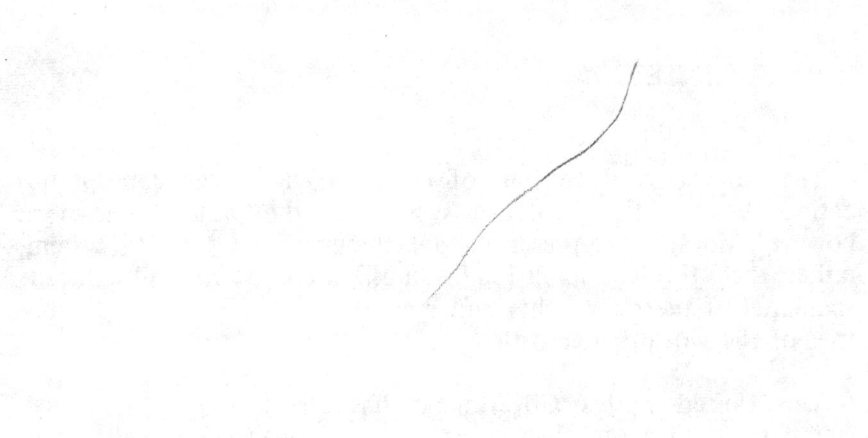

CONTENTS

ACKNOWLEDGMENT

All the basic data used in this book is derived from the translation approved by the International Bureau of Weights and Measures of its publication "Le Système International d'Unités." Grateful acknowledgment is made to the following for permission to include excerpts from their publications in this compendium.

Conférence Générale des Poids et Mesures (CGPM)
Consultative Committee for Units (CCU)
International Bureau of Weights and Measures (BIPM)
International Committee of Weights and Measures (CIPM)
International Organization for Standardization (ISO)
National Aeronautics and Space Administration (NASA)
National Bureau of Standards–U.S.A. (NBS)
National Physical Laboratory–U.K. (NPL)
South African Bureau of Standards (SABS)
South African Metrication Advisory Board (SAMAB)
United States Department of Commerce (USDC)

INTRODUCTION

The metric system of weights and measures has stood well the test of time and usage in many countries of the world which have used it since its introduction by the French many years ago. The Flemish mathematician, Simon Stevin, conceived the idea of a universal decimal system of measurement back in 1585 but despite his efforts to encourage its adoption it did not take hold until the latter part of the 18th century when the French Revolution brought in many radical changes.

The French revived the idea of creating a universal system of metric measurement in the 1790s when they proposed the *metre* as a unit of length and the *gramme* as a unit of mass. Although it attracted some attention in Europe, it did not jell until 1875 when the Treaty of the Metre was signed by the United States and sixteen other nations at Sevres, France, providing for an International Committee on Weights and Measures, and an international General Conference on Weights and Measures to guide its development and maintain uniform standards of measurement. Since that first step the metric system has progressed considerably, as its chronological history avers.

The expansion of trade between nations, the tremendous strides made in aeronautic and space technology, and the exchange of techniques between industrialized countries of the world have emphasized the need of a single universal system of weighing and measuring. This spurred the international authority on the metric system, known as the *Comférence Génèrale des Poids et Mesures* (CGPM), to adopt a universal system of metric weights and measures in 1960, and gave it the name *Le Système International d'Unitès* with the symbol (SI) as its abbreviation. This new standardized system has been favorably received throughout the world, and it will not be long before it will become the sole system of weighing and measuring used by all nations.

The transition that is to take place in our country from the present United States Customary to Metric will be a task of major proportion. The orientation of our people to the new system and the vast documentary changes which will have to be made in every sector of our national structure will, undoubtedly, create some confusion and inconvenience at the outset. Hence, it is imperative that the change-over is approached systematically, and pursued with orderly diligence to minimize any disruptive circumstances which might occur during the early stages when we will be in the process of discarding the old system we have used since colonial days and substitute a new one which is quite strange to many of us.

Metrication will call for a protracted period of patient and persevering work on the part of educators, students, engineers, scientists and technologists, as well as the legal fraternity and consumers. There is no royal road to attainment of this far reaching achievement. The work ahead will be trying, tedious, and burdensome.

This obvious situation actuated the compilation of this book to supply the immediate need of disseminating all the basic data concerning the International Metric System (SI) so as to make it possible for the individual to acquire quickly a working knowledge of the subject without having to wade through an enormous amount of tedious reading matter. During its preparation the authors had constantly before them a considerable quantity of data from various official sources here and abroad. This was carefully evaluated and screened to arrive at the best possible presentation in a condensed, comprehensive form without sacrificing thoroughness or omitting important material.

All pertinent up-to-the-minute data available to date has been included, and all of it is derived from the original translation approved by the International Bureau of Weights and Measures of its publication *Le Systѐme International d'Unitѐs* (SI). Our aim was to make a simple presentation of the underlying and unchanging principles of the (SI) metric system easy to understand, and to state the rudiments of the new system in such a manner as to provide the greatest amount of information at least cost of time and effort to the individual. While some rules and conversion factors may seem complex and intricate at the outset they will be found to be quite simple as one becomes more familiar with them.

We hope to have achieved the desired objective, and we trust this compendium will serve as an implement of good use during the period of transition to full metrication and for a long time into the future.

February 1973

A. L. Le Maraic
John P. Ciaramella

THE INTERNATIONAL SYSTEM OF UNITS

PART I

The International System of Units

Contents

I. INTRODUCTION

I.1 Historical note

In 1948 the 9th CGPM[1], by its Resolution 6, instructed the CIPM[1]: "to study the establishment of a complete set of rules for units of measurement"; "to find out for this purpose, by official inquiry, the opinion prevailing in scientific, technical, and educational circles in all countries" and "to make recommendations on the establishment of a *practical system of units of measurement* suitable for adoption by all signatories to the Metre Convention."

The same General Conference also laid down, by its Resolution 7, general principles for unit symbols (see II.1.2, page 6) and also gave a list of units with special names.

The 10th CGPM (1954), by its Resolution 6, and the 14th CGPM (1971) by its Resolution 3, adopted as base units of this "practical system of units", the units of the following seven quantities: length, mass, time, electric current, thermodynamic temperature, amount of substance, and luminous intensity (see II.1, page 3).

The 11th CGPM (1960), by its Resolution 12, adopted the name *International System of Units*, with the international abbreviation SI, for this practical system of units of measurement and laid down rules for the prefixes (see III.1, page 12), the derived and supplementary units (see II.2.2, page 10 and II.3, page 11) and other matters, thus establishing a comprehensive specification for units of measurement.

In the present document the expressions "SI units", "SI prefixes", "supplementary units" are used in accordance with Recommendation 1 (1969) of the CIPM.

[1] For the meaning of these abbreviations, see page VII.

I.2 The three classes of SI units

SI units are divided into three classes:

base units,
derived units,
supplementary units.

From the scientific point of view division of SI units into these three classes is to a certain extent arbitrary, because it is not essential to the physics of the subject.

Nevertheless the General Conference, considering the advantages of a single, practical, worldwide system for international relations, for teaching and for scientific work, decided to base the International System on a choice of seven well-defined units which by convention are regarded as dimensionally independent: the metre, the kilogram, the second, the ampere, the kelvin, the mole, and the candela (see II.1, page 3). These SI units are called *base units.*[2]

The second class of SI units contains *derived units*, i.e., units that can be formed by combining base units according to the algebraic relations linking the corresponding quantities. Several of these algebraic expressions in terms of base units can be replaced by special names and symbols which can themselves be used to form other derived units (see II.2, page 6).

Although it might be thought that SI units can only be base units or derived units, the 11th CGPM (1960) admitted a third class of SI units, called *supplementary units*, for which it declined to state whether they were base units or derived units (see II.3, page 11).

The SI units of these three classes form a coherent set in the sense normally attributed to the expression "coherent system of units".

The decimal multiples and sub-multiples of SI units formed by means of SI prefixes must be given their full name *multiples and sub-multiples of SI units* when it is desired to make a distinction between them and the coherent set of SI units.

[2] *Translators' note.* The spellings "metre" and "kilogram" are used in this USA/UK translation in the hope of securing worldwide uniformity in the English spelling of the names of the units of the International System.

II. SI UNITS

II.1 Base units

1. *Definitions*

a) Unit of length.—The 11th CGPM (1960) replaced the definition of the metre based on the international prototype of platinum-iridium, in force since 1889 and amplified in 1927, by the following definition:

The metre is the length equal to 1 650 763.73 *wavelengths in vacuum of the radiation corresponding to the transition between the levels* $2p_{10}$ *and* $5d_5$ *of the krypton*-86 *atom.* (11th CGPM (1960), Resolution 6).

The old international prototype of the metre which was legalized by the 1st CGPM in 1889 is still kept at the International Bureau of Weights and Measures under the conditions specified in 1889.

b) Unit of mass.—The 1st CGPM (1889) legalized the international prototype of the kilogram and declared: *this prototype shall henceforth be considered to be the unit of mass.*

With the object of removing the ambiguity which still occurred in the common use of the word "weight", the 3rd CGPM (1901) declared: *the kilogram is the unit of mass* [and not of weight or of force]; *it is equal to the mass of the international prototype of the kilogram.*

This international prototype made of platinum-iridium is kept at the BIPM under conditions specified by the 1st CGPM in 1889.

c) Unit of time.—Originally the unit of time, the second, was defined as the fraction 1/86 400 of the mean solar day. The exact definition of "mean solar day" was left to astronomers, but their measurements have shown that on account of irregularities in the rotation of the Earth the mean solar day does not guarantee the desired accuracy. In order to define the unit of time more precisely the 11th CGPM (1960) adopted a definition given by the International Astronomical Union which was based on the tropical year. Experimental work had however already shown that an atomic standard of time-interval, based on a transition between two energy levels of an atom or a molecule, could be realized and reproduced much more accurately. Considering that a very precise definition of the unit of time of the International System, the second, is indispensable for the needs of advanced metrology, the 13th CGPM (1967) decided to replace the definition of the second by the following:

The second is the duration of 9 192 631 770 *periods of the radiation corresponding to the transition between the two hyperfine levels of the ground state of the cesium*-133 *atom.* (13th CGPM (1967), Resolution 1).

d) Unit of electric current.—Electric units, called "international", for current and resistance, had been introduced by the International Electrical Congress held in Chicago in 1893, and the definitions of the "international" ampere and the "international" ohm were confirmed by the International Conference of London in 1908.

Although it was already obvious on the occasion of the 8th CGPM (1933) that there was a unanimous desire to replace those "international" units by so-called "absolute" units, the official decision to abolish them was only taken by the 9th CGPM (1948), which adopted for the unit of electric current, the ampere, the following definition:

The ampere is that constant current which, if maintained in two straight parallel conductors of infinite length, of negligible circular cross section, and placed 1 *metre apart in vacuum, would produce between these conductors a force equal to* 2×10^{-7} *newton per metre of length.* (CIPM (1946), Resolution 2 approved by the 9th CGPM, 1948)

The expression "MKS unit of force" which occurs in the original text has been replaced here by "newton" adopted by the 9th CGPM (1948, Resolution 7).

e) Unit of thermodynamic temperature.—The definition of the unit of thermodynamic temperature was given in substance by the 10th CGPM (1954, Resolution 3) which selected the triple point of water as fundamental fixed point and assigned to it the temperature 273.16 °K by definition. The 13th CGPM (1967, Resolution 3) adopted the name *kelvin* (symbol K) instead of "degree Kelvin" (symbol °K) and in its Resolution 4 defined the unit of thermodynamic temperature as follows:

The kelvin, unit of thermodynamic temperature, is the fraction 1/273.16 *of the thermodynamic temperature of the triple point of water.* (13th CGPM (1967), Resolution 4).

The 13th CGPM (1967, Resolution 3) also decided that the unit kelvin and its symbol K should be used to express an interval or a difference of temperature.

Note.—In addition to the thermodynamic temperature (symbol T), expressed in kelvins, use is also made of Celsius temperature (symbol t) defined by the equation

$$t = T - T_0$$

where $T_0 = 273.15$ K by definition. The Celsius temperature is in general expressed in degrees Celsius (symbol °C). The unit "degree Celsius" is thus equal to the unit "kelvin" and an interval or a difference of Celsius temperature may also be expressed in degrees Celsius.

f) Unit of amount of substance.—Since the discovery of the fundamental laws of chemistry, units of amount of substance called, for instance, "gram-atom" and "gram-molecule", have been used to specify amounts of chemical elements or compounds. These units had a direct connection with "atomic weights" and "molecular weights", which were in fact relative masses. "Atomic weights" were originally referred to the atomic weight of oxygen (by general agreement taken as 16). But whereas physicists separated isotopes in the mass spectrograph and attributed the value 16 to one of the isotopes of oxygen, chemists attributed that same value to the (slightly variable) mixture of isotopes 16, 17, and 18, which was for them the naturally occurring element oxygen. Finally an agreement between the International Union of Pure and Applied Physics (IUPAP) and the International Union of Pure and Applied Chemistry (IUPAC) brought this duality to an end in 1959/60. Physicists and chemists have ever since agreed to assign the value 12 to the isotope 12 of carbon, The unified scale thus obtained gives values of "relative atomic mass".

It remained to define the unit of amount of substance by fixing the corresponding mass of carbon 12; by international agreement, this mass has been fixed at 0.012 kg, and the unit of the quantity, "amount of substance",[3] has been given the name *mole* (symbol mol).

Following proposals of IUPAP, IUPAC, and ISO, the CIPM gave in 1967, and confirmed in 1969, the following definition of the mole, adopted by the 14th CGPM (1971, Resolution 3):

The mole is the amount of substance of a system which contains as many elementary entities as there are atoms in 0.012 *kilogram of carbon* 12.

Note. *When the mole is used, the elementary entities must be specified and may be atoms, molecules, ions, electrons, other particles, or specified groups of such particles.*

g) *Unit of luminous intensity.*—The units of luminous intensity based on flame or incandescent filament standards in use in various countries were replaced in 1948 by the "new candle". This decision had been prepared by the International Commission on Illumination (CIE) and by the CIPM before 1937, and was promulgated by the CIPM at its meeting in 1946 in virtue of powers conferred on it in 1933 by the 8th CGPM. The 9th CGPM (1948) ratified the decision of the CIPM and gave a new international name, *candela* (symbol cd), to the unit of luminous intensity. The text of the definition of the candela, as amended in 1967, is as follows.

The candela is the luminous intensity, in the perpendicular direction, of a surface of 1/600 000 *square metre of a blackbody at the temperature of freezing platinum under a pressure of* 101 325 *newtons per square metre.* (13th CGPM (1967). Resolution 5).

---

[3] The name of this quantity, adopted by IUPAP, IUPAC, and ISO is in French "quantité de matière" and in English "amount of substance"; (the German and Russian translations are "Stoffmenge" and "количество вещества"). The French name recalls "quantitas materiae" by which in the past the quantity now called mass used to be known; we must forget this old meaning, for mass and amount of substance are entirely different quantities.

2. Symbols

The base units of the International System are collected in table 1 with their names and their symbols (10th CGPM (1954), Resolution 6; 11th CGPM (1960), Resolution 12; 13th CGPM (1967), Resolution 3; 14th CGPM (1971), Resolution 3.

TABLE 1

SI base units

Quantity	Name	Symbol
length	metre	m
mass	kilogram	kg
time	second	s
electric current	ampere	A
thermodynamic temperature*	kelvin	K
amount of substance	mole	mol
luminous intensity	candela	cd

*Celsius temperature is in general expressed in degrees Celsius (symbol °C) (see Note, p. 4).

The general principle governing the writing of unit symbols had already been adopted by the 9th CGPM (1958), Resolution 7, according to which:

Roman [upright] *type, in general lower case, is used for symbols of units; if however the symbols are derived from proper names, capital roman type is used* [for the first letter]. *These symbols are not followed by a full stop* [period].

Unit symbols do not change in the plural.

II.2 Derived units

1. Expressions

Derived units are expressed algebraically in terms of base units by means of the mathematical symbols of multiplication and division. Several derived units have been given special names and symbols which may themselves be used to express other derived units in a simpler way than in terms of the base units.

Derived units may therfore be classified under three headings Some of them are given in tables 2, 3, and 4.

Table 2

Examples of SI derived units expressed in terms of base units

Quantity	SI unit	
	Name	Symbol
area	square metre	m^2
volume	cubic metre	m^3
speed, velocity	metre per second	m/s
acceleration	metre per second squared	m/s^2
wave number	1 per metre	m^{-1}
density, mass density	kilogram per cubic metre	kg/m^3
concentration (of amount of substance)	mole per cubic metre	mol/m^3
activity (radioactive)	1 per second	s^{-1}
specific volume	cubic metre per kilogram	m^3/kg
luminance	candela per square metre	cd/m^2

Table 3

SI derived units with special names

Quantity	SI unit			
	Name	Symbol	Expression in terms of other units	Expression in terms of SI base units
frequency	hertz	Hz		s^{-1}
force	newton	N		$m \cdot kg \cdot s^{-2}$
pressure	pascal	Pa	N/m^2	$m^{-1} \cdot kg \cdot s^{-2}$
energy, work, quantity of heat	joule	J	$N \cdot m$	$m^2 \cdot kg \cdot s^{-2}$
power, radiant flux	watt	W	J/s	$m^2 \cdot kg \cdot s^{-3}$
quantity of electricity, electric charge	coulomb	C	$A \cdot s$	$s \cdot A$
electric potential, potential difference, electromotive force	volt	V	W/A	$m^2 \cdot kg \cdot s^{-3} \cdot A^{-1}$
capacitance	farad	F	C/V	$m^{-2} \cdot kg^{-1} \cdot s^4 \cdot A^2$
electric resistance	ohm	Ω	V/A	$m^2 \cdot kg \cdot s^{-3} \cdot A^{-2}$
conductance	siemens	S	A/V	$m^{-2} \cdot kg^{-1} \cdot s^3 \cdot A^2$
magnetic flux	weber	Wb	$V \cdot s$	$m^2 \cdot kg \cdot s^{-2} \cdot A^{-1}$
magnetic flux density	tesla	T	Wb/m^2	$kg \cdot s^{-2} \cdot A^{-1}$
inductance	henry	H	Wb/A	$m^2 \cdot kg \cdot s^{-2} \cdot A^{-2}$
luminous flux	lumen	lm		$cd \cdot sr$ (a)
illuminance	lux	lx		$m^{-2} \cdot cd \cdot sr$ (a)

(a) In this expression the steradian (sr) is treated as a base unit.

TABLE 4

Examples of SI derived units expressed by means of special names

Quantity	SI unit		
	Name	Symbol	Expression in terms of SI base units
dynamic viscosity	pascal second	Pa·s	$m^{-1} \cdot kg \cdot s^{-1}$
moment of force	metre newton	N·m	$m^{2} \cdot kg \cdot s^{-2}$
surface tension	newton per metre	N/m	$kg \cdot s^{-2}$
heat flux density, irradiance	watt per square metre	W/m^2	$kg \cdot s^{-3}$
heat capacity, entropy	joule per kelvin	J/K	$m^{2} \cdot kg \cdot s^{-2} \cdot K^{-1}$
specific heat capacity, specific entropy	joule per kilogram kelvin	J/(kg·K)	$m^{2} \cdot s^{-2} \cdot K^{-1}$
specific energy	joule per kilogram	J/kg	$m^{2} \cdot s^{-2}$
thermal conductivity	watt per metre kelvin	W/(m·K)	$m \cdot kg \cdot s^{-3} \cdot K^{-1}$
energy density	joule per cubic metre	J/m^3	$m^{-1} \cdot kg \cdot s^{-2}$
electric field strength	volt per metre	V/m	$m \cdot kg \cdot s^{-3} \cdot A^{-1}$
electric charge density	coulomb per cubic metre	C/m^3	$m^{-3} \cdot s \cdot A$
electric flux density	coulomb per square metre	C/m^2	$m^{-2} \cdot s \cdot A$
permittivity	farad per metre	F/m	$m^{-3} \cdot kg^{-1} \cdot s^{4} \cdot A^{2}$
current density	ampere per square metre	A/m^2	
magnetic field strength	ampere per metre	A/m	
permeability	henry per metre	H/m	$m \cdot kg \cdot s^{-2} \cdot A^{-2}$
molar energy	joule per mole	J/mol	$m^{2} \cdot kg \cdot s^{-2} \cdot mol^{-1}$
molar entropy, molar heat capacity	joule per mole kelvin	J/(mol·K)	$m^{2} \cdot kg \cdot s^{-2} \cdot K^{-1} \cdot mol^{-1}$

Note a—The values of certain so-called dimensionless quantities, as for example refractive index, relative permeability or relative permittivity, are expressed by pure numbers. In this case the corresponding SI unit is the ratio of the same two SI units and may be expressed by the number 1.

Note b—Although a derived unit can be expressed in several equivalent ways by using names of base units and special names of derived units, CIPM sees no objection to the preferential use of certain combinations or of certain special names in order to distinguish more easily between quantities of the same dimension; for example, the hertz is often used in preference to the reciprocal second for the frequency of a periodic phenomenon, and the newton-metre in preference to the joule for the moment of a force, although, rigorously, 1 Hz$=$1s^{-1}, and 1 N·m$=$1 J.

2. *Recommendations*

The International Organization for Standardization (ISO) has issued additional recommendations with the aim of securing uniformity in the use of units, in particular those of the International System (see the series of Recommendations R 31 and Recommendation R 1000 of Technical Committee ISO/TC 12 "Quantities, units, symbols, conversion factors and conversion tables").

According to these recommendations:

a) The product of two or more units is preferably indicated by a dot. The dot may be dispensed with when there is no risk of confusion with another unit symbol

for example: N•m or N m *but not:* mN

b) A solidus (oblique stroke, /), a horizontal line, or negative powers may be used to express a derived unit formed from two others by division

for example: m/s, $\frac{\text{m}}{\text{s}}$ or m•s^{-1}

c) The solidus must not be repeated on the same line unless ambiguity is avoided by parentheses. In complicated cases negative powers or parentheses should be used

for example: m/s^2 or m•s^{-2} *but not:* m/s/s

m•kg/(s^3•A) or m•kg•s^{-3}•A^{-1} *but not:* m•kg/s^3/A

II.3 Supplementary units

The General Conference has not yet classified certain units of the International System under either base units or derived units. These SI units are assigned to the third class called "supplementary units", and may be regarded either as base units or as derived units.

For the time being this class contains only two, purely geometrical, units: the SI unit of plane angle, the *radian*, and the SI unit of solid angle, the *steradian* (11th CGPM (1960), Resolution 12).

TABLE 5

SI supplementary units

Quantity	SI unit	
	Name	Symbol
plane angle	radian	rad
solid angle	steradian	sr

The radian is the plane angle between two radii of a circle which cut off on the circumference an arc equal in length to the radius.

The steradian is the solid angle which, having its vertex in the center of a sphere, cuts off an area of the surface of the sphere equal to that of a square with sides of length equal to the radius of the sphere.

(ISO Recommendation R 31, part 1, second edition, December 1965).

Supplementary units may be used to form derived units. Examples are given in table 6.

TABLE 6

Examples of SI derived units formed by using supplementary units

Quantity	SI unit	
	Name	Symbol
angular velocity	radian per second	rad/s
angular acceleration	radian per second squared	rad/s^2
radiant intensity	watt per steradian	W/sr
radiance	watt per square metre steradian	$W \cdot m^{-2} \cdot sr^{-1}$

III. DECIMAL MULTIPLES AND SUB-MULTIPLES OF SI UNITS

III.1 SI Prefixes

The 11th CGPM (1960, Resolution 12) adopted a first series of names and symbols of prefixes to form decimal multiples and sub-multiples of SI units. Prefixes for 10^{-15} and 10^{-18} were added by the 12th CGPM (1964, Resolution 8).

TABLE 7

SI prefixes

Factor	Prefix	Symbol	Factor	Prefix	Symbol
10^{12}	tera	T	10^{-1}	deci	d
10^{9}	giga	G	10^{-2}	centi	c
10^{6}	mega	M	10^{-3}	milli	m
10^{3}	kilo	k	10^{-6}	micro	μ
10^{2}	hecto	h	10^{-9}	nano	n
10^{1}	deka	da	10^{-12}	pico	p
			10^{-15}	femto	f
			10^{-18}	atto	a

III.2 Recommendations

ISO recommends the following rules for the use of SI prefixes:

a) Prefix symbols are printed in roman (upright) type without spacing between the prefix symbol and the unit symbol.

b) An exponent attached to a symbol containing a prefix indicates that the multiple or sub-multiple of the unit is raised to the power expressed by the exponent,

for example: $1 \text{ cm}^3 = 10^{-6} \text{ m}^3$
$1 \text{ cm}^{-1} = 10^2 \text{ m}^{-1}$

c) Compound prefixes, formed by the juxtaposition of two or more SI prefixes, are not to be used.

for example: 1 nm *but not:* 1 mμm

III.3 The kilogram

Among the base units of the International System, the unit of mass is the only one whose name, for historical reasons, contains a prefix. Names of decimal multiples and sub-multiples of the unit of mass are formed by attaching prefixes to the word "gram" (CIPM (1967), Recommendation 2).

IV. UNITS OUTSIDE THE INTERNATIONAL SYSTEM

IV.1 Units used with the International System

The CIPM (1969) recognized that users of SI will wish to employ with it certain units which, although not part of it, but which are important and are widely used. These units are given in table 8. The combination of units of this table with SI units to form compound units should, however, be authorized only in limited cases; in particular, the kilowatt-hour should eventually be abandoned.

TABLE 8

Units in use with the International System

Name	Symbol	Value in SI unit
minute	min	1 min = 60 s
hour[a]	h	1 h = 60 min = 3 600 s
day	d	1 d = 24 h = 86 400 s
degree	°	1° = $(\pi/180)$ rad
minute	′	1′ = $(1/60)°$ = $(\pi/10\ 800)$ rad
second	″	1″ = $(1/60)'$ = $(\pi/648\ 000)$ rad
litre[a]	l	1 l = 1 dm^3 = 10^{-3} m^3
tonne[a]	t	1 t = 10^3 kg

[a] The symbol of this unit is included in Resolution 7 of the 9th CGPM (1948). The litre is defined in Resolution 6 of the 12th CGPM (1964).

It is likewise necessary to recognize, outside the International System, some others units which are useful in specialized fields, because their values expressed in SI units must be obtained by experiment, and are therefore not known exactly (table 9).

TABLE 9

Units used with the International System whose values in SI units are obtained experimentally

Name	Symbol	Definition
electronvolt	eV	([a])
unified atomic mass unit	u	([b])
astronomical unit	(c)	([c])
parsec	pc	([d])

([a]) 1 electronvolt is the kinetic energy acquired by an electron in passing through a potential difference of 1 volt in vacuum; 1 eV = $1.602\ 19 \times 10^{-19}$ J approximately.

([b]) The unified atomic mass unit is equal to the fraction 1/12 of the mass of an atom of the nuclide ^{12}C; 1 u = $1.660\ 53 \times 10^{-27}$ kg approximately.

([c]) This unit does not have an international symbol; abbreviations are used, for example, AU in English, UA in French, AE in German, a.e.Д in Russian, etc. The astronomical unit of distance is the length of the radius of the unperturbed circular orbit of a body of negligible mass moving round the Sun with a sidereal angular velocity of 0.017 202 098 950 radian per day of 86 400 ephemeris seconds. In the system of astronomical constants of the International Astronomical Union the value adopted for it is: 1 AU = $149\ 600 \times 10^{6}$ m.

([d]) 1 parsec is the distance at which 1 astronomical unit subtends an angle of 1 second of arc; we thus have approximately, 1 pc = 206 265 AU = $30\ 857 \times 10^{12}$ m.

IV.2 Units accepted temporarily

In view of existing practice the CIPM (1969) considered it was preferable to keep for the time being, for use with those of the International System, the units listed in table 10.

TABLE 10

Units to be used with the International System for a limited time

Name	Symbol	Value in SI units
nautical mile[a]		1 nautical mile = 1 852 m
knot		1 nautical mile per hour = (1852/3600) m/s
ångström	Å	1 Å = 0.1 nm = 10^{-10} m
are[b]	a	1 a = 1 dam^2 = 10^2 m^2
hectare[b]	ha	1 ha = 1 hm^2 = 10^4 m^2
barn[c]	b	1 b = 100 fm^2 = 10^{-28} m^2
bar[d]	bar	1 bar = 0.1 MPa = 10^5 Pa
standard atmosphere[e]	atm	1 atm = 101 325 Pa
gal[f]	Gal	1 Gal = 1 cm/s^2 = 10^{-2} m/s^2
curie[g]	Ci	1 Ci = 3.7 × 10^{10} s^{-1}
röntgen[h]	R	1 R = 2.58 × 10^{-4} C/kg
rad[i]	rad	1 rad = 10^{-2} J/kg

(a) The nautical mile is a special unit employed for marine and aerial navigation to express distances. The conventional value given above was adopted by the First International Extraordinary Hydrographic Conference, Monaco, 1929, under the name "International nautical mile".

(b) This unit and its symbol were adopted by the CIPM in 1879 (*Procès-Verbaux CIPM*, 1879, p. 41).

(c) The barn is a special unit employed in nuclear physics to express effective cross sections.

(d) This unit and its symbol are included in Resolution 7 of the 9th CGPM (1948).

(e) Resolution 4 of 10th CGPM (1954).

(f) The gal is a special unit employed in geodesy and geophysics to express the acceleration due to gravity.

(g) The curie is a special unit employed in nuclear physics to express activity of radionuclides (12th CGPM (1964), Resolution 7).

(h) The röntgen is a special unit employed to express exposure of X or γ radiations.

(i) The rad is a special unit employed to express absorbed dose of ionizing radiations. When there is risk of confusion with the symbol for radian, rd may be used as symbol for rad.

IV.3 CGS units

The CIPM considers that it is in general preferable not to use, with the units of the International System, CGS units which have special names.[4] Such units are listed in table 11.

TABLE 11

CGS units with special names

Name	Symbol	Value in SI units
erg[(a)]	erg	1 erg $= 10^{-7}$ J
dyne[(a)]	dyn	1 dyn $= 10^{-5}$ N
poise[(a)]	P	1 P = 1 dyn•s/cm^2 = 0.1 Pa• s
stokes	St	1 St = 1 cm^2/s = 10^{-4} m^2/s
gauss[(b)]	Gs, G	1 Gs corresponds to 10^{-4} T
oersted[(b)]	Oe	1 Oe corresponds to $\frac{1000}{4\pi}$ A/m
maxwell[(b)]	Mx	1 Mx corresponds to 10^{-8} Wb
stilb[(a)]	sb	1 stilb = 1 cd/cm^2 = 10^4 cd/m^2
phot	ph	1 ph = 10^4 lx

(a) This unit and its symbol were included in Resolution 7 of the 9th CGPM (1948).
(b) This unit is part of the so-called "electromagnetic" 3-dimensional CGS system and cannot strictly speaking be compared to the corresponding unit of the International System, which has four dimensions when only electric quantities are considered.

[4] The aim of the International System of Units and of the recommendations contained in this document is to secure a greater degree of uniformity, hence a better mutual understanding of the general use of units. Nevertheless in certain specialized fields of scientific research, in particular in theoretical physics, there may sometimes be very good reasons for using other systems or other units.
Whichever units are used, it is important that the *symbols* employed for them follow current international recommendations.

IV.4 Other units

As regards units outside the International System which do not come under sections IV.1, 2, and 3, the CIPM considers that it is in general preferable to avoid them, and to use instead units of the International System. Some of those units are listed in table 12.

TABLE 12

Other units generally deprecated

Name	Value in SI units
fermi	1 fermi = 1 fm = 10^{-15} m
metric carat[a]	1 metric carat = 200 mg = 2×10^{-4} kg
torr	1 torr = $\frac{101\ 325}{760}$ Pa
kilogram-force (kgf)	1 kgf = 9.806 65 N
calorie (cal)	1 cal = 4.186 8 J[b]
micron (μ)[c]	1 μ = 1 μm = 10^{-6} m
X unit[d]	
stere (st)[e]	1 st = 1 m^3
gamma (γ)	1 γ = 1 nT = 10^{-9} T
γ[f]	1 γ = 1 μg = 10^{-9} kg
λ[g]	1 λ = 1 μl = 10^{-6} l

(a) This name was adopted by the 4th CGPM (1907, pp. 89-91) for commercial dealings in diamonds, pearls, and precious stones.
(b) This value is that of the so-called "IT" calorie (5th International Conference on Properties of Steam, London, 1956).
(c) The name of this unit and its symbol, adopted by the CIPM in 1879 (*Procès-Verbaux CIPM,* 1879, p. 41) and retained in Resolution 7 of the 9th CGPM (1948) were abolished by the 13th CGPM (1967, Resolution 7).
(d) This special unit was employed to express wavelengths of X rays; 1 X unit = 1.002×10^{-4} nm approximately.
(e) This special unit employed to measure firewood was adopted by the CIPM in 1879 with the symbol "s" (*Procès-Verbaux CIPM,* 1879, p. 41). The 9th CGPM (1948, Resolution 7) changed the symbol to "st".
(f) This symbol is mentioned in *Procès-Verbaux CIPM,* 1880, p. 56.
(g) This symbol is mentioned in *Procès-Verbaux CIPM,* 1880, p. 30.

APPENDIX I

Decisions of the CGPM and the CIPM

CR: *Comptes rendus des séances de la Conférence Générale des Poids et Mesures (CGPM)*

PV: *Procès-Verbaux des séances du Comité International des Poids et Mesures (CIPM)*

1st CGPM, 1889

Sanction of the international prototypes of the metre and the kilogram (CR, 34-38)

The General Conference

considering

the "Compte rendu of the President of the CIPM" and the "Report of the CIPM", which show that, by the collaboration of the French section of the international Metre Commission and of the CIPM, the fundamental measurements of the international and national prototypes of the metre and of the kilogram have been made with all the accuracy and reliability which the present state of science permits; that the international and national prototypes of the metre and the kilogram are made of an alloy of platinum with 10 per cent iridium, to within 0.000 1; the equality in length of the international Metre and the equality in mass of the international Kilogram with the length of the Metre and the mass of the Kilogram kept in the Archives of France;

that the differences between the national Metres and the international Metre lie within 0.01 millimetre and that these differences are based on a hydrogen thermometer scale which can always be reproduced thanks to the stability of hydrogen, provided identical conditions are secured;

that the differences between the national Kilograms and the international Kilogram lie within 1 milligram;

that the international Metre and Kilogram and the national Metres and Kilograms fulfil the requirements of the Metre Convention,

sanctions

A. As regards international prototypes:

1 The Prototype of the metre chosen by the CIPM.

This prototype, at the temperature of melting ice, shall henceforth represent the metric unit of length.

2 The Prototype of the kilogram adopted by the CIPM.

This prototype shall henceforth be considered as the unit of mass.

3 The hydrogen thermometer centigrade scale in terms of which the equations of the prototype Metres have been established.

B. As regards national prototypes:

. . . .

3rd CGPM, 1901

Declaration concerning the definition of the litre (CR, 38)

. . . .

The Conference declares:

1 The unit of volume, for high accuracy determinations, is the volume occupied by a mass of 1 kilogram of pure water, at its maximum density and at standard atmospheric pressure; this volume is called "litre".

2 . . . .

Declaration on the unit of mass and on the definition of weight; conventional value of g_n (CR, 70)

Taking into account the decision of the CIPM of the 15 October 1887, according to which the kilogram has been defined as a unit of mass [5];

Taking into account the decision contained in the sanction of the prototypes of the Metric System, unanimously accepted by the CGPM on the 26 September 1889;

Considering the necessity to put an end to the ambiguity which in current practice still subsists on the meaning of the word *weight*, used sometimes for *mass*, sometimes for *mechanical force;*

The Conference declares:

"1 The kilogram is the unit of mass; it is equal to the mass of the international prototype of the kilogram;

"2 The word *weight* denotes a quantity of the same nature as a *force;* the weight of a body is the product of its mass and the acceleration due to gravity; in particular, the standard weight of a body is the product of its mass and the standard acceleration due to gravity;

"3 The value adopted in the international Service of Weights and Measures for the standard acceleration due to gravity is 980.665 cm/s^2, value already stated in the laws of some countries." [6]

---

[5] "The mass of the international Kilogram is taken as unit for the international Service of Weights and Measures" (PV, 1887, 88).

[6] *Note of BIPM.* This conventional reference "standard value" (g_n = 9.806 65 m/s^2) to be used in the reduction to standard gravity of measurements made in some place on the Earth has been reconfirmed in 1913 by the 5th CGPM (CR, 44).

7th CGPM, 1927

Definition of the metre by the international Prototype (CR, 49)

The unit of length is the metre, defined by the distance, at 0°, between the axes of the two central lines marked on the bar of platinum-iridium kept at the BIPM, and declared Prototype of the metre by the 1st CGPM, this bar being subject to standard atmospheric pressure and supported on two cylinders of at least one centimetre diameter, symmetrically placed in the same horizontal plane at a distance of 571 mm from each other.

CIPM, 1946

Definitions of photometric units (PV, 20, 119)

RESOLUTION [7]

. . . .

4. The photometric units may be defined as follows:

New candle (unit of luminous intensity).—The value of the new candle is such that the brightness of the full radiator at the temperature of solidification of platinum is 60 new candles per square centimetre.

New lumen (unit of luminous flux).—The new lumen is the luminous flux emitted in unit solid angle (steradian) by a uniform point source having a luminous intensity of 1 new candle.

5. . . . .

Definitions of electric units (PV, 20, 131)

RESOLUTION 2 [8]

. . . .

4. A) Definitions of the mechanical units which enter the definitions of electric units:

Unit of force.—The unit of force [in the MKS (Metre, Kilogram, Second) system] is that force which gives to a mass of 1 kilogram an acceleration of 1 metre per second, per second (*).

*The name "newton" has been proposed for the MKS unit of force.

[7] The two definitions contained in this Resolution were ratified by the 9th CGPM (1948), which also approved the name *candela* given to the "new candle" (CR, 54). For the lumen the qualifier "new" was later abandoned.

[8] The definitions contained in this Resolution 2 were approved by the 9th CGPM (1948), (CR, 49), which moreover adopted the name *newton* (Resolution 7).

Joule (unit of energy or work).—The joule is the work done when the point of application of 1 MKS unit of force [newton] moves a distance of 1 metre in the direction of the force.

Watt (unit of power).—The watt is that power which in one second gives rise to energy of 1 joule.

B) Definitions of electric units. The CIPM accepts the following propositions which define the theoretical value of the electric units:

Ampere (unit of electric current).—The ampere is that constant current which, if maintained in two straight parallel conductors of infinite length, of negligible circular cross-section, and placed 1 metre apart in vacuum, would produce between these conductors a force equal to 2×10^{-7} MKS unit of force [newton] per metre of length.

Volt (unit of potential difference and of electromotive force).—The volt is the difference of electric potential between two points of a conducting wire carrying a constant current of 1 ampere, when the power dissipated between these points is equal to 1 watt.

Ohm (unit of electric resistance).—The ohm is the electric resistance between two points of a conductor when a constant potential difference of 1 volt, applied to these points, produces in the conductor a current of 1 ampere, the conductor not being the seat of any electromotive force.

Coulomb (unit of quantity of electricity).—The coulomb is the quantity of electricity carried in 1 second by a current of 1 ampere.

Farad (unit of electric capacitance).—The farad is the capacitance of a capacitor between the plates of which there appears a potential difference of 1 volt when it is charged by a quantity of electricity of 1 coulomb.

Henry (unit of electric inductance).—The henry is the inductance of a closed circuit in which an electromotive force of 1 volt is produced when the electric current in the circuit varies uniformly at the rate of 1 ampere per second.

Weber (unit of magnetic flux).—The weber is that magnetic flux which, linking a circuit of one turn, would produce in it an electromotive force of 1 volt if it were reduced to zero at a uniform rate in 1 second.

Triple point of water; thermodynamic scale with a single fixed point; unit of quantity of heat (joule) (CR, 55 and 63)

Resolution 3[9]

1. With present-day technique, the triple point of water is capable of providing a thermometric reference point with an accuracy higher than can be obtained from the melting point of ice.

In consequence the Consultative Committee [for Thermometry and Calorimetry] considers that the zero of the centesimal thermodynamic scale must be defined as the temperature 0.010 0 degree below that of the triple point of pure water.

2. The CCTC accepts the principle of an absolute thermodynamic scale with a single fundamental fixed point, at present provided by the triple point of pure water, the absolute temperature of which will be fixed at a later date.

The introduction of this new scale does not affect in any way the use of the International Scale, which remains the recommended practical scale.

3. The unit of quantity of heat is the joule.

Note.—It is requested that the results of calorimetric experiments be as far as possible expressed in joules.

If the experiments are made by comparison with the rise of temperature of water (and that, for some reason, it is not possible to avoid using the calorie), the information necessary for conversion to joules must be provided.

The CIPM, advised by the CCTC, should prepare a table giving, in joules per degree, the most accurate values that can be obtained from experiments on the specific heat of water.

Adoption of "degree Celsius"

From three names ("degree centigrade", "centesimal degree", "degree Celsius") proposed to denote the degree of temperature, the CIPM has chosen "degree Celsius" (PV, **21**, 1948, 88).

This name is also adopted by the General Conference (CR, 64).

[9] The three propositions contained in this Resolution 3 have been adopted by the General Conference.

Proposal for establishing a practical system of units of measurement (CR, 64).

RESOLUTION 6

The General Conference,

considering

that the CIPM has been requested by the International Union of Physics to adopt for international use a practical international system of units; that the International Union of Physics recommends the MKS system and one electric unit of the absolute practical system, but does not recommend that the CGS system be abandoned by physicists;

that the CGPM has itself received from the French Government a similar request, accompanied by a draft to be used as basis of discussion for the establishment of a complete specification of units of measurement;

instructs the CIPM:

to seek by an energetic, active, official enquiry the opinion of scientific, technical, and educational circles of all countries (offering them in effect the French document as basis);

to gather and study the answers;

to make recommendations for a single practical system of units of measurement, suitable for adoption by all countries adhering to the Metre Convention.

Writing and printing of unit symbols and of numbers

RESOLUTION 7

Principles

Roman (upright) type, in general lower case, is used for symbols of units; if however the symbols are derived from proper names, capital roman type is used. These symbols are not followed by a full stop.

In numbers, the comma (French practice) or the dot (British practice) are used only to separate the integral part of numbers from the decimal part. Numbers may be divided in groups of three in order to facilitate reading; neither dots nor commas are ever inserted in the spaces between groups.

Unit	Symbol	Unit	Symbol
.metre	m	ampere	A
.square metre	m^2	volt	V
.cubic metre	m^3	watt	W
.micron	μ	ohm	Ω
.litre	l	coulomb	C
.gram	g	farad	F
.tonne	t	henry	H
second	s	hertz	Hz
erg	erg	poise	P
dyne	dyn	newton	N
degree Celsius	°C	.candela ("new candle")	cd
.degree absolute	°K	lux	lx
calorie	cal	lumen	lm
bar	bar	stilb	sb
hour	h		

Notes

1. The symbols whose unit names are preceded by dots are those which had already been adopted by a decision of the CIPM.

2. The symbol for the stere, the unit of volume for firewood, shall be "st" and not "s", which had been previously assigned to it by the CIPM.

3. To indicate a temperature interval or difference, rather than a temperature, the word "degree" in full, or the abbreviation "deg", must be used.

10th CGPM, 1954

Definition of the thermodynamic temperature scale (CR, 79)

RESOLUTION 3

The 10th CGPM decides to define the thermodynamic temperature scale by choosing the triple point of water as the fundamental fixed point, and assigning to it the temperature 273.16 degrees Kelvin, exactly.

Definition of standard atmosphere (CR, 79)

RESOLUTION 4

The 10th CGPM, having noted that the definition of the standard atmosphere given by the 9th CGPM when defining the International Temperature Scale, led some physicists to believe that this definition of the standard atmosphere was valid only for accurate work in thermometry,

declares that it adopts, for general use, the definition:
1 standard atmosphere = 1 013 250 dynes per square centimetre,
i.e., 101 325 newtons per square metre.

Practical system of units (CR, 80)

RESOLUTION 6

In accordance with the wish expressed by the 9th CGPM in its Resolution 6 concerning the establishment of a practical system of units of measurement for international use, the 10th CGPM

decides to adopt as base units of the system, the following units:

length	metre
mass	kilogram
time	second
electric current	ampere
thermodynamic temperature	degree Kelvin
luminous intensity	candela

CIPM, 1956

Definition of the unit of time (PV, **25**, 77)

RESOLUTION 1

In virtue of the powers invested in it by Resolution 5 of the 10th CGPM, the CIPM

considering

1 that the 9th General Assembly of the International Astronomical Union (Dublin, 1955) declared itself in favor of linking the second to the tropical year;

2 that, according to the decisions of the 8th General Assembly of the International Astronomical Union (Rome, 1952), the second of ephemeris time (ET) is the fraction $\frac{12\ 960\ 276\ 813}{408\ 986\ 496} \times 10^{-9}$ of the tropical year for 1900 January 0 at 12 h ET,

decides

"The second is the fraction 1/31 556 925.974 7 of the tropical year for 1900 January 0 at 12 hours ephemeris time".

International System of Units (PV, **25**, 83)

RESOLUTION 3

The CIPM

considering

the task entrusted to it by Resolution 6 of the 9th CGPM concerning the establishment of a practical system of units of measurement suitable for adoption by all countries adhering to the Metre Convention, the documents received from twenty-one countries in reply to the enquiry requested by the 9th CGPM,

Resolution 6 of the 10th CGPM, fixing the base units of the system to be established

recommends

1 that the name "International System of Units" be given to the system founded on the base units adopted by the 10th CGPM, viz:

[here follows the list of the six base units with their symbols, reproduced in Resolution 12 of the 11th CGPM (1960)]

2 that the units listed in the table below be used, without excluding others which might be added later:

[here follows the table of units reproduced in paragraph 4 of Resolution 12 of the 11th CGPM (1960)]

11th CGPM, 1960

Definition of the metre (CR, 85)

Resolution 6

The 11th CGPM

considering

that the international Prototype does not define the metre with an accuracy adequate for the present needs of metrology,

that it is moreover desirable to adopt a natural and indestructible standard,

decides

1 The metre is the length equal to 1 650 763.73 wavelengths in vacuum of the radiation corresponding to the transition between the levels 2 p_{10} and 5 d_5 of the krypton-86 atom.

2 The definition of the metre in force since 1889, based on the international Prototype of platinum-iridium, is abrogated.

3 The international Prototype of the metre sanctioned by the 1st CGPM in 1889 shall be kept at the BIPM under the conditions specified in 1889.

Resolution 7

The 11th CGPM

requests the CIPM

1 to prepare specifications for the realization of the new definition of the metre [10];

2 to select secondary wavelength standards for measurement of length by interferometry, and to prepare specifications for their use;

3 to continue the work in progress on improvement of length standards.

---

[10] See Appendix 2, p 35, for the relevant Recommendation adopted by the CIPM.

Definition of the unit of time (CR, 86)

RESOLUTION 9

The 11th CGPM
considering
the powers given to the CIPM by the 10th CGPM, to define the fundamental unit of time,
the decision taken by the CIPM in 1956,
ratifies the following definition:
"The second is the fraction 1/31 556 925.974 7 of the tropical year for 1900 January 0 at 12 hours ephemeris time".
International System of Units (CR, 87)

RESOLUTION 12

The 11th CGPM
considering
Resolution 6 of the 10th CGPM, by which it adopted six base units on which to establish a practical system of measurement for international use:

length	metre	m
mass	kilogram	kg
time	second	s
electric current	ampere	A
thermodynamic temperature	degree Kelvin	°K
luminous intensity	candela	cd

Resolution 3 adopted by the CIPM in 1956,
The recommendations adopted by the CIPM in 1958 concerning an abbreviation for the name of the system, and prefixes to form multiples and sub-multiples of the units,

decides
1 the system founded on the six base units above is called "International System of Units";
2 the international abbreviation of the name of the system is: SI;

3 names of multiples and sub-multiples of the units are formed by means of the following prefixes:

Multiplying factor	Prefix	Symbol
1 000 000 000 000 = 10^{12}	tera	T
1 000 000 000 = 10^{9}	giga	G
1 000 000 = 10^{6}	mega	M
1 000 = 10^{3}	kilo	k
100 = 10^{2}	hecto	h
10 = 10^{1}	deka	da
0.1 = 10^{-1}	deci	d
0.01 = 10^{-2}	centi	c
0.001 = 10^{-3}	milli	m
0.000 001 = 10^{-6}	micro	μ
0.000 000 001 = 10^{-9}	nano	n
0.000 000 000 001 = 10^{-12}	pico	p

4 the units listed below are used in the system, without excluding others which might be added later

	Supplementary Units		
plane angle	radian	rad	
solid angles	steradian	sr	
	Derived Units		
area	square metre	m^2	
volume	cubic metre	m^3	
frequency	hertz	Hz	1/s
mass density (density)	kilogram per cubic metre	kg/m^3	
speed, velocity	metre per second	m/s	
angular velocity	radian per second	rad/s	
acceleration	metre per second squared	m/s^2	
angular acceleration	radian per second squared	rad/s^2	
force	newton	N	$kg\cdot m/s^2$
pressure (mechanical stress)	newton per square metre	N/m^2	
kinematic viscosity	square metre per second	m^2/s	
dynamic viscosity	newton-second per square metre	$N\cdot s/m^2$	
work, energy, quantity of heat	joule	J	$N\cdot m$
power	watt	W	J/s
quantity of electricity	coulomb	C	$A\cdot s$
potential difference, electromotive force	volt	V	W/A
electric field strength	volt per metre	V/m	
electric resistance	ohm	Ω	V/A
capacitance	farad	F	$A\cdot s/V$
magnetic flux	weber	Wb	$V\cdot s$
inductance	henry	H	$V\cdot s/A$
magnetic flux density	tesla	T	Wb/m^2
magnetic field strength	ampere per metre	A/m	
magnetomotive force	ampere	A	
luminous flux	lumen	lm	$cd\cdot sr$
luminance	candela per square metre	cd/m^2	
illuminance	lux	lx	lm/m^2

Cubic decimetre and litre (CR, 88)

RESOLUTION 13

The 11th CGPM,

considering

that the cubic decimetre and the litre are unequal and differ by about 28 parts in 10^6,

that determinations of physical quantities which involve measurements of volume are being made more and more accurately, thus increasing the risk of confusion between the cubic decimetre and the litre,

requests the CIPM to study the problem and submit its conclusions to the 12th CGPM.

CIPM, 1961

Cubic decimetre and litre (PV, 29, 34)

RECOMMENDATION

The CIPM recommends that the results of accurate measurements of volume be expressed in units of the International System and not in litres.

12th CGPM, 1964

Atomic standard of frequency (CR, 93)

RESOLUTION 5

The 12th CGPM,

considering

that the 11th CGPM noted in its Resolution 10 the urgency, in the interests of accurate metrology, of adopting an atomic or molecular standard of time interval, that, in spite of the results already obtained with cesium atomic frequency standards, the time has not yet come for the CGPM to adopt a new definition of the second, base unit of the International System of Units, because of the new and considerable improvements likely to be obtained from work now in progress,

considering also that it is not desirable to wait any longer before time measurements in physics are based on atomic or molecular frequency standards,

empowers the CIPM to name the atomic or molecular frequency standards to be employed for the time being,

requests the Organizations and Laboratories knowledgeable in this field to pursue work connected with a new definition of the second.

DECLARATION OF THE CIPM (1964) (PV, 32, 26 and CR, 93)

The CIPM,

empowered by Resolution 5 of the 12th CGPM to name atomic or molecular frequency standards for temporary use for time measurements in physics,

declares that the standard to be employed is the transition between the hyperfine levels F = 4, M = 0 and F = 3, M = 0 of the ground state $^2S_{1/2}$ of the cesium-133 atom, unperturbed by external fields, and that the frequency of this transition is assigned the value 9 192 631 770 hertz.

Litre (CR, 93)

RESOLUTION 6

The 12th CGPM,

considering Resolution 13 adopted by the 11th CGPM in 1960 and the Recommendation adopted by the CIPM in 1961,

1 *abrogates* the definition of the litre given in 1901 by the 3rd CGPM

2 *declares* that the word "litre" may be employed as a special name for the cubic decimetre,

3 *recommends* that the name litre should not be employed to give the results of high accuracy volume measurements.

Curie (CR, 94)

RESOLUTION 7

The 12th CGPM,

considering that the curie has been used for a long time in many countries as a unit of activity for radionuclides,

recognizing that in the International System of Units (SI), the unit of this activity is the second to the power of minus one (s^{-1}),

accepts that the curie be still retained outside SI as unit of activity, with the value 3.7×10^{10} s^{-1}. The symbol for this unit is Ci.

Prefixes femto and atto (CR, 94)

RESOLUTION 8

The 12 CGPM,

decides to add to the list of prefixes for the formation of names of multiples and sub-multiples of units, adopted by the 11th CGPM, Resolution 12, paragraph 3, the following two new prefixes:

Multiplying factor	Prefix	Symbol
10^{-15}	femto	f
10^{-18}	atto	a

13th CGPM, 1967-1968

SI unit of time (second) (CR, 103)

RESOLUTION 1

The 13th CGPM,

considering

that the definition of the second adopted by the CIPM in 1956 (Resolution 1) and ratified by Resolution 9 of the 11th CGPM (1960), later upheld by Resolution 5 of the 12th CGPM (1964), is inadequate for the present needs of metrology,

that at its meeting of 1964 the CIPM, empowered by Resolution 5 of the 12th CGPM (1964) recommended, in order to fulfil these requirements, a cesium atomic frequency standard for temporary use,

that this frequency standard has now been sufficiently tested and found sufficiently accurate to provide a definition of the second fulfilling present requirements,

that the time has now come to replace the definition now in force of the unit of time of the International System of Units by an atomic definition based on that standard,

decides

1 The unit of time of the International System of Units is the second defined as follows:

"The second is the duration of 9 192 631 770 periods of the radiation corresponding to the transition between the two hyperfine levels of the ground state of the cesium-133 atom".

2 Resolution 1 adopted by the CIPM at its meeting of 1956 and Resolution 9 of the 11th CGPM are now abrogated.

Unit of thermodynamic temperature (kelvin) (CR, 104)

RESOLUTION 3

The 13th CGPM

considering

the names "degree Kelvin" and "degree", the symbols "°K" and "deg" and the rules for their use given in Resolution 7 of the 9th CGPM (1948), in Resolution 12 of the 11th CGPM (1960) and the decision taken by the CIPM in 1962 (PV, 30, 27)[11], that the unit of thermodynamic temperature and the unit of temperature interval are one and the same unit, which ought to be denoted by a single name and single symbol,

[11] "1 The unit degree Kelvin (symbol °K) may be employed for a difference of two thermodynamic temperatures as well as for thermodynamic temperature itself.

"2 If it is found necessary to suppress the name Kelvin, the international symbol "deg" is recommended for the unit of difference of temperature. (The symbol "deg" is read, for example: "degré" in French, "degree" in English, "gradous" (градус) in Russian, "Grad" in German, "graad" in Dutch").

decides

1 the unit of thermodynamic temperature is denoted by the name "kelvin" and its symbol is "K";

2 the same name and the same symbol are used to express a temperature interval;

3 a temperature interval may also be expressed in degrees Celsius;

4 the decisions mentioned in the opening paragraph concerning the name of the unit of thermodynamic temperature, its symbol and the designation of the unit to express an interval or a difference of temperatures are abrogated, but the usages which derive from these decisions remain permissible for the time being.

RESOLUTION 4

The 13th CGPM,

considering that it is useful to formulate more explicitly the definition of the unit of thermodynamic temperature contained in Resolution 3 of the 10th CGPM (1954),

decides to express this definition as follows:

"The kelvin, unit of thermodynamic temperature, is the fraction 1/273.16 of the thermodynamic temperature of the triple point of water".

Unit of luminous intensity (candela) (CR, 104)

RESOLUTION 5

The 13th CGPM

considering

the definition of the unit of luminous intensity ratified by the 9th CGPM (1948) and contained in the "Resolution concerning the change of photometric units" adopted by the CIPM in 1946 (PV, **20**, 119) in virtue of the powers conferred by the 8th CGPM (1933),

that this definition fixes satisfactorily the unit of luminous intensity, but that its wording may be open to criticism,

decides to express the definition of the candela as follows:

"The candela is the luminous intensity, in the perpendicular direction, of a surface of 1/600 000 square metre of a blackbody at the temperature of freezing platinum under a pressure of 101 325 newtons per square metre."

Derived units (CR, 105)

RESOLUTION 6

The 13th CGPM

considering that it is useful to add some derived units to the list of paragraph 4 of Resolution 12 of the 11th CGPM (1960),

decides to add:

wave number	1 per metre	m^{-1}
entropy	joule per kelvin	J/K
specific heat capacity	joule per kilogram kelvin	J/(kg•K)
thermal conductivity	watt per metre kelvin	W/(m•K)
radiant intensity	watt per steradian	W/sr
activity (of a radioactive source)	1 per second	s^{-1}

Abrogation of earlier decisions (micron, new candle) (CR, 105)

RESOLUTION 7

The 13th CGPM,

considering that subsequent decisions of the General Conference concerning the International System of Units are incompatible with parts of Resolution 7 of the 9th CGPM (1948),

decides accordingly to remove from Resolution 7 of the 9th Conference:

1 the unit name "micron", and the symbol "μ" which had been given to that unit, but which has now become a prefix;

2 the unit name "new candle".

CIPM, 1967

Decimal multiples and sub-multiples of the unit of mass (PV **35**, 29)

RECOMMENDATION 2

The CIPM,

considering that the rule for forming names of decimal multiples and sub-multiples of the units of paragraph 3 of Resolution 12 of the 11th CGPM (1960) might be interpreted in different ways when applied to the unit of mass,

declares that the rules of Resolution 12 of the 11th CGPM apply to the kilogram in the following manner: the names of decimal multiples and sub-multiples of the unit of mass are formed by attaching prefixes to the word "gram".

CIPM, 1969

International System of Units: Rules for application of Resolution 12 of the 11th CGPM (1960) (PV, **37**, 30)

RECOMMENDATION 1 (1969)

The CIPM,

considering that Resolution 12 of the 11th CGPM (1960) concerning the International System of Units, has provoked discussions on certain of its aspects,

declares

1 the base units, the supplementary units, and the derived units, of the International System of Units, which form a coherent set, are denoted by the name "SI units";

2 the prefixes adopted by the CGPM for the formation of decimal multiples and sub-multiples of SI units are called "SI prefixes";

and *recommends*

3 the use of SI units, and of their decimal multiples and sub-multiples whose names are formed by means of SI prefixes.

Note.—The name "supplementary units", appearing in Resolution 12 of the 11th CGPM (and in the present Recommendation) is given to SI units for which the General Conference declines to state whether they are base units or derived units.

14th CGPM, 1971

Pascal; siemens (CR,)

The 14th CGPM adopted the special names "pascal" (symbol Pa), for the SI unit of pressure (newton per square metre), and "siemens" (symbol S), for the SI unit of electric conductance (reciprocal ohm).

Unit of amount of substance (mole) (CR,)

RESOLUTION 3

The 14th CGPM

considering the advice of the International Union of Pure and Applied Physics, of the International Union of Pure and Applied Chemistry, and of the International Organization for Standardization, concerning the need to define a unit of amount of substance,

decides

1 The mole is the amount of substance of a system which contains as many elementary entities as there are atoms in 0.012 kilogram of carbon 12; its symbol is "mol".

2 When the mole is used, the elementary entities must be specified and may be atoms, molecules, ions, electrons, other particles, or specified groups of such particles.

3 The mole is a base unit of the International System of Units.

APPENDIX II

Practical realization of the definitions of some important units

1. Length

The following recommendation was adopted by the CIPM in 1960 to specify the characteristics of the discharge lamp radiating the standard line of krypton 86:

> In accordance with paragraph 1 of Resolution 7 adopted by the 11th CGPM (October 1960) the CIPM recommends that the line of krypton 86 adopted as primary standard of length be realized by means of a hot cathode discharge lamp containing krypton 86 of purity not less than 99% in sufficient quantity to ensure the presence of solid krypton at a temperature of 64 °K. The lamp shall have a capillary of internal diameter 2 to 4 millimetres, and wall thickness approximately 1 millimetre.
>
> It is considered that, provided the conditions listed below are satisfied, the wavelength of the radiation emitted by the positive column is equal to the wavelength corresponding to the transition between the unperturbed levels to within 1 in 10^8:
>
> 1. the capillary is observed end-on in a direction such that the light rays used travel from the cathode end to the anode end;
> 2. the lower part of the lamp including the capillary is immersed in a bath maintained to within 1 degree of the temperature of the triple point of nitrogen;
> 3. the current density in the capillary is 0.3 ± 0.1 ampere per square centimetre.
>
> (*Procès-Verbaux CIPM*, 1960, 28, 71; *Comptes rendus* 11*th CGPM*, 1960, 85)

The ancillary apparatus comprises the stabilized current supply for the lamp, a vacuum-tight cryostat, a thermometer for use in the region of 63 K, a vacuum pump, and either a monochromator, to isolate the line or special interference filters.

The wavelength of the standard line, reproducible to 1 in 10^8 according to the above specifications, might be made reproducible to 1 in 10^9 approximately with more stringent specifications.

Other lines of krypton 86 and several lines of mercury 198 and of cadmium 114 are recommended as secondary standards (*Procès-Verbaux* CIPM, 1963, **31**, Recommendation 1, 26 and *Comptes rendus* 12*th CGPM*, 1964, 18).

The wavelength of these lines varies with pressure, temperature, and composition of the air in which the light travels; the refractive index of the air must therefore in general be measured *in situ*.

To measure end or line standards these radiations are used in an interference comparator a complicated instrument with mechanical, optical interference, and thermometric components.

2. Mass

The primary standard of the unit of mass is the international prototype of the kilogram kept at the BIPM. The mass of 1 kg secondary standards of platinum-iridium or of stainless steel is compared with the mass of the prototype by means of balances whose precision can reach 1 in 10^8 or better.

By an easy operation a series of masses can be standardized to obtain multiples and sub-multiples of the kilogram.

3. Time

Some laboratories are able to make the equipment required to produce electric oscillations at the frequency of vibration of the atom of cesium-133 which defines the second. This equipment includes a quartz oscillator, frequency multipliers and synthesizers, a klystron, phase-sensitive detectors, an apparatus for producing an atomic beam of cesium in vacuum, cavity resonators, uniform and non-uniform magnetic fields, and an ion detector.

Complete assemblies to produce this frequency are also commercially available.

By division it is possible to obtain pulses at the desired frequencies, for instance 1 Hz, 1 kHz, etc.

The stability and the reproducibility can exceed 1 in 10^{11}.

Radio stations broadcast waves whose frequencies are known to about the same accuracy.

There are other standards besides the cesium beam, among them the hydrogen maser, rubidium clocks, quartz frequency standards and clocks, etc. Their frequency is controlled by comparison with a cesium standard, either directly, or by means of radio transmissions.

Conforming to a decision of the 14th CGPM (1971, Resolution 1), the CIPM has given the following definition of International Atomic Time (IAT):

> "International Atomic Time is the time reference coordinate established by the Bureau International de l'Heure on the basis of the readings of atomic clocks operating in various establishments in accordance with the definition of the second, the time unit of the International system of Units.
>
> "Most time signals broadcast by radio waves are given in a time scale called Coordinated Universal Time (UTC). UTC is defined in such a manner that it differs from IAT by an exact whole number of seconds. The difference UTC–IAT was set equal to −10 s starting the first of January 1972, the date of application of the reformulation of UTC which previously involved a frequency offset; this difference can be modified by 1 second, preferably on the first of January, and in case of need on the first of July, to keep UTC in agreement with the time defined by the rotation of the Earth with an approximation better than 0.7 s. Furthermore, the legal times of most countries are offset by a whole number of hours (time zones and 'summer' time)."

4. Electric quantities

So-called "absolute" electrical measurements, i.e. those that realize the unit according to its definition, can be undertaken only by laboratories enjoying exceptional facilities.

Electric current is obtained in amperes by measuring the force between two coils, of measurable shape and size, that carry the current.

The ohm, the farad, and the henry are accurately linked by impedance measurements at a known frequency, and may be determined in absolute value by calculation (1) of the self-inductance of a coil, or the mutual inductance of two coils, in terms of their linear dimensions, or (2) of the change in capacitance of a capacitor in terms of the change in length of its electrodes (method of Thompson-Lampard).

The volt is deduced from the ampere and the ohm.

The accuracy of these measurements lies between 1 and 3 in 10^6.

The results of absolute measurements are obtained by means of secondary standards which are, for instance:

1. coils of manganin wire for resistance standards;
2. galvanic cells with cadmium sulphate electrolyte for standards of electromotive force;
3. capacitors for standards of capacitance (of 10 pF for example).

Application of recent techniques also provides means of checking the stability of the secondary standards which maintain the electric units: measurement of the gyromagnetic ratio of the proton γ_p' for the ampere, measurement of the ratio h/e by the Josephson effect for the volt.

5. Temperature

Absolute measurements of temperature in accordance with the definition of the unit of thermodynamic temperature, the kelvin, are related to thermodynamics, for example by the gas thermometer.

At 273.16 K accuracy is of the order of 1 in 10^6, but it is not as good at higher and at lower temperatures.

The International Practical Temperature Scale adopted by the CIPM in 1968 agrees with the best thermodynamic results to date. The text on this scale (which replaces the 1948 scale, amended in 1960) is published in *Comité Consultatif de Thermométrie*, 8th session, 1967, Annexe 18, and *Comptes rendus*, 13th *CGPM*, 1967-1968, Annexe 2; the English translation is published in *Metrologia*, **5**, 35, 1969.

The instruments employed to measure temperatures in the International Scale are the platinum resistance thermometer, the platinum-10% rhodium/platinum thermocouple and the monochromatic optical pyrometer. These instruments are calibrated at a number of reproducible temperatures, called "defining fixed points," the values of which are assigned by agreement.

6. Amount of substance

All quantitative results of chemical analysis or of dosages can be expressed in moles, in other words in units of amount of substance of the constituent particles. The principle of physical measurements based on the definition of this unit is explained below.

The simplest case is that of a sample of a pure substance that is considered to be formed of atoms; call X the chemical symbol of these atoms. A mole of atoms X contains by definition as many atoms as there are ^{12}C atoms in 0.012 kilogram of carbon 12. As neither the mass $m(^{12}C)$ of an atom of carbon 12 nor the mass $m(X)$ of an atom X can be measured accurately, we use the ratio of these masses, $m(X)/m(^{12}C)$, which can be accurately determined.[12] The mass corresponding to 1 mole of X is then $[m(X)/m(^{12}C)] \times 0.012$ kg, which is expressed by saying that the molar mass $M(X)$ of X (quotient of mass by amount of substance) is

$$M(X) = [m(X)/m(^{12}C)] \times 0.012 \text{ kg/mol.}$$

For example, the atom of fluorine ^{19}F and the atom of carbon ^{12}C have masses which are in the ratio 18.9984/12. The molar mass of the molecular gas F_2 is:

$$M(F_2) = \frac{2 \times 18.9984}{12} \times 0.012 \text{ kg/mol} = 0.037\,996\,8 \text{ kg/mol.}$$

The amount of substance corresponding to a given mass of gas F_2, 0.05 kg for example, is:

$$\frac{0.05 \text{ kg}}{0.037\,996\,8 \text{ kg}\cdot\text{mol}^{-1}} = 1.315\,90 \text{ mol.}$$

In the case of a pure substance that is supposed made up of molecules B, which are combinations of atoms X, Y, . . . according to the chemical formula $B = X_\alpha Y_\beta \ldots$, the mass of one molecule is $m(B) = \alpha m(X) + \beta m(Y) + \ldots$

This mass is not known with accuracy, but the ratio $m(B)/m(^{12}C)$ can be determined accurately. The molar mass of a molecular substance B is then

$$M(B) = \frac{m(B)}{m(^{12}C)} \times 0.012 \text{ kg/mol} = \left\{ \alpha \frac{m(X)}{m(^{12}C)} + \beta \frac{m(Y)}{m(^{12}C)} + \cdots \right\} \times 0.012 \text{ kg/mol.}$$

[12] There are many methods of measuring this ratio, the most direct one being by the mass spectrograph.

The same procedure is used in the more general case when the composition of the substance B is specified as $X_\alpha Y_\beta$ . . . even if α, β, . . . are not integers. If we denote the mass ratios $m(X)/m(^{12}C)$, $m(Y)/m(^{12}C)$, . . . by $r(X)$, $r(Y)$, . . ., the molar mass of the substance B is given by the formula:

$$M(B) = [\alpha r(X) + \beta r(Y) + \ldots] \times 0.012 \text{ kg/mol}.$$

There are other methods based on the laws of physics and physical chemistry for measuring amounts of substance; three examples are given below.

With perfect gases, 1 mole of particles of any gas occupies the same volume at a temperature T and a pressure p (approximately 0.022 4 m^3 at T = 273.16 K and p = 101 325 Pa); hence a method of measuring the ratio of amounts of substance for any two gases (the corrections to apply if the gases are not perfect are well known).

For quantitative electrolytic reactions the ratio of amounts of substance can be obtained by measuring quantities of electricity. For example, 1 mole of Ag and 1 mole of (1/2) Cu are deposited on a cathode by the same quantity of electricity (approximately 96 487 C).

Application of the laws of Raoult is yet another method of determining ratios of amounts of substance in extremely dilute solutions.

7. Photometric quantities

Absolute photometric measurements by comparison with the luminance of a blackbody at the temperature of freezing platinum can only be undertaken by a few well-equipped laboratories. The accuracy of these measurements is somewhat better than 1%.

The results of these measurements are maintained by means of incandescent lamps fed with d.c. in a specified manner. These lamps constitute standards of luminous intensity and of luminous flux.

The method approved by CIPM in 1937 (Procès-Verbaux CIPM, 18, 237) for determining the value of photometric quantities for luminous sources having a color other than that of the primary standard, utilizes a procedure taking account of the "spectral luminous efficiencies" $V(\lambda)$ adopted by it in 1933. They were at that time known as "relative visibility (or luminosity) factors", and had been recommended by the CIE in 1924.

Photometric quantities are thereby defined in purely physical terms as quantities proportional to the sum or integral of a spectral power distribution, weighted according to a specified function of wavelength.

APPENDIX III

Organs of the Metre Convention

BIPM, CIPM, CGPM

The International Bureau of Weights and Measures (BIPM) was set up by the *Metre Convention* signed in Paris on 20 May 1875 by seventeen States during the final session of the Diplomatic Conference of the Metre. This Convention was amended in 1921.

BIPM has its headquarters near Paris, in the grounds of the Pavillon de Breteuil (Parc de Saint-Cloud), placed at its disposal by the French Government; its upkeep is financed jointly by the Member States of the Metre Convention.*

The task of BIPM is to ensure worldwide unification of physical measurements; it is responsible for:
—establishing the fundamental standards and scales for measurement of the principal physical quantities and maintaining the international prototypes;
—carrying out comparisons of national and international standards;
—ensuring the co-ordination of corresponding measuring techniques;
—carrying out and co-ordinating the determinations relating to the fundamental physical constants.

BIPM operates under the exclusive supervision of an *International Committee of Weights and Measures* (CIPM), which itself comes under the authority of a *General Conference of Weights and Measures* (CGPM).

The General Conference consists of delegates from all the Member States of the Metre Convention and meets at least once every six years. At each meeting it receives the Report of the International Committee on the work accomplished, and it is responsible for:
—discussing and instigating the arrangements required to ensure the propagation and improvement of the International System of Units (SI), which is the modern form of the metric system;
—confirming the results of new fundamental metrological determinations and the various scientific resolutions of international scope;

---

*As at 31 December 1971, forty-one States were members of this Convention: America (United States of), Argentina (Rep. of), Australia, Austria, Brazil, Belgium, Bulgaria, Cameroon, Canada, Chile, Czechoslovakia, Denmark, Dominican Republic, Finland, France, Germany, Hungary, India, Indonesia, Ireland, Italy, Japan, Korea, Mexico, the Netherlands, Norway, Poland, Portugal, Rumania, Spain, South Africa, Sweden, Switzerland, Thailand, Turkey, U.S.S.R., United Arab Republic, United Kingdom, Uruguay, Venezuela, Yugoslavia.

—adopting the important decisions concerning the organization and development of BIPM.

The International Committee consists of eighteen members each belonging to a different State; it meets at least once every two years. The officers of this Committee issue an *Annual Report* on the administrative and financial position of BIPM to the Governments of the Member States of the Metre Convention.

The activities of BIPM, which in the beginning were limited to the measurements of length and mass and to metrological studies in relation to these quantities, have been extended to standards of measurement for electricity (1927), photometry (1937) and ionizing radiations (1960). To this end the original laboratories, built in 1876–1878, were enlarged in 1929 and two new buildings were constructed in 1963–1964 for the ionizing radiation laboratories. Some thirty physicists or technicians work in the laboratories of BIPM. They do metrological research, and also undertake measurement and certification of material standards of the above-mentioned quantities. BIPM's annual budget is of the order of 3 000 000 gold francs, approximately 1 000 000 U.S. dollars.

In view of the extension of the work entrusted to BIPM, CIPM has set up since 1927, under the name of *Consultative Committees,* bodies designed to provide it with information on matters which it refers to them for study and advice. These Consultative Committees, which may form temporary or permanent "Working Groups" to study special subjects, are responsible for co-ordinating the international work carried out in their respective fields and proposing recommendations concerning the amendments to be made to the definitions and values of units. In order to ensure worldwide uniformity in units of measurement, the International Committee accordingly acts directly or submits proposals for sanction by the General Conference.

The Consultative Committees have common regulations (*Procès-Verbaux CIPM,* 1963, 31, 97). Each Consultative Committee, the chairman of which is normally a member of CIPM, is composed of a delegate from each of the large Metrology Laboratories and specialized Institutes, a list of which is drawn up by CIPM, as well as individual members also appointed by CIPM. These Committees hold their meetings at irregular intervals; at present there are seven of them in existence.

1. *The Consultative Committee for Electricity* (C.C.E.), set up in 1927.
2. *The Consultative Committee for Photometry* (C.C.P.), set up in 1933 (between 1930 and 1933 the preceding Committee dealt with matters concerning Photometry).
3. *The Consultative Committee for Thermometry* (C.C.T.), set up in 1937.
4. *The Consultative Committee for the Definition of the Metre* (C.C.D.M.), set up in 1952.

5. *The Consultative Committee for the Definition of the Second* (C.C.D.S.), set up in 1956.
6. *The Consultative Committee for the Standards of Measurement of Ionining Radiations* (C.C.E.M.R.I.), set up in 1958.

Since 1969 this Consultative Committee has consisted of four sections: Section I (measurement of x and γ rays); Section II (measurement of radionuclides); Section III (Newton measurements); Section IV (α-energy standards).

7. *The Consultative Committee for Units* (C.C.U.), set up in 1964.

The proceedings of the General Conference, the International Committee, the Consultative Committees, and the International Bureau are published under the auspices of the latter in the following series:

—*Comptes rendus des séances de la Conférence Générale des Poids et Mesures;*

—*Procès-Verbaux des séances du Comité International des Poids et Mesures;*

—*Sessions des Comités Consultatifs;*

—*Recueil de Travaux du Bureau International des Poids et Mesures* (this compilation brings together articles published in scientific and technical journals and books, as well as certain work published in the form of duplicated reports).

The collection of the *Travaux et Memoires du Bureau International des Poids et Mesures* (22 volumes published between 1881 and 1966) ceased in 1966 by a decision of CIPM.

From time to time BIPM publishes a report on the development of the Metric System throughout the world, entitled *Les récents progrès du Système Métrique.*

Since 1965 the international journal *Metrologia*, edited under the auspices of CIPM, has published articles on the more important work on scientific metrology carried out throughout the world, on the improvement in measuring methods and standards, of units, etc., as well as reports concerning the activities, decisions and recommendations of the various bodies created under the Metre Convention.

PHYSICAL CONSTANTS and CONVERSION FACTORS

PART II

The International System of Units: Physical Constants and Conversion Factors (NASA SP-7012) was first published in October 1964. Subsequently, the Twelfth and Thirteenth General Conferences convened, and it became desirable to revise SP-7012 to include resolutions adopted by these two more recent general conferences. The present edition not only incorporates material from the records of the Twelfth and Thirteenth General Conferences, but it also includes a new and more accurate set of values for the constants as derived from the recent work of Parker, Taylor, and Langenberg (1969).

NAMES OF INTERNATIONAL UNITS

Physical Quantity	*Name of Unit*	*Symbol*	
	BASIC UNITS		
Length	meter	m	
Mass	kilogram	kg	
Time	second	s	
Electric current	ampere	A	
Temperature	kelvin	K	
Luminous intensity	candela	cd	
	DERIVED UNITS		
Area	square meter	m^2	
Volume	cubic meter	m^3	
Frequency	hertz	Hz	(s^{-1})
Density	kilogram per cubic meter	kg/m^3	
Velocity	meter per second	m/s	
Angular velocity	radian per second	rad/s	
Acceleration	meter per second squared	m/s^2	
Angular acceleration	radian per second squared	rad/s^2	
Force	newton	N	$(kg{\cdot}m/s^2)$
Pressure	newton per sq meter	N/m^2	
Kinematic viscosity	sq meter per second	m^2/s	
Dynamic viscosity	newton-second per sq meter	$N{\cdot}s/m^2$	
Work, energy, quantity of heat	joule	J	(N·m)
Power	watt	W	(J/s)
Electric charge	coulomb	C	(A·s)
Voltage, potential difference, electromotive force	volt	V	(W/A)
Electric field strength	volt per meter	V/m	
Electric resistance	ohm	Ω	(V/A)
Electric capacitance	farad	F	(A·s/V)
Magnetic flux	weber	Wb	(V·s)
Inductance	henry	H	(V·s/A)
Magnetic flux density	tesla	T	(Wb/m^2)
Magnetic field strength	ampere per meter	A/m	
Magnetomotive force	ampere	A	
Luminous flux	lumen	lm	(cd·sr)
Luminance	candela per sq meter	cd/m^2	
Illumination	lux	lx	(lm/m^2)
Wave number	1 per meter	m^{-1}	
Entropy	joule per kelvin	J/K	
Specific heat	joule per kilogram kelvin	$J\ kg^{-1}\ K^{-1}$	
Thermal conductivity	watt per meter kelvin	$W\ m^{-1}\ K^{-1}$	
Radiant intensity	watt per steradian	W/sr	
Activity (of a radioactive source)	1 per second	s^{-1}	
	SUPPLEMENTARY UNITS		
Plane angle	radian	rad	
Solid angle	steradian	sr	

PREFIXES

The names of multiples and submultiples of SI Units may be formed by application of the prefixes:

Factor by which unit is multiplied	Prefix	Symbol
10^{12}	tera	T
10^{9}	giga	G
10^{6}	mega	M
10^{3}	kilo	k
10^{2}	hecto	h
10	deka	da
10^{-1}	deci	d
10^{-2}	centi	c
10^{-3}	milli	m
10^{-6}	micro	μ
10^{-9}	nano	n
10^{-12}	pico	p
10^{-15}	femto	f
10^{-18}	atto	a

PHYSICAL CONSTANTS

The following lists of physical constants are from the work of B. N. Taylor, W. H. Parker, and D. N. Langenberg (*Reviews of Modern Physics*, July 1969). Their least-squares adjustment of values of the constants depends strongly on a new and highly accurate (2.4 ppm) determination of e/h from the ac Josephson effect in superconductors, and is believed to be more accurate than the 1963 adjustment which appears to suffer from the use of an incorrect value of the fine structure constant as an input datum.

Quantity	Symbol	Value	Error ppm	Prefix	Unit
Speed of light in vacuum	c	2. 997 925 0	0. 33	$\times 10^{8}$	$m\ s^{-1}$
Gravitational constant	G	6. 673 2	460	10^{-11}	$N\ m^2\ kg^{-2}$
Avogadro constant	N_A	6. 022 169	6. 6	10^{26}	$kmole^{-1}$
Boltzmann constant	k	1. 380 622	43	10^{-23}	$J\ K^{-1}$
Gas constant	R	8. 314 34	42	10^{3}	$J\ kmole^{-1}\ K^{-1}$
Volume of ideal gas, standard conditions	V_0	2. 241 36	--------	10^{1}	$m^3\ kmole^{-1}$
Faraday constant	F	9. 648 670	5. 5	10^{7}	$C\ kmole^{-1}$
Unified atomic mass unit	u	1. 660 531	6. 6	10^{-27}	kg
Planck constant	h	6. 626 196	7. 6	10^{-34}	J s
	$h/2\pi$	1. 054 591 9	7. 6	10^{-34}	J s
Electron charge	e	1. 602 191 7	4. 4	10^{-19}	C
Electron rest mass	m_e	9. 109 558	6. 0	10^{-31}	kg
		5. 485 930	6. 2	10^{-4}	u
Proton rest mass	m_p	1. 672 614	6. 6	10^{-27}	kg
		1. 007 276 61	. 08	----------	u
Neutron rest mass	m_n	1. 674 920	6. 6	10^{-27}	kg
		1. 008 665 20	. 10	----------	u
Electron charge to mass ratio	e/m_e	1. 758 802 8	3. 1	10^{11}	$C\ kg^{-1}$
Stefan-Boltzmann constant	σ	5. 669 61	170	10^{-8}	$W\ m^{-2}\ K^4$
First radiation constant	$8\pi hc$	4. 992 579	7. 6	10^{-24}	J m
Second radiation constant	hc/k	1. 438 833	43	10^{-2}	m K
Rydberg constant	R_∞	1. 097 373 12	. 10	10^{7}	m^{-1}
Fine structure constant	α	7. 297 351	1. 5	10^{-3}	
	α^{-1}	1. 370 360 2	1. 5	10^{+2}	
Bohr radius	a_0	5. 291 771 5	1. 5	10^{-11}	m
Classical electron radius	r_e	2. 817 939	4. 6	10^{-15}	m
Compton wavelength of electron	λ_C	2. 426 309 6	3. 1	10^{-12}	m
	$\lambda_C/2\pi$	3. 861 592	3. 1	10^{-13}	m
Compton wavelength of proton	$\lambda_{C,p}$	1. 321 440 9	6. 8	10^{-15}	m
	$\lambda_{C,p}/2\pi$	2. 103 139	6. 8	10^{-16}	m
Compton wavelength of neutron	$\lambda_{C,n}$	1. 319 621 7	6. 8	10^{-15}	m
	$\lambda_{C,n}/2\pi$	2. 100 243	6. 8	10^{-16}	m
Electron magnetic moment	μ_e	9. 284 851	7. 0	10^{-24}	$J\ T^{-1}$
Proton magnetic moment	μ_p	1. 410 620 3	7. 0	10^{-26}	$J\ T^{-1}$
Bohr magneton	μ_B	9. 274 096	7. 0	10^{-24}	$J\ T^{-1}$
Nuclear magneton	μ_n	5. 050 951	10	10^{-27}	$J\ T^{-1}$

Quantity	Symbol	Value	Error ppm	Prefix	Unit
Gyromagnetic ratio of protons in H_2O	γ'_p	2. 675 127 0	3. 1	10^8	rad s^{-1} T^{-1}
	$\gamma'_p/2\pi$	4. 257 597	3. 1	10^7	Hz T^{-1}
Gyromagnetic ratio of protons in H_2O	γ_p	2. 675 196 5	3. 1	10^8	rad s^{-1} T^{-1}
corrected for diamagnetism of H_2O.	$\gamma_p/2\pi$	4. 257 707	3. 1	10^7	Hz T^{-1}
Magnetic flux quantum	Φ_0	2. 067 853 8	3. 3	10^{-15}	Wb
Quantum of circulation	$h/2m_e$	3. 636 947	3. 1	10^{-4}	J s kg^{-1}
	h/m_e	7. 273 894	3. 1	10^{-4}	J s kg^{-1}

"Dimensionless" combination	Value	Error ppm	Prefix
kg/eV	5. 609 538	4. 4	10^{35}
u/eV	9. 314 812	5. 5	10^8
u/kg	1. 660 531	6. 6	10^{-27}
m_e/eV	5. 110 041	3. 1	10^5
m_p/eV	9. 382 592	5. 5	10^8
m_n/eV	9. 395 527	5. 5	10^8
eV/J	1. 602 191 7	4. 4	10^{-19}
eV/Hz	2. 417 965 9	3. 3	10^{14}
eV m	8. 065 465	3. 3	10^5
eV/K	1. 160 485	42	10^4
$(\text{eV m})^{-1}$	1. 239 854 1	3. 3	10^{-6}
R_∞/J	2. 179 914	7. 6	10^{-18}
R_∞/eV	1. 360 582 6	3. 3	10^1
R_∞/Hz	3. 289 842 3	. 35	10^{15}
R_∞/K	1. 578 936	43	10^5
m_p/m_e	1. 836 109	6. 2	10^3
μ_e/μ_B	1. 001 159 638 9	. 0031	
μ'_p/μ_B	1. 520 993 12	. 066	10^{-3}
μ_p/μ_B	1. 521 032 64	. 30	10^{-3}
μ'_p/μ_n	2. 792 709	6. 2	
μ_p/μ_n	2. 792 782	6. 2	

CONVERSION FACTORS

The following tables express the definitions of miscellaneous units of measure as exact numerical multiples of coherent SI Units, and provide multiplying factors for converting numbers and miscellaneous units to corresponding new numbers and SI Units.

The first two digits of each numerical entry represent a power of 10. An asterisk follows each number which expresses an exact definition. For example, the entry "−02 2.54*" expresses the fact that 1 inch=2.54×10^{-2} meter, exactly, by definition. Most of the definitions are extracted from National Bureau of Standards documents. Numbers not followed by an asterisk are only approximate representations of definitions, or are the results of physical measurements.

The conversion factors are listed alphabetically and by physical quantity.

The Listing by Physical Quantity includes only relationships which are frequently encountered and deliberately omits the great multiplicity of combinations of units which are used for more specialized purposes. Conversion factors for combinations of units are easily generated from numbers given in the Alphabetical Listing by the technique of direct substitution or by other well-known rules for manipulating units. These rules are adequately discussed in many science and engineering textbooks and are not repeated here.

ALPHABETICAL LISTING

To convert from	*to*	*multiply by*
abampere	ampere	+01 1.00*
abcoulomb	coulomb	+01 1.00*
abfarad	farad	+09 1.00*
abhenry	henry	−09 1.00*
abmho	mho	+09 1.00*
abohm	ohm	−09 1.00*
abvolt	volt	−08 1.00*
acre	meter^2	+03 4.046 856 422 4*
ampere (international of 1948)	ampere	−01 9.998 35
angstrom	meter	−10 1.00*
are	meter^2	+02 1.00*
astronomical unit	meter	+11 1.495 978 9
atmosphere	newton/meter^2	+05 1.013 25*
bar	newton/meter^2	+05 1.00*
barn	meter^2	−28 1.00*
barrel (petroleum, 42 gallons)	meter^3	−01 1.589 873
barye	newton/meter^2	−01 1.00*
British thermal unit (ISO/TC 12)	joule	+03 1.055 06
British thermal unit (International Steam Table)	joule	+03 1.055 04
British thermal unit (mean)	joule	+03 1.055 87
British thermal unit (thermochemical)	joule	+03 1.054 350 264 488
British thermal unit (39° F)	joule	+03 1.059 67
British thermal unit (60° F)	joule	+03 1.054 68
bushel (U.S.)	meter^3	−02 3.523 907 016 688*

To convert from	*to*	*multiply by*
cable	meter	+02 2.194 56*
caliber	meter	−04 2.54*
calorie (International Steam Table)	joule	+00 4.1868
calorie (mean)	joule	+00 4.190 02
calorie (thermochemical)	joule	+00 4.184*
calorie (15° C)	joule	+00 4.185 80
calorie (20° C)	joule	+00 4.181 90
calorie (kilogram, International Steam Table)	joule	+03 4.1868
calorie (kilogram, mean)	joule	+03 4.190 02
calorie (kilogram, thermochemical)	joule	+03 4.184*
carat (metric)	kilogram	−04 2.00*
Celsius (temperature)	kelvin	$t_K = t_C + 273.15$
centimeter of mercury (0° C)	newton/meter^2	+03 1.333 22
centimeter of water (4° C)	newton/meter^2	+01 9.806 38
chain (engineer or ramden)	meter	+01 3.048*
chain (surveyor or gunter)	meter	+01 2.011 68*
circular mil	meter^2	−10 5.067 074 8
cord	meter^3	+00 3.624 556 3
coulomb (international of 1948)	coulomb	−01 9.998 35
cubit	meter	−01 4.572*
cup	meter^3	−04 2.365 882 365*
curie	disintegration/second	+10 3.70*
day (mean solar)	second (mean solar)	+04 8.64*
day (sidereal)	second (mean solar)	+04 8.616 409 0
degree (angle)	radian	−02 1.745 329 251 994 3
denier (international)	kilogram/meter	−07 1.00*
dram (avoirdupois)	kilogram	−03 1.771 845 195 312 5*
dram (troy or apothecary)	kilogram	−03 3.887 934 6*
dram (U.S. fluid)	meter^3	−06 3.696 691 195 312 5*
dyne	newton	−05 1.00*
electron volt	joule	−19 1.602 10
erg	joule	−07 1.00*
Fahrenheit (temperature)	kelvin	$t_K = (5/9)\ (t_F + 459.67)$
Fahrenheit (temperature)	Celsius	$t_C = (5/9)\ (t_F - 32)$
farad (international of 1948)	farad	−01 9.995 05
faraday (based on carbon 12)	coulomb	+04 9.648 70
faraday (chemical)	coulomb	+04 9.649 57
faraday (physical)	coulomb	+04 9.652 19
fathom	meter	+00 1.828 8*
fermi (femtometer)	meter	−15 1.00*
fluid ounce (U.S.)	meter^3	−05 2.957 352 956 25*
foot	meter	−01 3.048*
foot (U.S. survey)	meter	+00 1200/3937*
foot (U.S. survey)	meter	−01 3.048 006 096
foot of water (39.2° F)	newton/meter^2	+03 2.988 98
foot-candle	lumen/meter^2	+01 1.076 391 0
foot-lambert	candela/meter^2	+00 3.426 259
furlong	meter	+02 2.011 68*
gal (galileo)	meter/second^2	−02 1.00*
gallon (U.K. liquid)	meter^3	−03 4.546 087
gallon (U.S. dry)	meter^3	−03 4.404 883 770 86*
gallon (U.S. liquid)	meter^3	−03 3.785 411 784*

To convert from	*to*	*multiply by*
gamma	tesla	−09 1.00*
gauss	tesla	−04 1.00*
gilbert	ampere turn	−01 7.957 747 2
gill (U.K.)	meter³	−04 1.420 652
gill (U.S.)	meter³	−04 1.182 941 2
grad	degree (angular)	−01 9.00*
grad	radian	−02 1.570 796 3
grain	kilogram	−05 6.479 891*
gram	kilogram	−03 1.00*
hand	meter	−01 1.016*
hectare	meter²	+04 1.00*
henry (international of 1948)	henry	+00 1.000 495
hogshead (U.S.)	meter³	−01 2.384 809 423 92*
horsepower (550 foot lbf/second)	watt	+02 7.456 998 7
horsepower (boiler)	watt	+03 9.809 50
horsepower (electric)	watt	+02 7.46*
horsepower (metric)	watt	+02 7.354 99
horsepower (U.K.)	watt	+02 7.457
horsepower (water)	watt	+02 7.460 43
hour (mean solar)	second (mean solar)	+03 3.60*
hour (sidereal)	second (mean solar)	+03 3.590 170 4
hundredweight (long)	kilogram	+01 5.080 234 544*
hundredweight (short)	kilogram	+01 4.535 923 7*
inch	meter	−02 2.54*
inch of mercury (32° F)	newton/meter²	+03 3.386 389
inch of mercury (60° F)	newton/meter²	+03 3.376 85
inch of water (39.2° F)	newton/meter²	+02 2.490 82
inch of water (60° F)	newton/meter²	+02 2.4884
joule (international of 1948)	joule	+00 1.000 165
kayser	1/meter	+02 1.00*
kilocalorie (International Steam Table)	joule	+03 4.186 74
kilocalorie (mean)	joule	+03 4.190 02
kilocalorie (thermochemical)	joule	+03 4.184*
kilogram mass	kilogram	+00 1.00*
kilogram force (kgf)	newton	+00 9.806 65*
kilopond force	newton	+00 9.806 65*
kip	newton	+03 4.448 221 615 260 5*
knot (international)	meter/second	−01 5.144 444 444
lambert	candela/meter²	+04 1/π*
lambert	candela/meter²	+03 3.183 098 8
langley	joule/meter²	+04 4.184*
lbf (pound force, avoirdupois)	newton	+00 4.448 221 615 260 5*
lbm (pound mass, avoirdupois)	kilogram	−01 4.535 923 7*
league (British nautical)	meter	+03 5.559 552*
league (international nautical)	meter	+03 5.556*
league (statute)	meter	+03 4.828 032*
light year	meter	+15 9.460 55
link (engineer or ramden)	meter	−01 3.048*
link (surveyor or gunter)	meter	−01 2.011 68*
liter	meter³	−03 1.00*
lux	lumen/meter²	+00 1.00*
maxwell	weber	−08 1.00*

To convert from	*to*	*multiply by*
meter	wavelengths Kr 86	+06 1.650 763 73*
micron	meter	−06 1.00*
mil	meter	−05 2.54*
mile (U.S. statute)	meter	+03 1.609 344*
mile (U.K. nautical)	meter	+03 1.853 184*
mile (international nautical)	meter	+03 1.852*
mile (U.S. nautical)	meter	+03 1.852*
millibar	newton/meter^2	+02 1.00*
millimeter of mercury (0° C)	newton/meter^2	+02 1.333 224
minute (angle)	radian	−04 2.908 882 086 66
minute (mean solar)	second (mean solar)	+01 6.00*
minute (sidereal)	second (mean solar)	+01 5.983 617 4
month (mean calendar)	second (mean solar)	+06 2.628*
nautical mile (international)	meter	+03 1.852*
nautical mile (U.S.)	meter	+03 1.852*
nautical mile (U.K.)	meter	+03 1.853 184*
oersted	ampere/meter	+01 7.957 747 2
ohm (international of 1948)	ohm	+00 1.000 495
ounce force (avoirdupois)	newton	−01 2.780 138 5
ounce mass (avoirdupois)	kilogram	−02 2.834 952 312 5*
ounce mass (troy or apothecary)	kilogram	−02 3.110 347 68*
ounce (U.S. fluid)	meter^3	−05 2.957 352 956 25*
pace	meter	−01 7.62*
parsec	meter	+16 3.083 74
pascal	newton/meter^2	+00 1.00*
peck (U.S.)	meter^3	−03 8.809 767 541 72*
pennyweight	kilogram	−03 1.555 173 84*
perch	meter	+00 5.0292*
phot	lumen/meter^2	+04 1.00
pica (printers)	meter	−03 4.217 517 6*
pint (U.S. dry)	meter^3	−04 5.506 104 713 575*
pint (U.S. liquid)	meter^3	−04 4.731 764 73*
point (printers)	meter	−04 3.514 598*
poise	newton second/meter^2	−01 1.00*
pole	meter	+00 5.0292*
pound force (lbf avoirdupois)	newton	+00 4.448 221 615 260 5*
pound mass (lbm avoirdupois)	kilogram	−01 4.535 923 7*
pound mass (troy or apothecary)	kilogram	−01 3.732 417 216*
poundal	newton	−01 1.382 549 543 76*
quart (U.S. dry)	meter^3	−03 1.101 220 942 715*
quart (U.S. liquid)	meter^3	−04 9.463 529 5
rad (radiation dose absorbed)	joule/kilogram	−02 1.00*
Rankine (temperature)	kelvin	$t_K=(5/9)t_R$
rayleigh (rate of photon emission)	1/second meter^2	+10 1.00*
rhe	meter^2/newton second	+01 1.00*
rod	meter	+00 5.0292*
roentgen	coulomb/kilogram	−04 2.579 76*
rutherford	disintegration/second	+06 1.00*
second (angle)	radian	−06 4.848 136 811
second (ephemeris)	second	+00 1.000 000 000
second (mean solar)	second (ephemeris)	Consult American Ephemeris and Nautical Almanac

To convert from	*to*	*multiply by*
second (sidereal)	second (mean solar)	−01 9.972 695 7
section	meter²	+06 2.589 988 110 336*
scruple (apothecary)	kilogram	−03 1.295 978 2*
shake	second	−08 1.00
skein	meter	+02 1.097 28*
slug	kilogram	+01 1.459 390 29
span	meter	−01 2.286*
statampere	ampere	−10 3.335 640
statcoulomb	coulomb	−10 3.335 640
statfarad	farad	−12 1.112 650
stathenry	henry	+11 8.987 554
statmho	mho	−12 1.112 650
statohm	ohm	+11 8.987 554
statute mile (U.S.)	meter	+03 1.609 344*
statvolt	volt	+02 2.997 925
stere	meter³	+00 1.00*
stilb	candela/meter²	+04 1.00
stoke	meter²/second	−04 1.00*
tablespoon	meter³	−05 1.478 676 478 125*
teaspoon	meter³	−06 4.928 921 593 75*
ton (assay)	kilogram	−02 2.916 666 6
ton (long)	kilogram	+03 1.016 046 908 8*
ton (metric)	kilogram	+03 1.00*
ton (nuclear equivalent of TNT)	joule	+09 4.20
ton (register)	meter³	+00 2.831 684 659 2*
ton (short, 2000 pound)	kilogram	+02 9.071 847 4*
tonne	kilogram	+03 1.00*
torr (0° C)	newton/meter²	+02 1.333 22
township	meter²	+07 9.323 957 2
unit pole	weber	−07 1.256 637
volt (international of 1948)	volt	+00 1.000 330
watt (international of 1948)	watt	+00 1.000 165
yard	meter	−01 9.144*
year (calendar)	second (mean solar)	+07 3.1536*
year (sidereal)	second (mean solar)	+07 3.155 815 0
year (tropical)	second (mean solar)	+07 3.155 692 6
year 1900, tropical, Jan., day 0, hour 12	second (ephemeris)	+07 3.155 692 597 47*
year 1900, tropical, Jan., day 0, hour 12	second	+07 3.155 692 597 47

LISTING BY PHYSICAL QUANTITY

ACCELERATION

foot/second²	meter/second²	−01 3.048*
free fall, standard	meter/second²	+00 9.806 65*
gal (galileo)	meter/second²	−02 1.00*
inch/second²	meter/second²	−02 2.54*

AREA

acre	meter²	+03 4.046 856 422 4*
are	meter²	+02 1.00*
barn	meter²	−28 1.00*
circular mil	meter²	−10 5.067 074 8

To convert from	*to*	*multiply by*
foot²	meter²	−02 9.290 304*
hectare	meter²	+04 1.00*
inch²	meter²	−04 6.4516*
mile² (U.S. statute)	meter²	+06 2.589 988 110 336*
section	meter²	+06 2.589 988 110 336*
township	meter²	+07 9.323 957 2
yard²	meter²	−01 8.361 273 6*

DENSITY

gram/centimeter³	kilogram/meter³	+03 1.00*
lbm/inch³	kilogram/meter³	+04 2.767 990 5
lbm/foot³	kilogram/meter³	+01 1.601 846 3
slug/foot³	kilogram/meter³	+02 5.153 79

ENERGY

British thermal unit (ISO/TC 12)	joule	+03 1.055 06
British thermal unit (International Steam Table)	joule	+03 1.055 04
British thermal unit (mean)	joule	+03 1.055 87
British thermal unit (thermochemical)	joule	+03 1.054 350 264 488
British thermal unit (39° F)	joule	+03 1.059 67
British thermal unit (60° F)	joule	+03 1.054 68
calorie (International Steam Table)	joule	+00 4.1868
calorie (mean)	joule	+00 4.190 02
calorie (thermochemical)	joule	+00 4.184*
calorie (15° C)	joule	+00 4.185 80
calorie (20° C)	joule	+00 4.181 90
calorie (kilogram, International Steam Table)	joule	+03 4.1868
calorie (kilogram, mean)	joule	+03 4.190 02
calorie (kilogram, thermochemical)	joule	+03 4.184*
electron volt	joule	−19 1.602 10
erg	joule	−07 1.00*
foot lbf	joule	+00 1.355 817 9
foot poundal	joule	−02 4.214 011 0
joule (international of 1948)	joule	+00 1.000 165
kilocalorie (International Steam Table)	joule	+03 4.1868
kilocalorie (mean)	joule	+03 4.190 02
kilocalorie (thermochemical)	joule	+03 4.184*
kilowatt hour	joule	+06 3.60*
kilowatt hour (international of 1948)	joule	+06 3.600 59
ton (nuclear equivalent of TNT)	joule	+09 4.20
watt hour	joule	+03 3.60*

ENERGY/AREA TIME

Btu (thermochemical)/foot² second	watt/meter²	+04 1.134 893 1
Btu (thermochemical)/foot² minute	watt/meter²	+02 1.891 488 5
Btu (thermochemical)/foot² hour	watt/meter²	+00 3.152 480 8
Btu (thermochemical)/inch² second	watt/meter²	+06 1.634 246 2
calorie (thermochemical)/cm² minute	watt/meter²	+02 6.973 333 3
erg/centimeter² second	watt/meter²	−03 1.00*
watt/centimeter²	watt/meter²	+04 1.00*

FORCE

dyne	newton	−05 1.00*
kilogram force (kgf)	newton	+00 9.806 65*

To convert from	*to*	*multiply by*
kilopond force	newton	+00 9.806 65*
kip	newton	+03 4.448 221 615 260 5*
lbf (pound force, avoirdupois)	newton	+00 4.448 221 615 260 5*
ounce force (avoirdupois)	newton	−01 2.780 138 5
pound force, lbf (avoirdupois)	newton	+00 4.448 221 615 260 5*
poundal	newton	−01 1.382 549 543 76*

LENGTH

To convert from	*to*	*multiply by*
angstrom	meter	−10 1.00*
astronomical unit	meter	+11 1.495 978 9
cable	meter	+02 2.194 56*
caliber	meter	−04 2.54*
chain (surveyor or gunter)	meter	+01 2.011 68*
chain (engineer or ramden)	meter	+01 3.048*
cubit	meter	−01 4.572*
fathom	meter	+00 1.8288*
fermi (femtometer)	meter	−15 1.00*
foot	meter	−01 3.048*
foot (U.S. survey)	meter	+00 1200/3937*
foot (U.S. survey)	meter	−01 3.048 006 096
furlong	meter	+02 2.011 68*
hand	meter	−01 1.016*
inch	meter	−02 2.54*
league (U.K. nautical)	meter	+03 5.559 552*
league (international nautical)	meter	+03 5.556*
league (statute)	meter	+03 4.828 032*
light year	meter	+15 9.460 55
link (engineer or ramden)	meter	−01 3.048*
link (surveyor or gunter)	meter	−01 2.011 68*
meter	wavelengths Kr 86	+06 1.650 763 73*
micron	meter	−06 1.00*
mil	meter	−05 2.54*
mile (U.S. statute)	meter	+03 1.609 344*
mile (U.K. nautical)	meter	+03 1.853 184*
mile (international nautical)	meter	+03 1.852*
mile (U.S. nautical)	meter	+03 1.852*
nautical mile (U.K.)	meter	+03 1.853 184*
nautical mile (international)	meter	+03 1.852*
nautical mile (U.S.)	meter	+03 1.852*
pace	meter	−01 7.62*
parsec	meter	+16 3.083 74
perch	meter	+00 5.0292*
pica (printers)	meter	−03 4.217 517 6*
point (printers)	meter	−04 3.514 598*
pole	meter	+00 5.0292*
rod	meter	+00 5.0292*
skein	meter	+02 1.097 28*
span	meter	−01 2.286*
statute mile (U.S.)	meter	+03 1.609 344*
yard	meter	−01 9.144*

MASS

To convert from	*to*	*multiply by*
carat (metric)	kilogram	−04 2.00*
dram (avoirdupois)	kilogram	−03 1.771 845 195 312 5*
dram (troy or apothecary)	kilogram	−03 3.887 934 6*
grain	kilogram	−05 6.479 891*
gram	kilogram	−03 1.00*

To convert from	*to*	*multiply by*
hundredweight (long)	kilogram	+01 5.080 234 544*
hundredweight (short)	kilogram	+01 4.535 923 7*
kgf second^2 meter (mass)	kilogram	+00 9.806 65*
kilogram mass	kilogram	+00 1.00*
lbm (pound mass, avoirdupois)	kilogram	−01 4.535 923 7*
ounce mass (avoirdupois)	kilogram	−02 2.834 952 312 5*
ounce mass (troy or apothecary)	kilogram	−02 3.110 347 68*
pennyweight	kilogram	−03 1.555 173 84*
pound mass, lbm (avoirdupois)	kilogram	−01 4.535 923 7*
pound mass (troy or apothecary)	kilogram	−01 3.732 417 216*
scruple (apothecary)	kilogram	−03 1.295 978 2*
slug	kilogram	+01 1.459 390 29
ton (assay)	kilogram	−02 2.916 666 6
ton (long)	kilogram	+03 1.016 046 908 8*
ton (metric)	kilogram	+03 1.00*
ton (short, 2000 pound)	kilogram	+02 9.071 847 4*
tonne	kilogram	+03 1.00*

POWER

Btu (thermochemical)/second	watt	+03 1.054 350 264 488
Btu (thermochemical)/minute	watt	+01 1.757 250 4
calorie (thermochemical)/second	watt	+00 4.184*
calorie (thermochemical)/minute	watt	−02 6.973 333 3
foot lbf/hour	watt	−04 3.766 161 0
foot lbf/minute	watt	−02 2.259 696 6
foot lbf/second	watt	+00 1.355 817 9
horsepower (550 foot lbf/second)	watt	+02 7.456 998 7
horsepower (boiler)	watt	+03 9.809 50
horsepower (electric)	watt	+02 7.46*
horsepower (metric)	watt	+02 7.354 99
horsepower (U.K.)	watt	+02 7.457
horsepower (water)	watt	+02 7.460 43
kilocalorie (thermochemical)/minute	watt	+01 6.973 333 3
kilocalorie (thermochemical)/second	watt	+03 4.184*
watt (international of 1948)	watt	+00 1.000 165

PRESSURE

atmosphere	newton/meter^2	+05 1.013 25*
bar	newton/meter^2	+05 1.00*
barye	newton/meter^2	−01 1.00*
centimeter of mercury (0° C)	newton/meter^2	+03 1.333 22
centimeter of water (4° C)	newton/meter^2	+01 9.806 38
dyne/centimeter^2	newton/meter^2	−01 1.00*
foot of water (39.2° F)	newton/meter^2	+03 2.988 98
inch of mercury (32° F)	newton/meter^2	+03 3.386 389
inch of mercury (60° F)	newton/meter^2	+03 3.376 85
inch of water (39.2° F)	newton/meter^2	+02 2.490 82
inch of water (60° F)	newton/meter^2	+02 2.4884
kgf/centimeter^2	newton/meter^2	+04 9.806 65*
kgf/meter^2	newton/meter^2	+00 9.806 65*
lbf/foot^2	newton/meter^2	+01 4.788 025 8
lbf/inch^2 (psi)	newton/meter^2	+03 6.894 757 2
millibar	newton/meter^2	+02 1.00*
millimeter of mercury (0° C)	newton/meter^2	+02 1.333 224
pascal	newton/meter^2	+00 1.00*
psi (lbf/inch^2)	newton/meter^2	+03 6.894 757 2
torr (0° C)	newton/meter^2	+02 1.333 22

To convert from	*to*	*multiply by*
SPEED		
foot/hour	meter/second	−05 8.466 666 6
foot/minute	meter/second	−03 5.08*
foot/second	meter/second	−01 3.048*
inch/second	meter/second	−02 2.54*
kilometer/hour	meter/second	−01 2.777 777 8
knot (international)	meter/second	−01 5.144 444 444
mile/hour (U.S. statute)	meter/second	−01 4.4704*
mile/minute (U.S. statute)	meter/second	+01 2.682 24*
mile/second (U.S. statute)	meter/second	+03 1.609 344*
TEMPERATURE		
Celsius	kelvin	$t_K = t_C + 273.15$
Fahrenheit	kelvin	$t_K = (5/9)(t_F + 459.67)$
Fahrenheit	Celsius	$t_C = (5/9)(t_F - 32)$
Rankine	kelvin	$t_K = (5/9)t_R$
TIME		
day (mean solar)	second (mean solar)	+04 8.64*
day (sidereal)	second (mean solar)	+04 8.616 409 0
hour (mean solar)	second (mean solar)	+03 3.60*
hour (sidereal)	second (mean solar)	+03 3.590 170 4
minute (mean solar)	second (mean solar)	+01 6.00*
minute (sidereal)	second (mean solar)	+01 5.983 617 4
month (mean calendar)	second (mean solar)	+06 2.628*
second (ephemeris)	second	+00 1.000 000 000
second (mean solar)	second (ephemeris)	Consult American Ephemeris and Nautical Almanac
second (sidereal)	second (mean solar)	−01 9.972 695 7
year (calendar)	second (mean solar)	+07 3.1536*
year (sidereal)	second (mean solar)	+07 3.155 815 0
year (tropical)	second (mean solar)	+07 3.155 692 6
year 1900, tropical, Jan., day 0, hour 12	second (ephemeris)	+07 3.155 692 597 47*
year 1900, tropical, Jan., day 0, hour 12	second	+07 3.155 692 597 47
VISCOSITY		
centistoke	meter²/second	−06 1.00*
stoke	meter²/second	−04 1.00*
foot²/second	meter²/second	−02 9.290 304*
centipoise	newton second/meter²	−03 1.00*
lbm/foot second	newton second/meter²	+00 1.488 163 9
lbf second/foot²	newton second/meter²	+01 4.788 025 8
poise	newton second/meter²	−01 1.00*
poundal second/foot²	newton second/meter²	+00 1.488 163 9
slug/foot second	newton second/meter²	+01 4.788 025 8
rhe	meter²/newton second	+01 1.00*
VOLUME		
acre foot	meter³	+03 1.233 481 9
barrel (petroleum, 42 gallons)	meter³	−01 1.589 873
board foot	meter³	−03 2.359 737 216*
bushel (U.S.)	meter³	−02 3.523 907 016 688*
cord	meter³	+00 3.624 556 3

To convert from	*to*	*multiply by*
cup	$meter^3$	−04 2.365 882 365*
dram (U.S. fluid)	$meter^3$	−06 3.696 691 195 312 5*
fluid ounce (U.S.)	$meter^3$	−05 2.957 352 956 25*
$foot^3$	$meter^3$	−02 2.831 684 659 2*
gallon (U.K. liquid)	$meter^3$	−03 4.546 087
gallon (U.S. dry)	$meter^3$	−03 4.404 883 770 86*
gallon (U.S. liquid)	$meter^3$	−03 3.785 411 784*
gill (U.K.)	$meter^3$	−04 1.420 652
gill (U.S.)	$meter^3$	−04 1.182 941 2
hogshead (U.S.)	$meter^3$	−01 2.384 809 423 92*
$inch^3$	$meter^3$	−05 1.638 706 4*
liter	$meter^3$	−03 1.00*
ounce (U.S. fluid)	$meter^3$	−05 2.957 352 956 25*
peck (U.S.)	$meter^3$	−03 8.809 767 541 72*
pint (U.S. dry)	$meter^3$	−04 5.506 104 713 575*
pint (U.S. liquid)	$meter^3$	−04 4.731 764 73*
quart (U.S. dry)	$meter^3$	−03 1.101 220 942 715*
quart (U.S. liquid)	$meter^3$	−04 9.463 529 5
stere	$meter^3$	+00 1.00*
tablespoon	$meter^3$	−05 1.478 676 478 125*
teaspoon	$meter^3$	−06 4.928 921 593 75*
ton (register)	$meter^3$	+00 2.831 684 659 2*
$yard^3$	$meter^3$	−01 7.645 548 579 84*

UNITS OF MEASURE

Definitions

In its original conception, the meter was the fundamental unit of the Metric System, and all units of length and capacity were to be derived directly from the meter which was intended to be equal to one ten-millionth of the earth's quadrant. Furthermore, it was originally planned that the unit of mass, the kilogram, should be identical with the mass of a cubic decimeter of water at its maximum density. The units of length and mass are now defined independently of these conceptions.

In October 1960 the Eleventh General (International) Conference on Weights and Measures redefined the meter as equal to 1 650 763.73 wavelengths of the orange-red radiation in vacuum of krypton 86 corresponding to the unperturbed transition between the $2p_{10}$ and $5d_5$ levels.

The kilogram is independently defined as the mass of a particular platinum-iridium standard, the International Prototype Kilogram, which is kept at the International Bureau of Weights and Measures in Sèvres, France.

The liter has been defined, since October 1964, as being equal to a cubic decimeter. The meter is thus a unit on which is based all metric standards and measurements of length, area, and volume.

Definitions of Units

Length

A *meter* is a unit of length equal to 1 650 763.73 wavelengths in a vacuum of the orange-red radiation of krypton 86.

A *yard* is a unit of length equal to 0.914 4 meter.

Mass

A *kilogram* is a unit of mass equal to the mass of the International Prototype Kilogram.

An *avoirdupois pound* is a unit of mass equal to 0.453 592 37 kilogram.

Capacity, or Volume

A *cubic meter* is a unit of volume equal to a cube the edges of which are 1 meter.

A *liter* is a unit of volume equal to a cubic decimeter.

A *cubic yard* is a unit of volume equal to a cube the edges of which are 1 yard.

A *gallon* is a unit of volume equal to 231 cubic inches. It is used for measuring liquids only.

A *bushel* is a unit of volume equal to 2 150.42 cubic inches. It is used for measuring dry commodities only.

Area

A *square meter* is a unit of area equal to the area of a square the sides of which are 1 meter.

A *square yard* is a unit of area equal to the area of a square the sides of which are 1 yard.

Spelling and Symbols for Units

The spelling of the names of units as adopted by the National Bureau of Standards is that given in the list below. The spelling of the metric units is in accordance with that given in the law of July 28, 1866, legalizing the Metric System in the United States.

Following the name of each unit in the list below is given the symbol that the Bureau has adopted. Attention is particularly called to the following principles:

1. No period is used with symbols for units. Whenever "in" for inch might be confused with the preposition "in", "inch" should be spelled out.
2. The exponents "²" and "³" are used to signify "square" and "cubic," respectively, instead of the symbols "sq" or "cu," which are, however, frequently used in technical literature for the U. S. Customary units.
3. The same symbol is used for both singular and plural.

Some Units and Their Symbols

Unit	Symbol
acre	acre
are	a
barrel	bbl
board foot	fbm
bushel	bu
carat	c
Celsius, degree	°C
centare	ca
centigram	cg
centiliter	cl
centimeter	cm
chain	ch
cubic centimeter	cm^3
cubic decimeter	dm^3
cubic dekameter	dam^3
cubic foot	ft^3
cubic hectometer	hm^3
cubic inch	in^3
cubic kilometer	km^3
cubic meter	m^3
cubic mile	mi^3
cubic millimeter	mm^3
cubic yard	yd^3
decigram	dg
deciliter	dl
decimeter	dm
dekagram	dag
dekaliter	dal
dekameter	dam
dram, avoirdupois	dr avdp
fathom	fath
foot	ft
furlong	furlong
gallon	gal
grain	grain
gram	g
hectare	ha
hectogram	hg
hectoliter	hl
hectometer	hm
hogshead	hhd
hundredweight	cwt
inch	in
International Nautical Mile	INM
kelvin	K
kilogram	kg
kiloliter	kl
kilometer	km
link	link
liquid	liq
liter	liter
meter	m
microgram	μg
microinch	μin
microliter	μl
mile	mi
milligram	mg
milliliter	ml
millimeter	mm
minim	minim
ounce	oz
ounce, avoirdupois	oz avdp
ounce, liquid	liq oz
ounce, troy	oz tr
peck	peck
pennyweight	dwt
pint, liquid	liq pt
pound	lb
pound, avoirdupois	lb avdp
pound, troy	lb tr
quart, liquid	liq qt
rod	rod
second	s
square centimeter	cm^2
square decimeter	dm^2
square dekameter	dam^2
square foot	ft^2
square hectometer	hm^2
square inch	in^2
square kilometer	km^2
square meter	m^2
square mile	mi^2
square millimeter	mm^2
square yard	yd^2
stere	stere
ton, long	long ton
ton, metric	t
ton, short	short ton
yard	yd

Units of Measurement—Conversion Factors*

Units of Length

To Convert from **Centimeters** To	Multiply by
Inches	0.393 700 8
Feet	0.032 808 40
Yards	0.010 936 13
Meters	**0.01**

To Convert from **Meters** To	Multiply by
Inches	39.370 08
Feet	3.280 840
Yards	1.093 613
Miles	0.000 621 37
Millimeters	**1 000**
Centimeters	**100**
Kilometers	**0.001**

To Convert from **Inches** To	Multiply by
Feet	0.083 333 33
Yards	0.027 777 78
Centimeters	**2.54**
Meters	**0.025 4**

To Convert from **Feet** To	Multiply by
Inches	**12**
Yards	0.333 333 3
Miles	0.000 189 39
Centimeters	**30.48**
Meters	**0.304 8**
Kilometers	**0.000 304 8**

* All boldface figures are exact; the others generally are given to seven significant figures.

In using conversion factors, it is possible to perform division as well as the multiplication process shown here. Division may be particularly advantageous where more than the significant figures published here are required. Division may be performed in lieu of multiplication by using the reciprocal of any indicated multiplier as divisor. For example, to convert from centimeters to inches by division, refer to the table headed "To Convert from *Inches*" and use the factor listed at "centimeters" (*2.54*) as divisor.

To Convert from **Yards** To	Multiply by
Inches	**36**
Feet	**3**
Miles	0.000 568 18
Centimeters	**91.44**
Meters	**0.914 4**

To Convert from **Miles** To	Multiply by
Inches	**63 360**
Feet	**5 280**
Yards	**1 760**
Centimeters	**160 934.4**
Meters	**1 609.344**
Kilometers	**1.609 344**

Units of Mass

To Convert from Grams

To	Multiply by
Grains	15.432 36
Avoirdupois Drams	0.564 383 4
Avoirdupois Ounces	0.035 273 96
Troy Ounces	0.032 150 75
Troy Pounds	0.002 679 23
Avoirdupois Pounds	0.002 204 62
Milligrams	**1 000**
Kilograms	**0.001**

To Convert from Metric Tons

To	Multiply by
Avoirdupois Pounds	2 204.623
Short Hundredweights	22.046 23
Short Tons	1.102 311 3
Long Tons	0.984 206 5
Kilograms	**1 000**

To Convert from Kilograms

To	Multiply by
Grains	15 432.36
Avoirdupois Drams	564.383 4
Avoirdupois Ounces	35.273 96
Troy Ounces	32.150 75
Troy Pounds	2.679 229
Avoirdupois Pounds	2.204 623
Grams	**1 000**
Short Hundredweights	0.022 046 23
Short Tons	0.001 102 31
Long Tons	0.000 984 2
Metric Tons	**0.001**

To Convert from Grains

To	Multiply by
Avoirdupois Drams	0.036 571 43
Avoirdupois Ounces	0.002 285 71
Troy Ounces	0.002 083 33
Troy Pounds	0.000 173 61
Avoirdupois Pounds	0.000 142 86
Milligrams	**64.798 91**
Grams	**0.064 798 91**
Kilograms	**0.000 064 798 91**

To Convert from Avoirdupois Pounds

To	Multiply by
Grains	**7 000**
Avoirdupois Drams	**256**
Avoirdupois Ounces	**16**
Troy Ounces	14.583 33
Troy Pounds	1.215 278
Grams	**453.592 37**
Kilograms	**0.453 592 37**
Short Hundredweights	**0.01**
Short Tons	**0.000 5**
Long Tons	0.000 446 428 6
Metric Tons	**0.000 453 592 37**

To Convert from Avoirdupois Ounces

To	Multiply by
Grains	**437.5**
Avoirdupois Drams	**16**
Troy Ounces	0.911 458 3
Troy Pounds	0.075 954 86
Avoirdupois Pounds	**0.062 5**
Grams	**28.349 523 125**
Kilograms	**0.028 349 523 125**

To Convert from Short Hundredweights

To	Multiply by
Avoirdupois Pounds	**100**
Short Tons	**0.05**
Long Tons	0.044 642 86
Kilograms	**45.359 237**
Metric Tons	**0.045 359 237**

To Convert from **Short Tons** To	Multiply by
Avoirdupois Pounds	**2 000**
Short Hundredweights	**20**
Long Tons	0.892 857 1
Kilograms	**907.184 74**
Metric Tons	**0.907 184 74**

To Convert from **Long Tons** To	Multiply by
Avoirdupois Ounces	**35 840**
Avoirdupois Pounds	**2 240**
Short Hundredweights	**22.4**
Short Tons	**1.12**
Kilograms	**1 016.046 908 8**
Metric Tons	**1.016 046 908 8**

To Convert from **Troy Ounces** To	Multiply by
Grains	**480**
Avoirdupois Drams	17.554 29
Avoirdupois Ounces	1.097 143
Troy Pounds	0.083 333 3
Avoirdupois Pounds	0.068 571 43
Grams	**31.103 476 8**

To Convert from **Troy Pounds** To	Multiply by
Grains	**5 760**
Avoirdupois Drams	210.651 4
Avoirdupois Ounces	13.165 71
Troy Ounces	**12**
Avoirdupois Pounds	0.822 857 1
Grams	**373.241 721 6**

Units of Capacity, or Volume, Liquid Measure

To Convert from **Milliliters** To	Multiply by
Minims	16.230 73
Liquid Ounces	0.033 814 02
Gills	0.008 453 5
Liquid Pints	0.002 113 4
Liquid Quarts	0.001 056 7
Gallons	0.000 264 17
Cubic Inches	0.061 023 74
Liters	**0.001**

To Convert from **Liters** To	Multiply by
Liquid Ounces	33.814 02
Gills	8.453 506
Liquid Pints	2.113 376
Liquid Quarts	1.056 688
Gallons	0.264 172 05
Cubic Inches	61.023 74
Cubic Feet	0.035 314 67
Milliliters	**1 000**
Cubic Meters	**0.001**
Cubic Yards	0.001 307 95

To Convert from **Cubic Meters** To	Multiply by
Gallons	264.172 05
Cubic Inches	61 023.74
Cubic Feet	35.314 67
Liters	**1 000**
Cubic Yards	1.307 950 6

To Convert from **Minims** To	Multiply by
Liquid Ounces	0.002 083 33
Gills	0.000 520 83
Cubic Inches	0.003 759 77
Milliliters	0.061 611 52

To Convert from
Gills

To	Multiply by
Minims	**1 920**
Liquid Ounces	**4**
Liquid Pints	**0.25**
Liquid Quarts	**0.125**
Gallons	**0.031 25**
Cubic Inches	**7.218 75**
Cubic Feet	0.004 177 517
Milliliters	**118.294 118 25**
Liters	**0.118 294 118 25**

To Convert from
Liquid Ounces

To	Multiply by
Minims	**480**
Gills	**0.25**
Liquid Pints	**0.062 5**
Liquid Quarts	**0.031 25**
Gallons	**0.007 812 5**
Cubic Inches	1.804 687 5
Cubic Feet	0.001 044 38
Milliliters	29.573 53
Liters	0.029 573 53

To Convert from
Cubic Inches

To	Multiply by
Minims	265.974 0
Liquid Ounces	0.554 112 6
Gills	0.138 528 1
Liquid Pints	0.034 632 03
Liquid Quarts	0.017 316 02
Gallons	0.004 329 0
Cubic Feet	0.000 578 7
Milliliters	**16.387 064**
Liters	**0.016 387 064**
Cubic Meters	**0.000 016 387 064**
Cubic Yards	0.000 021 43

To Convert from
Liquid Pints

To	Multiply by
Minims	**7 680**
Liquid Ounces	**16**
Gills	**4**
Liquid Quarts	**0.5**
Gallons	**0.125**
Cubic Inches	**28.875**
Cubic Feet	0.016 710 07
Milliliters	**473.176 473**
Liters	**0.473 176 473**

To Convert from
Cubic Feet

To	Multiply by
Liquid Ounces	957.506 5
Gills	239.376 6
Liquid Pints	59.844 16
Liquid Quarts	29.922 08
Gallons	7.480 519
Cubic Inches	**1 728**
Liters	**28.316 846 592**
Cubic Meters	**0.028 316 846 592**
Cubic Yards	0.037 037 04

To Convert from
Cubic Yards

To	Multiply by
Gallons	201.974 0
Cubic Inches	**46 656**
Cubic Feet	**27**
Liters	**764.554 857 984**
Cubic Meters	**0.764 554 857 984**

To	To Convert from **Liquid Quarts** Multiply by
Minims	**15 360**
Liquid Ounces	**32**
Gills	**8**
Liquid Pints	**2**
Gallons	**0.25**
Cubic Inches	**57.75**
Cubic Feet	0.033 420 14
Milliliters	**946.352 946**
Liters	**0.946 352 946**

To	To Convert from **Gallons** Multiply by
Minims	**61 440**
Liquid Ounces	**128**
Gills	**32**
Liquid Pints	**8**
Liquid Quarts	**4**
Cubic Inches	**231**
Cubic Feet	0.133 680 6
Milliliters	**3 785.411 784**
Liters	**3.785 411 784**
Cubic Meters	**0.003 785 411 784**
Cubic Yards	0.004 951 13

Units of Capacity, or Volume, Dry Measure

To	To Convert from **Liters** Multiply by
Dry Pints	1.816 166
Dry Quarts	0.908 082 98
Pecks	0.113 510 4
Bushels	0.028 377 59
Dekaliters	**0.1**

To	To Convert from **Dekaliters** Multiply by
Dry Pints	18.161 66
Dry Quarts	9.080 829 8
Pecks	1.135 104
Bushels	0.283 775 9
Cubic Inches	610.237 4
Cubic Feet	0.353 146 7
Liters	**10**

To	To Convert from **Cubic Meters** Multiply by
Pecks	113.510 4
Bushels	28.377 59

To	To Convert from **Dry Pints** Multiply by
Dry Quarts	**0.5**
Pecks	**0.062 5**
Bushels	**0.015 625**
Cubic Inches	**33.600 312 5**
Cubic Feet	0.019 444 63
Liters	0.550 610 47
Dekaliters	0.055 061 05

To Convert from **Dry Quarts** To	Multiply by
Dry Pints	**2**
Pecks	**0.125**
Bushels	**0.031 25**
Cubic Inches	**67.200 625**
Cubic Feet	0.038 889 25
Liters	1.101 221
Dekaliters	0.110 122 1

To Convert from **Pecks** To	Multiply by
Dry Pints	**16**
Dry Quarts	**8**
Bushels	**0.25**
Cubic Inches	**537.605**
Cubic Feet	0.311 114
Liters	8.809 767 5
Dekaliters	0.880 976 75
Cubic Meters	0.008 809 77
Cubic Yards	0.011 522 74

To Convert from **Bushels** To	Multiply by
Dry Pints	**64**
Dry Quarts	**32**
Pecks	**4**
Cubic Inches	**2 150.42**
Cubic Feet	1.244 456
Liters	35.239 07
Dekaliters	3.523 907
Cubic Meters	0.035 239 07
Cubic Yards	0.046 090 96

To Convert from **Cubic Inches** To	Multiply by
Dry Pints	0.029 761 6
Dry Quarts	0.014 880 8
Pecks	0.001 860 10
Bushels	0.000 465 025

To Convert from **Cubic Feet** To	Multiply by
Dry Pints	51.428 09
Dry Quarts	25.714 05
Pecks	3.214 256
Bushels	0.803 563 95

To Convert from **Cubic Yards** To	Multiply by
Pecks	86.784 91
Bushels	21.696 227

Units of Area

To Convert from Square Centimeters

To	Multiply by
Square Inches	0.155 000 3
Square Feet	0.001 076 39
Square Yards	0.000 119 599
Square Meters	**0.000 1**

To Convert from Hectares

To	Multiply by
Square Feet	107 639.1
Square Yards	11 959.90
Acres	2.471 054
Square Miles	0.003 861 02
Square Meters	**10 000**

To Convert from Square Feet

To	Multiply by
Square Inches	**144**
Square Yards	0.111 111 1
Acres	0.000 022 957
Square Centimeters	**929.030 4**
Square Meters	**0.092 903 04**

To Convert from Acres

To	Multiply by
Square Feet	**43 560**
Square Yards	**4 840**
Square Miles	**0.001 562 5**
Square Meters	**4 046.856 422 4**
Hectares	**0.404 685 642 24**

To Convert from Square Meters

To	Multiply by
Square Inches	1 550.003
Square Feet	10.763 91
Square Yards	1.195 990
Acres	0.000 247 105
Square Centimeters	**10 000**
Hectares	**0.000 1**

To Convert from Square Inches

To	Multiply by
Square Feet	0.006 944 44
Square Yards	0.000 771 605
Square Centimeters	**6.451 6**
Square Meters	**0.000 645 16**

To Convert from Square Yards

To	Multiply by
Square Inches	**1 296**
Square Feet	**9**
Acres	0.000 206 611 6
Square Miles	0.000 000 322 830 6
Square Centimeters	**8 361.273 6**
Square Meters	**0.836 127 36**
Hectares	**0.000 083 612 736**

To Convert from Square Miles

To	Multiply by
Square Feet	**27 878 400**
Square Yards	**3 097 600**
Acres	**640**
Square Meters	**2 589 988.110 336**
Hectares	**258.998 811 033 6**

Units of Weight and Measure
International (Metric) and U.S. Customary

With

"TWO-WAY" CONVERSION TABLES

PART III

CONTENTS

International (Metric) and U.S. Customary

In the construction of the tables in this publication, when the fundamental relation of the units furnished directly a reduction factor for use in determining the multiples of the units, this factor was used in its fundamental form, as, for example, that 1 meter=39.37 inches. Reduction factors which it was necessary to obtain, however, by multiplication, division, powers, or roots, etc., of the fundamental relations were usually carried out to a greater degree of accuracy than that to which it is usually possible to make measurements, for convenience in computing the multiples to the accuracy desired.

When the tables do not give the equivalent of any desired quantity directly and completely, the equivalent can usually be obtained, without the necessity of making a multiplication of these reduction factors, by using quantities from several tables, making a shift of decimal points, if necessary, and merely adding the results.

The supplementary metric units are formed by combining the words "meter", "gram", and "liter" with the six numerical prefixes, as in the following table:

Prefixes		Meaning			Units
milli-	=	*one-thousandth*	$\frac{1}{1000}$	.001	"meter" *for length.*
centi-	=	*one-hundredth*	$\frac{1}{100}$	.01	
deci-	=	*one-tenth*	$\frac{1}{10}$	.1	"gram" *for weight or mass.*
Unit	=	*one*		1	
deka-	=	*ten*	$\frac{10}{1}$	10	"liter" *for capacity.*
hecto-	=	*one hundred*	$\frac{100}{1}$	100	
kilo-	=	*one thousand*	$\frac{1000}{1}$	1000	

All lengths, areas, and cubic measures in the following tables are derived from the international meter, the basic relation between units of the customary and the metric systems being:

1 meter=39.37 inches,

contained in the law of July 28, 1866, and set forth in the Mendenhall Order of April 5, 1893, namely:

$$\frac{1 \text{ yard}}{1 \text{ meter}} = \frac{3600}{3937}.$$

In its original conception the meter was the fundamental unit of the metric system, and all units of length and capacity were to be derived directly from the meter which was intended to be equal to one ten-millionth of the earth's quadrant. Furthermore, it was originally planned that the unit of mass, the kilogram, should be identical with the mass of a cubic decimeter of water at its maximum density. At present, however, the units of length and mass are defined independently of these conceptions.

A supplementary definition of the meter in terms of the wave length of light was adopted provisionally by the Seventh General (International) Conference on Weights and Measures, in 1927. According to this definition, the relation for red cadmium light waves under specified conditions of temperature, pressure, and humidity is

$$1 \text{ meter} = 1\,553\,164.13 \text{ wave lengths.}$$

This corresponds to a wave length (cadmium red) of

$$6\,438.469\,6 \times 10^{-7} \text{ millimeters.}$$

The kilogram is the fundamental unit on which are based all metric standards of mass. The liter is a secondary or derived unit of capacity or volume. The liter is larger by about 27 parts per million than the cube of the tenth of the meter, i. e., the cubic decimeter—that is, 1 liter=1.000 027 cubic decimeters.

The conversion tables in this publication which involve the relative length of the yard and meter are based upon the relation:

$$1 \text{ meter} = 39.37 \text{ inches,}$$

contained in the act of Congress of 1866. From this relation it follows that 1 inch= 25.400 05 millimeters (nearly).

All capacities are based on the equivalent 1 liter equals 1.000 027 cubic decimeters. The decimeter is equal to 3.937 inches in accordance with the legal equivalent of the meter given above. The gallon referred to in the tables is the United States gallon of 231 cubic inches. The bushel is the United States bushel of 2 150.42 cubic inches. These units must not be confused with the British units of the same name which differ from those used in the United States. The British gallon (277.420 cubic inches) is approximately 20 percent larger, and the British bushel (2 219.36 cubic inches) is 3 percent larger than the corresponding units used in this country.

All weights are derived from the International Kilogram, as authorized in the Mendenhall Order of April 5, 1893. The relation used is 1 avoirdupois pound= 453.592 427 7 grams.

In recent years engineering and industrial interests the world over have urged the adoption of the simpler relation, 1 inch=25.4 millimeters exactly, which differs from the preceding value by only 2 parts in a million.

DEFINITIONS OF UNITS

1. LENGTH

Fundamental Units

A meter (m) is a unit of length equal to the distance between the defining lines on the International Prototype Meter when this standard is at the temperature of melting ice (0° C).

A yard (yd) is a unit of length equal to $\frac{3600}{3937}$ of a meter.

Multiples and Submultiples

1 kilometer (km)=1 000 meters.
1 hectometer (hm)=100 meters.
1 dekameter (dkm)=10 meters.
1 decimeter (dm)=0.1 meter.
1 centimeter (cm)=0.01 meter.
1 millimeter (mm)=0.001 meter.
1 micron (μ)=0.000 001 meter=0.001 millimeter.
1 millimicron (mμ)=0.000 000 001 meter=0.001 micron.
1 angstrom (A) {=0.000 000 1 millimeter. = .000 1 micron. = .1 millimicron.
1 statute mile {=8 furlongs=320 rods. =1 760 yards=5 280 feet.
1 furlong=⅛ mile=40 rods=220 yards=660 feet.
1 rod=5½ yards=16½ feet=25 links.
1 foot=⅓ yard=12 inches.
1 hand=4 inches.
1 inch=1/36 yard=1/12 foot.
1 line (button)=1/40 inch.
1 point (printers)=1/72 inch.
1 mil=1/1000 inch.
1 chain (Gunter's)=4 rods=22 yards=66 feet=100 links.
1 link (Gunter's)=1/100 chain=7.92 inches.
1 U. S. nautical mile
1 sea mile } =1 853.248 meters=6 080.20 feet.
1 geographical mile
1 international nautical mile=1 852 meters=6 076.10 feet.
1 fathom=6 feet=8 spans.
1 span=⅛ fathom=9 inches.

2. AREA

Fundamental Units

A square meter (m^2) is a unit of area equal to the area of a square the sides of which are 1 meter.

A square yard (sq yd) is a unit of area equal to the area of a square the sides of which are 1 yard.

Multiples and Submultiples

1 square kilometer (km^2)=1 000 000 square meters.
1 hectare (ha), or square hectometer (hm^2)=10 000 square meters.

1 are (a), or square dekameter (dkm^2)=100 square meters.
1 centare (ca)=1 square meter.
1 square decimeter (dm^2)=0.01 square meter.
1 square centimeter (cm^2)=0.000 1 square meter.
1 square millimeter (mm^2)=0.000 001 square meter.
1 square mile (sq mi) {=640 acres=102 400 square rods. =3 097 600 square yards=27 878 400 square feet.}
1 acre (acre)=10 square chains {=160 square rods=4 840 square yards. =43 560 square feet.}
1 square chain (sq ch) {=16 square rods=484 square yards=4 356 square feet. =10 000 square links.}
1 square link (sq li) {=0.000 1 square chain=0.048 4 square yard. =0.435 6 square foot=62.726 4 square inches.}
1 square rod (sq rd)=30.25 square yards=272.25 square feet=625 square links.
1 square foot (sq ft)=1/9 square yard=144 square inches.
1 square inch (sq in.)=1/1296 square yard=1/144 square foot.

3. VOLUME

Fundamental Units

A cubic meter (m^3) is a unit of volume equal to a cube the edges of which are 1 meter.

A cubic yard (cu yd) is a unit of volume equal to a cube the edges of which are 1 yard.

Multiples and Submultiples

1 cubic kilometer (km^3)=1 000 000 000 cubic meters.
1 cubic hectometer (hm^3)=1 000 000 cubic meters.
1 cubic dekameter (dkm^3)=1 000 cubic meters.
1 stere (s)=1 cubic meter.
1 cubic decimeter (dm^3)=0.001 cubic meter.
1 cubic centimeter (cm^3)=0.000 001 cubic meter=0.001 cubic decimeter.
1 cubic millimeter (mm^3)=0.000 000 001 cubic meter=0.001 cubic centimeter.
1 cubic foot (cu ft)=1/27 cubic yard.
1 cubic inch (cu in.)=1/466 56 cubic yard=1/1728 cubic foot.
1 board foot (fbm)=144 cubic inches=1/12 cubic foot.
1 cord (cd)=128 cubic feet.

4. CAPACITY

Fundamental Units

A liter (liter) is a unit of capacity equal to the volume occuped by the mass of 1 kilogram of pure water at its maximum density (at a temperature of 4° C, practically) and under the standard atmospheric pressure (of 760 mm). It is equivalent in volume to 1.000 027 cubic decimeters.

A gallon (gal) is a unit of capacity equal to the volume of 231 cubic inches. It is used for the measurement of liquid commodities only.

A bushel (bu) is a unit of capacity equal to the volume of 2 150.42 cubic inches. It is used in the measurement of dry commodities only.[7]

Multiples and Submultiples

1 hectoliter (hl)=100 liters.
1 dekaliter (dkl)=10 liters.
1 deciliter (dl)=0.1 liter.
1 centiliter (cl)=0.01 liter.

[7] This is the so-called stricken or struck bushel. A heaped bushel for apples of 2 747.715 cubic inches was established by the U. S. Court of Customs Appeals on Feb. 15, 1912, in United States v. Weber (no. 757). A heaped bushel, equivalent to 1¼ stricken bushels, is also recognized.

1 milliliter (ml)=0.001 liter=1.000 027 cubic centimeters.
1 liquid quart (liq qt)=¼ gallon=57.75 cubic inches.
1 liquid pint (liq pt)=⅛ gallon=½ liquid quart=28.875 cubic inches.
1 gill (gi)=1/32 gallon=¼ liquid pint=7.218 75 cubic inches.
1 fluid ounce (fl oz)=1/128 gallon=1/16 liquid pint.
1 fluid dram (fl dr)=⅛ fluid ounce=1/128 liquid pint.
1 minim (min *or* ♏)=1/60 fluid dram=1/480 fluid ounce.
1 peck (pk)=¼ bushel=537.605 cubic inches.
1 dry quart (dry qt)=1/32 bushel=⅛ peck=67.200 625 cubic inches.
1 dry pint (dry pt)=1/64 bushel=½ dry quart=33.600 312 5 cubic inches.
1 barrel, for fruits, vegetables, and other dry commodities, other than cranberries, =7 056 cubic inches=105 dry quarts.
1 barrel for cranberries=5 826 cubic inches.

5. MASS

Fundamental Units

A kilogram (kg) is a unit of mass equal to the mass of the International Prototype Kilogram.

A gram (g) is a unit of mass equal to one-thousandth of the mass of the International Prototype Kilogram.

An avoirdupois pound (lb avdp) is a unit of mass equal to 0.453 592 427 7 kilogram.

A troy pound (lb t) is a unit of mass equal to $\frac{5760}{7000}$ of that of the avoirdupois pound.

Multiples and Submultiples

1 metric ton (t)=1 000 kilograms.
1 hectogram (hg)=100 grams.
1 dekagram (dkg)=10 grams.
1 decigram (dg)=0.1 gram.
1 centigram (cg)=0.01 gram.
1 milligram (mg)=0.001 gram.
1 metric carat (c)=200 milligrams=0.2 gram.
1 avoirdupois ounce (oz avdp)=1/16 avoirdupois pound=437.5 grains.
1 avoirdupois dram (dr avdp)=1/256 avoirdupois pound=1/16 avoirdupois ounce.
1 grain (grain)=1/7000 avoirdupois pound=1/5760 troy pound.
1 apothecaries' pound (lb ap)=1 troy pound=$\frac{5760}{7000}$ avoirdupois pound.
1 apothecaries' or troy ounce (oz ap *or* ℥, oz t)=1/12 troy pound=$\frac{480}{7000}$ avoirdupois pound=480 grains.
1 apothecaries' dram (dr ap *or* ʒ)=1/96 apothecaries' pound=⅛ apothecaries' ounce=60 grains.
1 pennyweight (dwt)=1/20 troy ounce=24 grains.
1 apothecaries' scruple (s ap *or* ℈)=⅓ apothecaries' dram=20 grains.
1 short hundredweight (sh cwt)=100 avoirdupois pounds.
1 long hundredweight (l cwt)=112 avoirdupois pounds.
1 short ton (sh tn)=2 000 avoirdupois pounds.
1 long ton (l tn)=2 240 avoirdupois pounds.

STANDARDS OF MEASUREMENT

Units of measurement should be distinguished from standards of measurement. Units of length are fixed distances, independent of any other consideration, while length standards are affected by the expansion and contraction resulting from changes of temperature of the material of which the standard is composed. It is therefore necessary to fix upon some temperature at which the distance between the defining lines or end surfaces of the standards shall be equal to the unit. The same is true of standards of capacity, which at some definite temperature contain a given number of units of volume.

The recommended standard temperature for commercial and industrial length standards is 20° C (68° F). Some metric standards, especially those made in Europe until recently, are intended to be correct at 0° C. In the past some length standards graduated in the customary units have been made to be correct at 62° F (16.67° C).

For measurements of high precision it is also necessry to specify the manner of support of the standards, whether at certain points only or throughout their entire length, and in the case of tapes it is also necessary to give the tension applied to the tape when in use.

In the United States the capacity standards, both metric and customary, are made to hold the specified volumes at 4° C. Standards of capacity are usually made of brass and the capacity at any other temperature may be computed by the use of the coefficient of cubical expansion of that material usually assumed to be 0.000 054 per degree centigrade. In the purchase and sale of liquids a more important consideration than the temperature of the measures is the temperature of the liquid when measured, for the reason that the large coefficient of expansion of many liquids makes the actual mass of a given volume delivered vary considerably with temperature.

While the temperature of a weight does not affect its mass, it is nevertheless important that when two weights are compared in air they both be at the same temperature as the air. If there is a difference between the temperature of the air and the weights, convection currents will be set up and the readings of the balance will be thereby affected. Also, since weights are buoyed up by the surrounding air by amounts dependent upon their volumes, it is desirable that the weights of any set be of the same material. If two weights of the same density balance in air of a certain density they will balance in vacuo or in air of a different density.

Brass is the material most widely used for standard weights, although platinum and aluminum are quite commonly used for weights of 1 gram or less. In the absence of any knowledge as to the actual density of weights, those made of brass are assumed to have a density of 8.4 grams per cm^3 at 0° C, while those of platinum and of aluminum are assumed to have densities of 21.5 and 2.7 grams per cm^3, respectively.

SPELLING AND ABBREVIATION OF UNITS

The spelling of the names of units as adopted by the National Bureau of Standards is that given in the list below. The spelling of the metric units is in accordance with that given in the law of July 28, 1866, legalizing the metric system in the United States.

Following the name of each unit in the list below is given the abbreviation which the Bureau has adopted. Attention is particularly called to the following principles:

1. The period is omitted after all abbreviations of units, except where the abbreviation forms an English word.

2. The exponents "^{2}" and "^{3}" are used to signify "square" and "cubic", respectively, in the case of the metric units, instead of the abbreviations "sq" or "cu". In conformity with this principle the abbreviation for cubic centimeter is "cm^3" (instead of "cc" or "c cm"). The term "cubic centimeter", as used in chemical work, is, in fact, a misnomer, since the unit actually used is the "milliliter", of which the correct abbreviation is "ml".

3. The use of the same abbreviation for both singular and plural is recommended. This practice is already established in expressing metric units and is in accordance with the spirit and chief purpose of abbreviations.

4. It is also suggested that, unless all the text is printed in capital letters, only small letters be used for abbreviations, except in such case as A for angstrom, etc., where the use of capital letters is general.

LIST OF THE MOST COMMON UNITS OF WEIGHT AND MEASURE AND THEIR ABBREVIATIONS

Unit	Abbreviation	Unit	Abbreviation	Unit	Abbreviation
acre	acre	dram, apothacaries'	dr ap *or* ʒ	ounce, avoirdupois	oz avdp
angstrom	A	dram, avoirdupois	dr avdp	ounce, fluid	fl oz
are	a	dram, fluid	fl dr	ounce, troy	oz t
avoirdupois	avdp	fathom	fath	peck	pk
barrel	bbl	foot	ft	pennyweight	dwt
board foot	fbm	furlong	fur.	pint	pt
bushel	bu	gallon	gal	pound	lb
carat	c	grain	grain	pound, apothecaries'	lb ap
centare	ca	gram	g	pound, avoirdupois	lb avdp
centigram	cg	hectare	ha	pound, troy	lb t
centiliter	cl	hectogram	hg	quart	qt
centimeter	cm	hectoliter	hl	rod	rd
chain	ch	hectometer	hm	scruple, apothecaries	s ap *or* ℈
cubic centimeter	cm^3	hogshead	hhd	square centimeter	cm^2
cubic decimeter	dm^3	hundredweight	cwt	square chain	sq ch
cubic dekameter	dkm^3	inch	in.	square decimeter	dm^2
cubic foot	cu ft	kilogram	kg	square dekameter	dkm^2
cubic hectometer	hm^3	kiloliter	kl	square foot	sq ft
cubic inch	cu in.	kilometer	km	square hectometer	hm^2
cubic kilometer	km^3	link	li	square inch	sq in.
cubic meter	m^3	liquid	liq	square kilometer	km^2
cubic mile	cu mi	liter	liter	square link	sq li
cubic millimeter	mm^3	meter	m	square meter	m^2
cubic yard	cu yd	metric ton	t	square mile	sq mi
decigram	dg	micron	μ	square millimeter	mm^2
deciliter	dl	mile	mi	square rod	sq rd
decimeter	dm	milligram	mg	square yard	sq yd
decistere	ds	milliliter	ml	stere	s
dekagram	dkg	millimeter	mm	ton	tn
dekaliter	dkl	millimicron	$m\mu$	ton, metric	t
dekameter	dkm	minim	min *or* ♏	troy	t
dekastere	dks	ounce	oz	yard	yd
dram	dr	ounce, apothecaries'	oz ap *or* ℥		

TABLES OF INTERRELATION

1. UNITS OF

Units	Inches	Links	Feet	Yards	Rods
1 inch =	1	0.126 263	0.083 333 3	0.027 777 8	0.005 050 51
1 link =	7.92	1	0.66	0.22	0.04
1 foot =	12	1.515 152	1	0.333 333	0.060 606 1
1 yard =	36	4.545 45	3	1	0.181 818
1 rod =	198	25	16.5	5.5	1
1 chain =	792	100	66	22	4
1 mile =	63 360	8000	5280	1760	320
1 centimeter=	0.3937	0.049 709 60	0.032 808 33	0.010 936 111	0.001 988 384
1 meter =	39.37	4.970 960	3.280 833	1.093 611 1	0.198 838 4

2. UNITS OF

Units	Square inches	Square links	Square feet	Square yards	Square rods	Square chains
1 square inch =	1	0.015 942 3	0.006 944 44	0.000 771 605	0.000 025 507 6	0.000 001 594 23
1 square link =	62.7264	1	0.4356	0.0484	0.0016	0.0001
1 square foot =	144	2.295 684	1	0.111 111 1	0.003 673 09	0.000 229 568
1 square yard =	1296	20.6612	9	1	0.033 057 85	0.002 066 12
1 square rod =	39 204	625	272.25	30.25	1	0.0625
1 square chain =	627 264	10 000	4356	484	16	1
1 acre =	6 272 640	100 000	43 560	4840	160	10
1 square mile =	4 014 489 600	64 000 000	27 878 400	3 097 600	102 400	6400
1 square centimeter=	0.154 999 7	0.002 471 04	0.001 076 387	0.000 119 598 5	0.000 003 953 67	0.000 000 247 104
1 square meter =	1549.9969	24.7104	10.763 87	1.195 985	0.039 536 7	0.002 471 04
1 hectare =	15 499 969	247 104	107 638.7	11 959.85	395.367	24.7104

3. UNITS OF

Units	Cubic inches	Cubic feet	Cubic yards
1 cubic inch =	1	0.000 578 704	0.000 021 433 47
1 cubic foot =	1728	1	0.037 037 0
1 cubic yard =	46 656	27	1
1 cubic centimeter=	0.061 023 38	0.000 035 314 45	0.000 001 307 94
1 cubic decimeter =	61.023 38	0.035 314 45	0.001 307 943
1 cubic meter =	61 023.38	35.314 45	1.307 942 8

4. UNITS OF CAPACITY

Units	Minims	Fluid drams	Fluid ounces	Gills	Liquid pints
1 minim =	1	0.016 666 7	0.002 083 33	0.000 520 833	0.000 130 208
1 fluid dram =	60	1	0.125	0.031 25	0.007 812 5
1 fluid ounce=	480	8	1	0.25	0.0625
1 gill =	1920	32	4	1	0.25
1 liquid pint =	7680	128	16	4	1
1 liquid quart=	15 360	256	32	8	2
1 gallon =	61 440	1024	128	32	8
1 milliliter =	16.2311	0.270 518	0.033 814 7	0.008 453 68	0.002 113 42
1 liter =	16 231.1	270.518	33.8147	8.453 68	2.113 42
1 cubic inch =	265.974	4.432 90	0.554 113	0.138 528	0.034 632 0

OF UNITS OF MEASUREMENT

LENGTH

Chains	Miles	Centimeters	Meters	Units
0.001 262 63	0.000 015 782 8	2.540 005	0.025 400 05	=1 inch
0.01	**0.000 125**	20.116 84	0.201 168 4	=1 link
0.015 151 5	0.000 189 393 9	30.480 06	0.304 800 6	=1 foot
0.045 454 5	0.000 568 182	91.440 18	0.914 401 8	=1 yard
0.25	**0.003 125**	502.9210	5.029 210	=1 rod
1	**0.0125**	2011.684	20.116 84	=1 chain
80	**1**	160 934.72	1609.3472	=1 mile
0.000 497 096 0	0.000 006 213 699	**1**	**0.01**	=1 centimeter
0.049 709 60	0.000 621 369 9	**100**	**1**	=1 meter

AREA

Acres	Square miles	Square centimeters	Square meters	Hectares	Units
0.000 000 159 423	0.000 000 000 249 1	6.451 626	0.000 645 162 6	0.000 000 064 516	=1 square inch
0.000 01	**0.000 000 015 625**	404.6873	0.040 468 73	0.000 004 046 87	=1 square link
0.000 022 956 8	0.000 000 035 870 1	929.0341	0.092 903 41	0.000 009 290 34	=1 square foot
0.000 206 612	0.000 000 322 831	8361.307	0.836 130 7	0.000 083 613 1	=1 square yard
0.006 25	**0.000 009 765 625**	252 929.5	25.292 95	0.002 529 295	=1 square rod
0.1	**0.000 156 25**	4 046 873	404.6873	0.040 468 7	=1 square chain
1	**0.001 562 5**	40 468 726	4046.873	0.404 687	=1 acre
640	**1**	25 899 984 703	2 589 998	258.9998	=1 square mile
0.000 000 024 710 4	0.000 000 000 038 610 06	**1**	**0.0001**	**0.000 000 01**	=1 square centimeter
0.000 247 104	0.000 000 386 100 6	**10 000**	**1**	**0.0001**	=1 square meter
2.471 04	0.003 861 006	**100 000 000**	**10 000**	**1**	=1 hectare

VOLUME

Cubic centimeters	Cubic decimeters	Cubic meters	Units
16.387 162	0.016 387 16	0.000 016 387 16	=1 cubic inch
28 317.016	28.317 016	0.028 317 016	=1 cubic foot
764 559.4	764.5594	0.764 559 4	=1 cubic yard
1	**0.001**	**0.000 001**	=1 cubic centimeter
1 000	**1**	**0.001**	=1 cubic decimeter
1 000 000	**1000**	**1**	=1 cubic meter

LIQUID MEASURE

Liquid quarts	Gallons	Milliliters	Liters	Cubic inches	Units
0.000 065 104	0.000 016 276	0.061 610 2	0.000 061 610 2	0.003 759 77	=1 minim
0.003 906 25	0.000 976 562	3.696 61	0.003 696 61	0.225 586	=1 fluid dram
0.031 25	**0.007 812 5**	29.5729	0.029 572 9	1.804 69	=1 fluid ounce
0.125	**0.031 25**	118.292	0.118 292	**7.218 75**	=1 gill
0.5	**0.125**	473.167	0.473 167	**28.875**	=1 liquid pint
1	**0.25**	946.333	0.946 333	**57.75**	=1 liquid quart
4	**1**	3785.332	3.785 332	**231**	=1 gallon
0.001 056 71	0.000 264 178	**1**	**0.001**	0.061 025 0	=1 milliliter
1.056 71	0.264 178	**1000**	**1.**	61.0250	=1 liter
0.017 316 0	0.004 329 00	16.3867	0.016 386 7	**1**	=1 cubic inch

5. UNITS OF CAPACITY

Units	Dry pints	Dry quarts	Pecks	Bushels
1 dry pint =	**1**	**0.5**	**0.0625**	**0.015 625**
1 dry quart =	**2**	**1**	**0.125**	**0.031 25**
1 peck =	**16**	**8**	**1**	**0.25**
1 bushel =	**64**	**32**	**4**	**1**
1 liter =	1.816 20	0.908 102	0.113 513	0.028 378
1 dekaliter =	18.1620	9.081 02	1.135 13	0.283 78
1 cubic inch =	0.029 761 6	0.014 880 8	0.001 860 10	0.000 465 025

6. UNITS OF MASS LESS

Units *	Grains	Apothecaries' scruples	Pennyweights	Avoirdupois drams	Apothecaries' drams	Avoirdupois ounces
1 grain =	**1**	**0.05**	0.041 666 67	0.036 571 43	0.016 666 7	0.002 285 71
1 apoth. scruple =	**20**	**1**	0.833 333 3	0.731 428 6	0.333 333	0.045 714 3
1 pennyweight =	**24**	**1.2**	**1**	0.877 714 3	**0.4**	0.054 857 1
1 avoir. dram =	**27.343 75**	**1.367 187 5**	1.139 323	**1**	0.455 729 2	**0.0625**
1 apoth. dram =	**60**	**3**	**2.5**	2.194 286	**1**	0.137 142 9
1 avoir. ounce =	**437.5**	**21.875**	18.229 17	**16**	7.291 67	**1**
1 apoth. or troy ounce =	**480**	**24**	**20**	17.554 28	**8**	1.097 142 9
1 apoth. or troy pound =	**5760**	**288**	**240**	210.6514	**96**	13.165 714
1 avoir. pound =	**7000**	**350**	291.6667	**256**	116.6667	**16**
1 milligram =	0.015 432 356	0.000 771 618	0.000 643 014 8	0.000 564 383 3	0.000 257 205 9	0.000 035 273 96
1 gram =	15.432 356	0.771 618	0.643 014 85	0.564 383 3	0.257 205 9	0.035 273 96
1 kilogram =	15 432.356	771.6178	643.014 85	564.383 32	257.205 94	35.273 96

7. UNITS OF MASS

Units	Avoirdupois ounces	Avoirdupois pounds	Short hundred-weights	Short tons
1 avoirdupois ounce =	**1**	**0.0625**	**0.000 625**	**0.000 031 25**
1 avoirdupois pound =	**16**	**1**	**0.01**	**0.0005**
1 short hundredweight =	**1600**	**100**	**1**	**0.05**
1 short ton =	**32 000**	**2000**	**20**	**1**
1 long ton =	**35 840**	**2240**	**22.4**	**1.12**
1 kilogram =	35.273 957	2.204 622 34	0.022 046 223	0.001 102 311 2
1 metric ton =	35 273.957	2204.622 34	22.046 223	1.102 311 2

* "Avoir." is now abbreviated "avdp".

DRY MEASURE

Liters	Dekaliters	Cubic inches	Units
0.550 599	0.055 060	**33.600 312 5**	=1 dry pint
1.101 198	0.110 120	**67.200 625**	=1 dry quart
8.809 58	0.880 958	**537.605**	=1 peck
35.2383	3.523 83	**2150.42**	=1 bushel
1	**0.1**	61.0250	=1 liter
10	**1**	610.250	=1 dekaliter
0.016 386 7	0.001 638 67	**1**	=1 cubic inch

THAN POUNDS AND KILOGRAMS

Apothecaries' or troy ounces	Apothecaries' or troy pounds	Avoirdupois pounds	Milligrams	Grams	Kilograms	Units
0.002 083 33	0.000 173 611 1	0.000 142 857 1	64.798 918	0.064 798 918	0.000 064 798 9	=1 grain
0.041 666 7	0.003 472 222	0.002 857 143	1295.9784	1.295 978 4	0.001 295 978	=1 apoth. scruple
0.05	0.004 166 667	0.003 428 571	1555.1740	1.555 174 0	0.001 555 174	=1 pennyweight
0.056 966 146	0.004 747 178 8	**0.003 906 25**	1771.8454	1.771 845 4	0.001 771 845	=1 avoir. dram
0.125	0.010 416 667	0.008 571 429	3887.9351	3.887 935 1	0.003 887 935	=1 apoth. dram
0.911 458 3	0.075 954 861	**0.0625**	28 349.527	28.349 527	0.028 349 53	=1 avoir. ounce
1	0.083 333 33	0.068 571 43	31 103.481	31.103 481	0.031 103 48	=1 apoth. or troy ounce
12	**1**	0.822 857 1	373 241.77	373.241 77	0.373 241 77	=1 apoth. or troy pound
14.583 333	1.215 277 8	**1**	**453 592.4277**	**453.592 4277**	**0.453 592 427 7**	=1 avoir. pound
0.000 032 150 74	0.000 002 679 23	0.000 002 204 62	**1**	**0.001**	**0.000 001**	=1 milligram
0.032 150 74	0.002 679 23	0.002 204 62	**1000**	**1**	**0.001**	=1 gram
32.150 742	2.679 228 5	2.204 622 341	**1 000 000**	**1000**	**1**	=1 kilogram

GREATER THAN AVOIRDUPOIS OUNCES

Long tons	Kilograms	Metric tons	Units
0.000 027 901 79	0.028 349 53	0.000 028 349 53	=1 avoirdupois ounce
0.000 446 428 6	**0.453 592 427 7**	0.000 453 592 43	=1 avoirdupois pound
0.044 642 86	45.359 243	0.045 359 243	=1 short hundredweight
0.892 857 1	907.184 86	0.907 184 86	=1 short ton
1	1016.047 04	1.016 047 04	=1 long ton
0.000 984 206 4	**1**	**0.001**	=1 kilogram
0.984 206 40	**1000**	**1**	=1 metric ton

COMPARISON OF METRIC AND CUSTOMARY UNITS FROM 1 TO 9

1. LENGTH

Inches (in.) = Millimeters (mm)	Feet (ft) = Meters (m)	Yards (yd) = Meters (m)	Rods (rd) = Meters (m)	U.S. miles (mi) = Kilometers (km)
1= 25.4001	1=0.304 801	1=0.914 402	1= 5.029 21	1= 1.609 347
2= 50.8001	2=0.609 601	2=1.828 804	2=10.058 42	2= 3.218 694
3= 76.2002	3=0.914 402	3=2.743 205	3=15.087 63	3= 4.828 042
4=101.6002	4=1.219 202	4=3.657 607	4=20.116 84	4= 6.437 389
5=127.0003	5=1.524 003	5=4.572 009	5=25.146 05	5= 8.046 736
6=152.4003	6=1.828 804	6=5.486 411	6=30.175 26	6= 9.656 083
7=177.8004	7=2.133 604	7=6.400 813	7=35.204 47	7=11.265 431
8=203.2004	8=2.438 405	8=7.315 215	8=40.233 68	8=12.874 778
9=228.6005	9=2.743 205	9=8.229 616	9=45.262 89	9=14.484 125
0.039 37=1	3.280 83=1	1.093 611=1	0.198 838=1	0.621 370=1
0.078 74=2	6.561 67=2	2.187 222=2	0.397 677=2	1.242 740=2
0.118 11=3	9.842 50=3	3.280 833=3	0.596 515=3	1.864 110=3
0.157 48=4	13.123 33=4	4.374 444=4	0.795 354=4	2.485 480=4
0.196 85=5	16.404 17=5	5.468 056=5	0.994 192=5	3.106 850=5
0.236 22=6	19.685 00=6	6.561 667=6	1.193 030=6	3.728 220=6
0.275 59=7	22.965 83=7	7.655 278=7	1.391 869=7	4.349 590=7
0.314 96=8	26.246 67=8	8.748 889=8	1.590 707=8	4.970 960=8
0.354 33=9	29.527 50=9	9.842 500=9	1.789 545=9	5.592 330=9

2. AREA

Square inches (sq in.) = Square centimeters (cm^2)	Square feet (sq ft) = Square meters (m^2)	Square yards (sq yd) = Square meters (m^2)	Acres (acre) = Hectares (ha)	Square miles (sq mi) = Square kilometers (km^2)
1= 6.452	1=0.092 90	1=0.8361	1=0.4047	1= 2.5900
2=12.903	2=0.185 81	2=1.6723	2=0.8094	2= 5.1800
3=19.355	3=0.278 71	3=2.5084	3=1.2141	3= 7.7700
4=25.807	4=0.371 61	4=3.3445	4=1.6187	4=10.3600
5=32.258	5=0.464 52	5=4.1807	5=2.0234	5=12.9500
6=38.710	6=0.557 42	6=5.0168	6=2.4281	6=15.5400
7=45.161	7=0.650 32	7=5.8529	7=2.8328	7=18.1300
8=51.613	8=0.743 23	8=6.6890	8=3.2375	8=20.7200
9=58.065	9=0.836 13	9=7.5252	9=3.6422	9=23.3100
0.155 00=1	10.764=1	1.1960=1	2.471=1	0.3861=1
0.310 00=2	21.528=2	2.3920=2	4.942=2	0.7722=2
0.465 00=3	32.292=3	3.5880=3	7.413=3	1.1583=3
0.620 00=4	43.055=4	4.7839=4	9.884=4	1.5444=4
0.775 00=5	53.819=5	5.9799=5	12.355=5	1.9305=5
0.930 00=6	64.583=6	7.1759=6	14.826=6	2.3166=6
1.085 00=7	75.347=7	8.3719=7	17.297=7	2.7027=7
1.240 00=8	86.111=8	9.5679=8	19.768=8	3.0888=8
1.395 00=9	96.875=9	10.7639=9	22.239=9	3.4749=9

3. VOLUME

Cubic inches (cu in.) = Cubic centimeters (cm^3)	Cubic feet (cu ft) = Cubic meters (m^3)	Cubic yards (cu yd) = Cubic meters (m^3)	Cubic inches (cu in.) = Liters (liter)	Cubic feet (cu ft) = Liters (liter)
1= 16.3872	1=0.028 317	1=0.7646	1=0.016 386 7	1= 28.316
2= 32.7743	2=0.056 634	2=1.5291	2=0.032 773 4	2= 56.633
3= 49.1615	3=0.084 951	3=2.2937	3=0.049 160 2	3= 84.949
4= 65.5486	4=0.113 268	4=3.0582	4=0.065 546 9	4=113.265
5= 81.9358	5=0.141 585	5=3.8228	5=0.081 933 6	5=141.581
6= 98.3230	6=0.169 902	6=4.5874	6=0.098 320 3	6=169.898
7=114.7101	7=0.198 219	7=5.3519	7=0.114 707 0	7=198.214
8=131.0973	8=0.226 536	8=6.1165	8=0.131 093 8	8=226.530
9=147.4845	9=0.254 853	9=6.8810	9=0.147 480 5	9=254.846
0.061 02=1	35.314=1	1.3079=1	61.025=1	0.035 315=1
0.122 05=2	70.629=2	2.6159=2	122.050=2	0.070 631=2
0.183 07=3	105.943=3	3.9238=3	183.075=3	0.105 946=3
0.244 09=4	141.258=4	5.2318=4	244.100=4	0.141 262=4
0.305 12=5	176.572=5	6.5397=5	305.125=5	0.176 577=5
0.366 14=6	211.887=6	7.8477=6	366.150=6	0.211 892=6
0.427 16=7	247.201=7	9.1556=7	427.175=7	0.247 208=7
0.488 19=8	282.516=8	10.4635=8	488.200=8	0.282 523=8
0.549 21=9	317.830=9	11.7715=9	549.225=9	0.317 839=9

4. CAPACITY—LIQUID MEASURE

U. S. fluid drams (fl dr) / Milliliters (ml)	U. S. fluid ounces (fl oz) / Milliliters (ml)	U. S. liquid pints (pt) / Liters (liter)	U. S. liquid quarts (qt) / Liters (liter)	U. S. gallons (gal) / Liters (liter)
1= 3.6966	1= 29.573	1=0.473 17	1=0.946 33	1= 3.785 33
2= 7.3932	2= 59.146	2=0.946 33	2=1.892 67	2= 7.570 66
3=11.0898	3= 88.719	3=1.419 50	3=2.839 00	3=11.356 00
4=14.7865	4=118.292	4=1.892 67	4=3.785 33	4=15.141 33
5=18.4831	5=147.865	5=2.365 83	5=4.731 67	5=18.926 66
6=22.1797	6=177.437	6=2.839 00	6=5.678 00	6=22.711 99
7=25.8763	7=207.010	7=3.312 17	7=6.624 33	7=26.497 33
8=29.5729	8=236.583	8=3.785 33	8=7.570 66	8=30.282 66
9=33.2695	9=266.156	9=4.258 50	9=8.517 00	9=34.067 99
0.270 52=1	0.033 815=1	2.1134=1	1.056 71=1	0.264 18=1
0.541 04=2	0.067 629=2	4.2268=2	2.113 42=2	0.528 36=2
0.811 55=3	0.101 444=3	6.3403=3	3.170 13=3	0.792 53=3
1.082 07=4	0.135 259=4	8.4537=4	4.226 84=4	1.056 71=4
1.352 59=5	0.169 074=5	10.5671=5	5.283 55=5	1.320 89=5
1.623 11=6	0.202 888=6	12.6805=6	6.340 26=6	1.585 07=6
1.893 63=7	0.236 703=7	14.7939=7	7.396 97=7	1.849 24=7
2.164 14=8	0.270 518=8	16.9074=8	8.453 68=8	2.113 42=8
2.434 66=9	0.304 333=9	19.0208=9	9.510 39=9	2.377 60=9

5. CAPACITY—DRY MEASURE

U. S. dry quarts (qt) / Liters (liter)	U. S. pecks (pk) / Liters (liter)	U. S. pecks (pk) / Dekaliters (dkl)	U. S. bushels (bu) / Hectoliters (hl)	U. S. bushels per acre / Hectoliters per hectare
1=1.1012	1= 8.810	1=0.8810	1=0.352 38	1=0.8708
2=2.2024	2=17.619	2=1.7619	2=0.704 77	2=1.7415
3=3.3036	3=26.429	3=2.6429	3=1.057 15	3=2.6123
4=4.4048	4=35.238	4=3.5238	4=1.409 53	4=3.4830
5=5.5060	5=44.048	5=4.4048	5=1.761 92	5=4.3538
6=6.6072	6=52.857	6=5.2857	6=2.114 30	6=5.2245
7=7.7084	7=61.667	7=6.1667	7=2.466 68	7=6.0953
8=8.8096	8=70.477	8=7.0477	8=2.819 07	8=6.9660
9=9.9108	9=79.286	9=7.9286	9=3.171 45	9=7.8368
0.9081=1	0.113 51=1	1.1351=1	2.8378=1	1.1484=1
1.8162=2	0.227 03=2	2.2703=2	5.6756=2	2.2969=2
2.7243=3	0.340 54=3	3.4054=3	8.5135=3	3.4453=3
3.6324=4	0.454 05=4	4.5405=4	11.3513=4	4.5937=4
4.5405=5	0.567 56=5	5.6756=5	14.1891=5	5.7421=5
5.4486=6	0.681 08=6	6.8108=6	17.0269=6	6.8906=6
6.3567=7	0.794 59=7	7.9459=7	19.8647=7	8.0390=7
7.2648=8	0.908 10=8	9.0810=8	22.7026=8	9.1874=8
8.1729=9	1.021 61=9	10.2161=9	25.5404=9	10.3359=9

6. MASS

Grains (grain) / Grams (g)	Apothecaries' drams (dr ap or ʒ) / Grams (g)	Troy ounces (oz t) / Grams (g)	Avoirdupois ounces (oz avdp) / Grams (g)	Avoirdupois pounds (lb avdp) / Kilograms (kg)
1=0.064 799	1= 3.8879	1= 31.103	1= 28.350	1=0.453 59
2=0.129 598	2= 7.7759	2= 62.207	2= 56.699	2=0.907 18
3=0.194 397	3=11.6638	3= 93.310	3= 85.049	3=1.360 78
4=0.259 196	4=15.5517	4=124.414	4=113.398	4=1.814 37
5=0.323 995	5=19.4397	5=155.517	5=141.748	5=2.267 96
6=0.388 794	6=23.3276	6=186.621	6=170.097	6=2.721 55
7=0.453 592	7=27.2155	7=217.724	7=198.447	7=3.175 15
8=0.518 391	8=31.1035	8=248.828	8=226.796	8=3.628 74
9=0.583 190	9=34.9914	9=279.931	9=255.146	9=4.082 33
15.4324=1	0.257 21=1	0.032 151=1	0.035 274=1	2.204 62=1
30.8647=2	0.514 41=2	0.064 301=2	0.070 548=2	4.409 24=2
46.2971=3	0.771 62=3	0.096 452=3	0.105 822=3	6.613 87=3
61.7294=4	1.028 82=4	0.128 603=4	0.141 096=4	8.818 49=4
77.1618=5	1.286 03=5	0.160 754=5	0.176 370=5	11.023 11=5
92.5941=6	1.543 24=6	0.192 904=6	0.211 644=6	13.227 73=6
108.0265=7	1.800 44=7	0.225 055=7	0.246 918=7	15.432 36=7
123.4589=8	2.057 65=8	0.257 206=8	0.282 192=8	17.636 98=8
138.8912=9	2.314 85=9	0.289 357=9	0.317 466=9	19.841 60=9

COMPARISON OF THE VARIOUS TONS AND POUNDS IN USE IN THE UNITED STATES (FROM 1 TO 9 UNITS)

Troy pounds	Avoirdupois pounds	Kilograms	Short tons	Long tons	Metric tons
1	0.822 857	0.373 24	0.000 411 43	0.000 367 35	0.000 373 24
2	1.645 71	0.746 48	0.000 822 86	0.000 734 69	0.000 746 48
3	2.468 57	1.119 73	0.001 234 29	0.001 102 04	0.001 119 73
4	3.291 43	1.492 97	0.001 645 71	0.001 469 39	0.001 492 97
5	4.114 29	1.866 21	0.002 057 14	0.001 836 73	0.001 866 21
6	4.937 14	2.239 45	0.002 468 57	0.002 204 08	0.002 239 45
7	5.760 00	2.612 69	0.002 880 00	0.002 571 43	0.002 612 69
8	6.582 86	2.985 93	0.003 291 43	0.002 938 78	0.002 985 93
9	7 405 71	3.359 18	0.003 702 86	0.003 306 12	0.003 359 18
1.215 28	**1**	0.453 59	0.0005	0.000 446 43	0.000 453 59
2.430 56	**2**	0.907 18	0.0010	0.000 892 86	0.000 907 18
3.645 83	**3**	1.360 78	0.0015	0.001 339 29	0.001 360 78
4.861 11	**4**	1.814 37	0.0020	0.001 785 71	0.001 814 37
6.076 39	**5**	2.267 96	0.0025	0.002 232 14	0.002 267 96
7.291 67	**6**	2.721 55	0.0030	0.002 678 57	0.002 721 55
8.506 94	**7**	3.175 15	0.0035	0.003 125 00	0.003 175 15
9.722 22	**8**	3.628 74	0.0040	0.003 571 43	0.003 628 74
10.937 50	**9**	4.082 33	0.0045	0.004 017 86	0.004 082 33
2.679 23	2.204 62	**1**	0.001 102 31	0.000 984 21	0.001
5.358 46	4.409 24	**2**	0.002 204 62	0.001 968 41	0.002
8.037 69	6.613 87	**3**	0.003 306 93	0.002 952 62	0.003
10.716 91	8.818 49	**4**	0.004 409 24	0.003 936 83	0.004
13.396 14	11.023 11	**5**	0.005 511 56	0.004 921 03	0.005
16.075 37	13.227 73	**6**	0.006 613 87	0.005 905 24	0.006
18.754 60	15.432 36	**7**	0.007 716 18	0.006 889 44	0.007
21.433 83	17.636 98	**8**	0.008 818 49	0.007 873 65	0.008
24.113 06	19.841 60	**9**	0.009 920 80	0.008 857 86	0.009
2430.56	2000	907.18	**1**	0.892 86	0.907 18
4861.11	4000	1814.37	**2**	1.785 71	1.814 37
7291.67	6000	2721.55	**3**	2.678 57	2.721 55
9722.22	8000	3628.74	**4**	3.571 43	3.628 74
12 152.78	10 000	4535.92	**5**	4.464 29	4.535 92
14 583.33	12 000	5443.11	**6**	5.357 14	5.443 11
17 013.89	14 000	6350.29	**7**	6.250 00	6.350 29
19 444.44	16 000	7257.48	**8**	7.142 86	7.257 48
21 875.00	18 000	8164.66	**9**	8.035 71	8.164 66
2722.22	2240	1016.05	1.12	**1**	1.016 05
5444.44	4480	2032.09	2.24	**2**	2.032 09
8166.67	6720	3048.14	3.36	**3**	3.048 14
10 888.89	8960	4064.19	4.48	**4**	4.064 19
13 611.11	11 200	5080.24	5.60	**5**	5.080 24
16 333.33	13 440	6096.28	6.72	**6**	6.096 28
19 055.56	15 680	7112.32	7.84	**7**	7.112 32
21 777.78	17 920	8128.38	8.96	**8**	8.128 38
24 500.00	20 160	9144.42	10.08	**9**	9.144 42
2679.23	2204.62	1000	1.102 31	0.984 21	**1**
5358.46	4409.24	2000	2.204 62	1.968 41	**2**
8037.69	6613.87	3000	3.306 93	2.952 62	**3**
10 716.91	8818.49	4000	4.409 24	3.936 83	**4**
13 396.14	11 023.11	5000	5.511 56	4.921 03	**5**
16 075.37	13 227.73	6000	6.613 87	5.905 24	**6**
18 754.60	15 432.36	7000	7.716 18	6.889 44	**7**
21 433.83	17 636.98	8000	8.818 49	7.873 65	**8**
24 113.06	19 841.60	9000	9.920 80	8.857 86	**9**

LENGTH—INCHES AND MILLIMETERS—EQUIVALENTS OF DECIMAL AND BINARY FRACTIONS OF AN INCH IN MILLIMETERS

From 1/64 to 1 Inch

½'s	¼'s	8ths	16ths	32ds	64ths	Millimeters	Decimals of an inch	Inch	½'s	¼'s	8ths	16ths	32ds	64ths	Millimeters	Decimals of an inch
					1	= 0.397	0.015625							33	=13.097	0.515625
				1	2	= .794	.03125						17	34	=13.494	.53125
					3	= 1.191	.046875							35	=13.891	.546875
			1	2	4	= 1.588	.0625					9	18	36	=14.288	.5625
					5	= 1.984	.078125							37	=14.684	.578125
				3	6	= 2.381	.09375						19	38	=15.081	.59375
					7	= 2.778	.109375							39	=15.478	.609375
		1	2	4	8	= 3.175	.1250				5	10	20	40	=15.875	.625
					9	= 3.572	.140625							41	=16.272	.640625
				5	10	= 3.969	.15625						21	42	=16.669	.65625
					11	= 4.366	.171875							43	=17.066	.671875
			3	6	12	= 4.763	.1875					11	22	44	=17.463	.6875
					13	= 5.159	.203125							45	=17.859	.703125
				7	14	= 5.556	.21875						23	46	=18.256	.71875
					15	= 5.953	.234375							47	=18.653	.734375
	1	2	4	8	16	= 6.350	.2500			3	6	12	24	48	=19.050	.75
					17	= 6.747	.265625							49	=19.447	.765625
				9	18	= 7.144	.28125						25	50	=19.844	.78125
					19	= 7.541	.296875							51	=20.241	.796875
			5	10	20	= 7.938	.3125					13	26	52	=20.638	.8125
					21	= 8.334	.328125							53	=21.034	.828125
				11	22	= 8.731	.34375						27	54	=21.431	.84375
					23	= 9.128	.359375							55	=21.828	.859375
		3	6	12	24	= 9.525	.3750				7	14	28	56	=22.225	.875
					25	= 9.922	.390625							57	=22.622	.890625
				13	26	=10.319	.40625						29	58	=23.019	.90625
					27	=10.716	.421875							59	=23.416	.921875
			7	14	28	=11.113	.4375					15	30	60	=23.813	.9375
					29	=11.509	.453125							61	=24.209	.953125
				15	30	=11.906	.46875						31	62	=24.606	.96875
					31	=12.303	.484375							63	=25.003	.984375
1	2	4	8	16	32	=12.700	.5	1	2	4	8	16	32	64	=25.400	1.000

LENGTH—HUNDREDTHS OF AN INCH TO MILLIMETERS

From 1 to 99 Hundredths

Hundredths of an inch	0	1	2	3	4	5	6	7	8	9
	0	0.254	0.508	0.762	1.016	1.270	1.524	1.778	2.032	2.286
10	2.540	2.794	3.048	3.302	3.556	3.810	4.064	4.318	4.572	4.826
20	5.080	5.334	5.588	5.842	6.096	6.350	6.604	6.858	7.112	7.366
30	7.620	7.874	8.128	8.382	8.636	8.890	9.144	9.398	9.652	9.906
40	10.160	10.414	10.668	10.922	11.176	11.430	11.684	11.938	12.192	12.446
50	12.700	12.954	13.208	13.462	13.716	13.970	14.224	14.478	14.732	14.986
60	15.240	15.494	15.748	16.002	16.256	16.510	16.764	17.018	17.272	17.526
70	17.780	18.034	18.288	18.542	18.796	19.050	19.304	19.558	19.812	20.066
80	20.320	20.574	20.828	21.082	21.336	21.590	21.844	22.098	22.352	22.606
90	22.860	23.114	23.368	23.622	23.876	24.130	24.384	24.638	24.892	25.146

LENGTH—MILLIMETERS TO DECIMALS OF AN INCH

From 1 to 99 Units

Millimeters	0	1	2	3	4	5	6	7	8	9
	0	0.03937	0.07874	0.11811	0.15748	0.19685	0.23622	0.27559	0.31496	0.35433
10	0.39370	.43307	.47244	.51181	.55118	.59055	.62992	.66929	.70866	.74803
20	.78740	.82677	.86614	.90551	.94488	.98425	1.02362	1.06299	1.10236	1.14173
30	1.18110	1.22047	1.25984	1.29921	1.33858	1.37795	1.41732	1.45669	1.49606	1.53543
40	1.57480	1.61417	1.65354	1.69291	1.73228	1.77165	1.81102	1.85039	1.88976	1.92913
50	1.96850	2.00787	2.04724	2.08661	2.12598	2.16535	2.20472	2.24409	2.28346	2.32283
60	2.36220	2.40157	2.44094	2.48031	2.51968	2.55905	2.59842	2.63779	2.67716	2.71653
70	2.75590	2.79527	2.83464	2.87401	2.91338	2.95275	2.99212	3.03149	3.07086	3.11023
80	3.14960	3.18897	3.22834	3.26771	3.30708	3.34645	3.38582	3.42519	3.46456	3.50393
90	3.54330	3.58267	3.62204	3.66141	3.70078	3.74015	3.77952	3.81889	3.85826	3.89763

LENGTH—UNITED STATES NAUTICAL MILES, INTERNATIONAL NAUTICAL MILES, AND KILOMETERS

Basic relations:
- 1 U. S. nautical mile=1.853 248 kilometers.
- 1 International nautical mile=1.852 kilometers.
- 1 U. S. nautical mile=1.000 673 9 int. nautical miles.

U. S. nautical miles	Int. nautical miles	Kilometers	U. S. nautical miles	Int. nautical miles	Kilometers	U. S. nautical miles	Int. nautical miles	Kilometers
0				0				0
1	1.0007	1.8532	0.9993	1	1.8520	0.5396	0.5400	1
2	2.0013	3.7065	1.9987	2	3.7040	1.0792	1.0799	2
3	3.0020	5.5597	2.9980	3	5.5560	1.6188	1.6199	3
4	4.0027	7.4130	3.9973	4	7.4080	2.1584	2.1598	4
5	5.0034	9.2662	4.9966	5	9.2600	2.6980	2.6998	5
6	6.0040	11.1195	5.9960	6	11.1120	3.2376	3.2397	6
7	7.0047	12.9727	6.9953	7	12.9640	3.7772	3.7797	7
8	8.0054	14.8260	7.9946	8	14.8160	4.3167	4.3197	8
9	9.0061	16.6792	8.9939	9	16.6680	4.8563	4.8596	9
10	10.0067	18.5325	9.9933	10	18.5200	5.3959	5.3996	10
1	11.0074	20.3857	10.9926	1	20.3720	5.9355	5.9395	1
2	12.0081	22.2390	11.9919	2	22.2240	6.4751	6.4795	2
3	13.0088	24.0922	12.9912	3	24.0760	7.0147	7.0194	3
4	14.0094	25.9455	13.9906	4	25.9280	7.5543	7.5594	4
5	15.0101	27.7987	14.9899	5	27.7800	8.0939	8.0994	5
6	16.0108	29.6520	15.9892	6	29.6320	8.6335	8.6393	6
7	17.0115	31.5052	16.9886	7	31.4840	9.1731	9.1793	7
8	18.0121	33.3585	17.9879	8	33.3360	9.7127	9.7192	8
9	19.0128	35.2117	18.9872	9	35.1880	10.2523	10.2592	9
20	20.0135	37.0650	19.9865	20	37.0400	10.7919	10.7991	20
1	21.0142	38.9182	20.9859	1	38.8920	11.3315	11.3391	1
2	22.0148	40.7715	21.9852	2	40.7440	11.8711	11.8790	2
3	23.0155	42.6247	22.9845	3	42.5960	12.4106	12.4190	3
4	24.0162	44.4780	23.9838	4	44.4480	12.9502	12.9590	4
5	25.0168	46.3312	24.9832	5	46.3000	13.4898	13.4989	5
6	26.0175	48.1844	25.9825	6	48.1520	14.0294	14.0389	6
7	27.0182	50.0377	26.9818	7	50.0040	14.5690	14.5788	7
8	28.0189	51.8909	27.9811	8	51.8560	15.1086	15.1188	8
9	29.0195	53.7442	28.9805	9	53.7080	15.6482	15.6587	9
30	30.0202	55.5974	29.9798	30	55.5600	16.1878	16.1987	30
1	31.0209	57.4507	30 9791	1	57.4120	16.7274	16.7387	1
2	32.0216	59.3039	31.9785	2	59.2640	17.2670	17.2786	2
3	33.0222	61.1572	32.9778	3	61.1160	17.8066	17.8186	3
4	34.0229	63.0104	33.9771	4	62.9680	18.3462	18.3585	4
5	35.0236	64.8637	34.9764	5	64.8200	18.8858	18.8985	5
6	36.0243	66.7169	35.9758	6	66.6720	19.4254	19.4384	6
7	37.0249	68.5702	36.9751	7	68.5240	19.9649	19.9784	7
8	38.0256	70.4234	37.9744	8	70.3760	20.5045	20.5184	8
9	39.0263	72.2767	38.9737	9	72.2280	21.0441	21.0583	9
40	40.0270	74.1299	39.9731	40	74.0800	21.5837	21.5983	40
1	41.0276	75.9832	40.9724	1	75.9320	22.1233	22.1382	1
2	42.0283	77.8364	41.9717	2	77.7840	22.6629	22.6782	2
3	43.0290	79.6897	42.9710	3	79.6360	23.2025	23.2181	3
4	44.0297	81.5429	43.9704	4	81.4880	23.7421	23.7581	4
5	45.0303	83.3962	44.9697	5	83.3400	24.2817	24.2981	5
6	46.0310	85.2494	45.9690	6	85.1920	24.8213	24.8380	6
7	47.0317	87.1027	46.9683	7	87.0440	25.3609	25.3780	7
8	48.0323	88.9559	47.9677	8	88.8960	25.9005	25.9179	8
9	49.0330	90.8092	48.9670	9	90.7480	26.4401	26.4579	9

50	50.0337	92.6624	49.9663	**50**	92.6000	26.9797	26.9978	**50**
1	51.0344	94.5156	50.9657	**1**	94.4520	27.5193	27.5378	**1**
2	52.0350	96.3689	51.9650	**2**	96.3040	28.0588	28.0778	**2**
3	53.0357	98.2221	52.9643	**3**	98.1560	28.5984	28.6177	**3**
4	54.0364	100.0754	53.9636	**4**	100.0080	29.1380	29.1577	**4**
5	55.0371	101.9286	54.9630	**5**	101.8600	29.6776	29.6976	**5**
6	56.0377	103.7819	55.9623	**6**	103.7120	30.2172	30.2376	**6**
7	57.0384	105.6351	56.9616	**7**	105.5640	30.7568	30.7775	**7**
8	58.0391	107.4884	57.9609	**8**	107.4160	31.2964	31.3175	**8**
9	59.0398	109.3416	58.9603	**9**	109.2680	31.8360	31.8575	**9**
60	60.0404	111.1949	59.9596	**60**	111.1200	32.3756	32.3974	**60**
1	61.0411	113.0481	60.9589	**1**	112.9720	32.9152	32.9374	**1**
2	62.0418	114.9014	61.9582	**2**	114.8240	33.4548	33.4773	**2**
3	63.0425	116.7546	62.9576	**3**	116.6760	33.9944	34.0173	**3**
4	64.0431	118.6079	63.9569	**4**	118.5280	34.5340	34.5572	**4**
5	65.0438	120.4611	64.9562	**5**	120.3800	35.0736	35.0972	**5**
6	66.0445	122.3144	65.9556	**6**	122.2320	35.6132	35.6371	**6**
7	67.0452	124.1676	66.9549	**7**	124.0840	36.1527	36.1771	**7**
8	68.0458	126.0209	67.9542	**8**	125.9360	36.6923	36.7171	**8**
9	69.0465	127.8741	68.9535	**9**	127.7880	37.2319	37.2570	**9**
70	70.0472	129.7274	69.9529	**70**	129.6400	37.7715	37.7970	**70**
1	71.0478	131.5806	70.9522	**1**	131.4920	38.3111	38.3369	**1**
2	72.0485	133.4339	71.9515	**2**	133.3440	38.8507	38.8769	**2**
3	73.0492	135.2871	72.9508	**3**	135.1960	39.3903	39.4168	**3**
4	74.0499	137.1404	73.9502	**4**	137.0480	39.9299	39.9568	**4**
5	75.0505	138.9936	74.9495	**5**	138.9000	40.4695	40.4968	**5**
6	76.0512	140.8468	75.9488	**6**	140.7520	41.0091	41.0367	**6**
7	77.0519	142.7001	76.9481	**7**	142.6040	41.5487	41.5767	**7**
8	78.0526	144.5533	77.9475	**8**	144.4560	42.0883	42.1166	**8**
9	79.0532	146.4066	78.9468	**9**	146.3080	42.6279	42.6566	**9**
80	80.0539	148.2598	79.9461	**80**	148.1600	43.1675	43.1965	**80**
1	81.0546	150.1131	80.9455	**1**	150.0120	43.7070	43.7365	**1**
2	82.0553	151.9663	81.9448	**2**	151.8640	44.2466	44.2765	**2**
3	83.0559	153.8196	82.9441	**3**	153.7160	44.7862	44.8164	**3**
4	84.0566	155.6728	83.9434	**4**	155.5680	45.3258	45.3564	**4**
5	85.0573	157.5261	84.9428	**5**	157.4200	45.8654	45.8963	**5**
6	86.0580	159.3793	85.9421	**6**	159.2720	46.4050	46.4363	**6**
7	87.0586	161.2326	86.9414	**7**	161.1240	46.9446	46.9762	**7**
8	88.0593	163.0858	87.9407	**8**	162.9760	47.4842	47.5162	**8**
9	89.0600	164.9391	88.9401	**9**	164.8280	48.0238	48.0562	**9**
90	90.0607	166.7923	89.9394	**90**	166.6800	48.5634	48.5961	**90**
1	91.0613	168.6456	90.9387	**1**	168.5320	49.1030	49.1361	**1**
2	92.0620	170.4988	91.9380	**2**	170.3840	49.6426	49.6760	**2**
3	93.0627	172.3521	92.9374	**3**	172.2360	50.1822	50.2160	**3**
4	94.0633	174.2053	93.9367	**4**	174.0880	50.7218	50.7559	**4**
5	95.0640	176.0586	94.9360	**5**	175.9400	51.2614	51.2959	**5**
6	96.0647	177.9118	95.9354	**6**	177.7920	51.8009	51.8359	**6**
7	97.0654	179.7651	96.9347	**7**	179.6440	52.3405	52.3758	**7**
8	98.0660	181.6183	97.9340	**8**	181.4960	52.8801	52.9158	**8**
9	99.0667	183.4716	98.9333	**9**	183.3480	53.4197	53.4557	**9**
100	100.0674	185.3248	99.9327	**100**	185.2000	53.9593	53.9957	**100**

LENGTH—MILLIMETERS TO INCHES

[From 0.00 to 25.40 millimeters by 0.01 millimeter. 1 millimeter=0.03937 inch.]

Milli-meters	Hundredths of millimeters									
	0.00	0.01	0.02	0.03	0.04	0.05	0.06	0.07	0.08	0.09
	Inches	Inches	Inches	Inches	Inches	Inches	Inches	Inches	Inches	Inches
0.00	0. 000000	0. 000394	0. 000787	0. 001181	0. 001575	0. 001968	0. 002362	0. 002756	0. 003150	0. 003543
0. 10	. 003937	. 004331	. 004724	. 005118	. 005512	. 005906	. 006299	. 006693	. 007087	. 007480
0. 20	. 007874	. 008268	. 008661	. 009055	. 009449	. 009842	. 010236	. 010630	. 011024	. 011417
0. 30	. 011811	. 012205	. 012598	. 012992	. 013386	. 013780	. 014173	. 014567	. 014961	. 015354
0. 40	. 015748	. 016142	. 016535	. 016929	. 017323	. 017716	. 018110	. 018504	. 018898	. 019291
0.50	0. 019685	0. 020079	0. 020472	0. 020866	0. 021260	0. 021654	0. 022047	0. 022441	0. 022835	0. 023228
0. 60	. 023622	. 024016	. 024409	. 024803	. 025197	. 025590	. 025984	. 026378	. 026772	. 027165
0. 70	. 027559	. 027953	. 028346	. 028740	. 029134	. 029528	. 029921	. 030315	. 030709	. 031102
0. 80	. 031496	. 031890	. 032283	. 032677	. 033071	. 033464	. 033858	. 034252	. 034646	. 035039
0. 90	. 035433	. 035827	. 036220	. 036614	. 037008	. 037402	. 037795	. 038189	. 038583	. 038976
1.00	0. 03937	0. 03976	0. 04016	0. 04055	0. 04094	0. 04134	0. 04173	0. 04213	0. 04252	0. 04291
1. 10	. 04331	. 04370	. 04409	. 04449	. 04488	. 04528	. 04567	. 04606	. 04646	. 04685
1. 20	. 04724	. 04764	. 04803	. 04843	. 04882	. 04921	. 04961	. 05000	. 05039	. 05079
1. 30	. 05118	. 05157	. 05197	. 05236	. 05276	. 05315	. 05354	. 05394	. 05433	. 05472
1. 40	. 05512	. 05551	. 05591	. 05630	. 05669	. 05709	. 05748	. 05787	. 05827	. 05866
1.50	0. 05906	0. 05945	0. 05984	0. 06024	0. 06063	0. 06102	0. 06142	0. 06181	0. 06220	0. 06260
1. 60	. 06299	. 06339	. 06378	. 06417	. 06457	. 06496	. 06535	. 06575	. 06614	. 06654
1. 70	. 06693	. 06732	. 06772	. 06811	. 06850	. 06890	. 06929	. 06968	. 07008	. 07047
1. 80	. 07087	. 07126	. 07165	. 07205	. 07244	. 07283	. 07323	. 07362	. 07402	. 07441
1. 90	. 07480	. 07520	. 07559	. 07598	. 07638	. 07677	. 07717	. 07756	. 07795	. 07835
2.00	0. 07874	0. 07913	0. 07953	0. 07992	0. 08031	0. 08071	0. 08110	0. 08150	0. 08189	0. 08228
2. 10	. 08268	. 08307	. 08346	. 08386	. 08425	. 08465	. 08504	. 08543	. 08583	. 08622
2. 20	. 08661	. 08701	. 08740	. 08780	. 08819	. 08858	. 08898	. 08937	. 08976	. 09016
2. 30	. 09055	. 09094	. 09134	. 09173	. 09213	. 09252	. 09291	. 09331	. 09370	. 09409
2. 40	. 09449	. 09488	. 09528	. 09567	. 09606	. 09646	. 09685	. 09724	. 09764	. 09803

2.50	0. 09842	0. 09882	0. 09921	0. 09961	0. 10000	0. 10039	0. 10079	0. 10118	0. 10157	0. 10197
2. 60	. 10236	. 10276	. 10315	. 10354	. 10394	. 10433	. 10472	. 10512	. 10551	. 10591
2. 70	. 10630	. 10669	. 10709	. 10748	. 10787	. 10827	. 10866	. 10905	. 10945	. 10984
2. 80	. 11024	. 11063	. 11102	. 11142	. 11181	. 11220	. 11260	. 11299	. 11339	. 11378
2. 90	. 11417	. 11457	. 11496	. 11535	. 11575	. 11614	. 11654	. 11693	. 11732	. 11772
3.00	0. 11811	0. 11850	0. 11890	0. 11929	0. 11968	0. 12008	0. 12047	0. 12087	0. 12126	0. 12165
3. 10	. 12205	. 12244	. 12283	. 12323	. 12362	. 12402	. 12441	. 12480	. 12520	. 12559
3. 20	. 12598	. 12638	. 12677	. 12717	. 12756	. 12795	. 12835	. 12874	. 12913	. 12953
3. 30	. 12992	. 13031	. 13071	. 13110	. 13150	. 13189	. 13228	. 13268	. 13307	. 13346
3. 40	. 13386	. 13425	. 13465	. 13504	. 13543	. 13583	. 13622	. 13661	. 13701	. 13740
3.50	0. 13780	0. 13819	0. 13858	0. 13898	0. 13937	0. 13976	0. 14016	0. 14055	0. 14094	0. 14134
3. 60	. 14173	. 14213	. 14252	. 14291	. 14331	. 14370	. 14409	. 14449	. 14488	. 14528
3. 70	. 14567	. 14606	. 14646	. 14685	. 14724	. 14764	. 14803	. 14842	. 14882	. 14921
3. 80	. 14961	. 15000	. 15039	. 15079	. 15118	. 15157	. 15197	. 15236	. 15276	. 15315
3. 90	. 15354	. 15394	. 15433	. 15472	. 15512	. 15551	. 15591	. 15630	. 15669	. 15709
4.00	0. 15748	0. 15787	0. 15827	0. 15866	0. 15905	0. 15945	0. 15984	0. 16024	0. 16063	0. 16102
4. 10	. 16142	. 16181	. 16220	. 16260	. 16299	. 16339	. 16378	. 16417	. 16457	. 16496
4. 20	. 16535	. 16575	. 16614	. 16654	. 16693	. 16732	. 16772	. 16811	. 16850	. 16890
4. 30	. 16929	. 16968	. 17008	. 17047	. 17087	. 17126	. 17165	. 17205	. 17244	. 17283
4. 40	. 17323	. 17362	. 17402	. 17441	. 17480	. 17520	. 17559	. 17598	. 17638	. 17677
4.50	0. 17716	0. 17756	0. 17795	0. 17835	0. 17874	0. 17913	0. 17953	0. 17992	0. 18031	0. 18071
4. 60	. 18110	. 18150	. 18189	. 18228	. 18268	. 18307	. 18346	. 18386	. 18425	. 18465
4. 70	. 18504	. 18543	. 18583	. 18622	. 18661	. 18701	. 18740	. 18779	. 18819	. 18858
4. 80	. 18898	. 18937	. 18976	. 19016	. 19055	. 19094	. 19134	. 19173	. 19213	. 19252
4. 90	. 19291	. 19331	. 19370	. 19409	. 19449	. 19488	. 19528	. 19567	. 19606	. 19646
5.00	0. 19685	0. 19724	0. 19764	0. 19803	0. 19842	0. 19882	0. 19921	0. 19961	0. 20000	0. 20039
5. 10	. 20079	. 20118	. 20157	. 20197	. 20236	. 20276	. 20315	. 20354	. 20394	. 20433
5. 20	. 20472	. 20512	. 20551	. 20591	. 20630	. 20669	. 20709	. 20748	. 20787	. 20827
5. 30	. 20866	. 20905	. 20945	. 20984	. 21024	. 21063	. 21102	. 21142	. 21181	. 21220
5. 40	. 21260	. 21299	. 21339	. 21378	. 21417	. 21457	. 21496	. 21535	. 21575	. 21614
5.50	0. 21654	0. 21693	0. 21732	0. 21772	0. 21811	0. 21850	0. 21890	0. 21929	0. 21968	0. 22008
5. 60	. 22047	. 22087	. 22126	. 22165	. 22205	. 22244	. 22283	. 22323	. 22362	. 22402
5. 70	. 22441	. 22480	. 22520	. 22559	. 22598	. 22638	. 22677	. 22716	. 22756	. 22795
5. 80	. 22835	. 22874	. 22913	. 22953	. 22992	. 23031	. 23071	. 23110	. 23150	. 23189
5. 90	. 23228	. 23268	. 23307	. 23346	. 23386	. 23425	. 23465	. 23504	. 23543	. 23583

LENGTH—MILLIMETERS TO INCHES—Continued

[From 0.00 to 25.40 millimeters by 0.01 millimeter. 1 millimeter=0.03937 inch.]

Milli-meters	Hundredths of millimeters									
	0.00	0.01	0.02	0.03	0.04	0.05	0.06	0.07	0.08	0.09
	Inches	Inches	Inches	Inches	Inches	Inches	Inches	Inches	Inches	Inches
6.00	0.23622	0.23661	0.23701	0.23740	0.23779	0.23819	0.23858	0.23898	0.23937	0.23976
6.10	.24016	.24055	.24094	.24134	.24173	.24213	.24252	.24291	.24331	.24370
6.20	.24409	.24449	.24488	.24528	.24567	.24606	.24646	.24685	.24724	.24764
6.30	.24803	.24842	.24882	.24921	.24961	.25000	.25039	.25079	.25118	.25157
6.40	.25197	.25236	.25276	.25315	.25354	.25394	.25433	.25472	.25512	.25551
6.50	0.25590	0.25630	0.25669	0.25709	0.25748	0.25787	0.25827	0.25866	0.25905	0.25945
6.60	.25984	.26024	.26063	.26102	.26142	.26181	.26220	.26260	.26299	.26339
6.70	.26378	.26417	.26457	.26496	.26535	.26575	.26614	.26653	.26693	.26732
6.80	.26772	.26811	.26850	.26890	.26929	.26968	.27008	.27047	.27087	.27126
6.90	.27165	.27205	.27244	.27283	.27323	.27362	.27402	.27441	.27480	.27520
7.00	0.27559	0.27598	0.27638	0.27677	0.27716	0.27756	0.27795	0.27835	0.27874	0.27913
7.10	.27953	.27992	.28031	.28071	.28110	.28150	.28189	.28228	.28268	.28307
7.20	.28346	.28386	.28425	.28465	.28504	.28543	.28583	.28622	.28661	.28701
7.30	.28740	.28779	.28819	.28858	.28898	.28937	.28976	.29016	.29055	.29094
7.40	.29134	.29173	.29213	.29252	.29291	.29331	.29370	.29409	.29449	.29488
7.50	0.29528	0.29567	0.29606	0.29646	0.29685	0.29724	0.29764	0.29803	0.29842	0.29882
7.60	.29921	.29961	.30000	.30039	.30079	.30118	.30157	.30197	.30236	.30276
7.70	.30315	.30354	.30394	.30433	.30472	.30512	.30551	.30590	.30630	.30669
7.80	.30709	.30748	.30787	.30827	.30866	.30905	.30945	.30984	.31024	.31063
7.90	.31102	.31142	.31181	.31220	.31260	.31299	.31339	.31378	.31417	.31457
8.00	0.31496	0.31535	0.31575	0.31614	0.31653	0.31693	0.31732	0.31772	0.31811	0.31850
8.10	.31890	.31929	.31968	.32008	.32047	.32087	.32126	.32165	.32205	.32244
8.20	.32283	.32323	.32362	.32402	.32441	.32480	.32520	.32559	.32598	.32638
8.30	.32677	.32716	.32756	.32795	.32835	.32874	.32913	.32953	.32992	.33031
8.40	.33071	.33110	.33150	.33189	.33228	.33268	.33307	.33346	.33386	.33425

8.50	0. 33464	0. 33504	0. 33543	0. 33583	0. 33622	0. 33661	0. 33701	0. 33740	0. 33779	0. 33819
8. 60	. 33858	. 33898	. 33937	. 33976	. 34016	. 34055	. 34094	. 34134	. 34173	. 34213
8. 70	. 34252	. 34291	. 34331	. 34370	. 34409	. 34449	. 34488	. 34527	. 34567	. 34606
8. 80	. 34646	. 34685	. 34724	. 34764	. 34803	. 34842	. 34882	. 34921	. 34961	. 35000
8. 90	. 35039	. 35079	. 35118	. 35157	. 35197	. 35236	. 35276	. 35315	. 35354	. 35394
9.00	0. 35433	0. 35472	0. 35512	0. 35551	0. 35590	0. 35630	0. 35669	0. 35709	0. 35748	0. 35787
9. 10	. 35827	. 35866	. 35905	. 35945	. 35984	. 36024	. 36063	. 36102	. 36142	. 36181
9. 20	. 36220	. 36260	. 36299	. 36339	. 36378	. 36417	. 36457	. 36496	. 36535	. 36575
9. 30	. 36614	. 36653	. 36693	. 36732	. 36772	. 36811	. 36850	. 36890	. 36929	. 36968
9. 40	. 37008	. 37047	. 37087	. 37126	. 37165	. 37205	. 37244	. 37283	. 37323	. 37362
9.50	0. 37402	0. 37441	0. 37480	0. 37520	0. 37559	0. 37598	0. 37638	0. 37677	0. 37716	0. 37756
9. 60	. 37795	. 37835	. 37874	. 37913	. 37953	. 37992	. 38031	. 38071	. 38110	. 38150
9. 70	. 38189	. 38228	. 38268	. 38307	. 38346	. 38386	. 38425	. 38464	. 38504	. 38543
9. 80	. 38583	. 38622	. 38661	. 38701	. 38740	. 38779	. 38819	. 38858	. 38898	. 38937
9. 90	. 38976	. 39016	. 39055	. 39094	. 39134	. 39173	. 39213	. 39252	. 39291	. 39331
10.00	0. 39370	0. 39409	0. 39449	0. 39488	0. 39527	0. 39567	0. 39606	0. 39646	0. 39685	0. 39724
10. 10	. 39764	. 39803	. 39842	. 39882	. 39921	. 39961	. 40000	. 40039	. 40079	. 40118
10. 20	. 40157	. 40197	. 40236	. 40276	. 40315	. 40354	. 40394	. 40433	. 40472	. 40512
10. 30	. 40551	. 40590	. 40630	. 40669	. 40709	. 40748	. 40787	. 40827	. 40866	. 40905
10. 40	. 40945	. 40984	. 41024	. 41063	. 41102	. 41142	. 41181	. 41220	. 41260	. 41299
10.50	0. 41338	0. 41378	0. 41417	0. 41457	0. 41496	0. 41535	0. 41575	0. 41614	0. 41653	0. 41693
10. 60	. 41732	. 41772	. 41811	. 41850	. 41890	. 41929	. 41968	. 42008	. 42047	. 42087
10. 70	. 42126	. 42165	. 42205	. 42244	. 42283	. 42323	. 42362	. 42401	. 42441	. 42480
10. 80	. 42520	. 42559	. 42598	. 42638	. 42677	. 42716	. 42756	. 42795	. 42835	. 42874
10. 90	. 42913	. 42953	. 42992	. 43031	. 43071	. 43110	. 43150	. 43189	. 43228	. 43268
11.00	0. 43307	0. 43346	0. 43386	0. 43425	0. 43464	0. 43504	0. 43543	0. 43583	0. 43622	0. 43661
11. 10	. 43701	. 43740	. 43779	. 43819	. 43858	. 43898	. 43937	. 43976	. 44016	. 44055
11. 20	. 44094	. 44134	. 44173	. 44213	. 44252	. 44291	. 44331	. 44370	. 44409	. 44449
11. 30	. 44488	. 44527	. 44567	. 44606	. 44646	. 44685	. 44724	. 44764	. 44803	. 44842
11. 40	. 44882	. 44921	. 44961	. 45000	. 45039	. 45079	. 45118	. 45157	. 45197	. 45236
11.50	0. 45276	0. 45315	0. 45354	0. 45394	0. 45433	0. 45472	0. 45512	0. 45551	0. 45590	0. 45630
11. 60	. 45669	. 45709	. 45748	. 45787	. 45827	. 45866	. 45905	. 45945	. 45984	. 46024
11. 70	. 46063	. 46102	. 46142	. 46181	. 46220	. 46260	. 46299	. 46338	. 46378	. 46417
11. 80	. 46457	. 46496	. 46535	. 46575	. 46614	. 46653	. 46693	. 46732	. 46772	. 46811
11. 90	. 46850	. 46890	. 46929	. 46968	. 47008	. 47047	. 47087	. 47126	. 47165	. 47205

LENGTH—MILLIMETERS TO INCHES—Continued.

[From 0.00 to 25.40 millimeters by 0.01 millimeter. 1 millimeter=0.03937 inch.]

Millimeters	Hundredths of millimeters									
	0.00	0.01	0.02	0.03	0.04	0.05	0.06	0.07	0.08	0.09
	Inches	Inches	Inches	Inches	Inches	Inches	Inches	Inches	Inches	Inches
12.00	0. 47244	0. 47283	0. 47323	0. 47362	0. 47401	0. 47441	0. 47480	0. 47520	0. 47559	0. 47598
12. 10	. 47638	. 47677	. 47716	. 47756	. 47795	. 47835	. 47874	. 47913	. 47953	. 47992
12. 20	. 48031	. 48071	. 48110	. 48150	. 48189	. 48228	. 48268	. 48307	. 48346	. 48386
12. 30	. 48425	. 48464	. 48504	. 48543	. 48583	. 48622	. 48661	. 48701	. 48740	. 48779
12. 40	. 48819	. 48858	. 48898	. 48937	. 48976	. 49016	. 49055	. 49094	. 49134	. 49173
12.50	0. 49212	0. 49252	0. 49291	0. 49331	0. 49370	0. 49409	0. 49449	0. 49488	0. 49527	0. 49567
12. 60	. 49606	. 49646	. 49685	. 49724	. 49764	. 49803	. 49842	. 49882	. 49921	. 49961
12. 70	. 50000	. 50039	. 50079	. 50118	. 50157	. 50197	. 50236	. 50275	. 50315	. 50354
12. 80	. 50394	. 50433	. 50472	. 50512	. 50551	. 50590	. 50630	. 50669	.50709	. 50748
12. 90	. 50787	. 50827	. 50866	. 50905	. 50945	. 50984	. 51024	. 51063	. 51102	. 51142
13.00	0. 51181	0. 51220	0. 51260	0. 51299	0. 51338	0. 51378	0. 51417	0. 51457	0. 51496	0. 51535
13. 10	. 51575	. 51614	. 51653	. 51693	. 51732	. 51772	. 51811	. 51850	. 51890	. 51929
13. 20	. 51968	. 52008	. 52047	. 52087	. 52126	. 52165	. 52205	. 52244	. 52283	. 52323
13. 30	. 52362	. 52401	. 52441	. 52480	. 52520	. 52559	. 52598	. 52638	. 52677	. 52716
13. 40	. 52756	. 52795	. 52835	. 52874	. 52913	. 52953	. 52992	. 53031	. 53071	. 53110
13.50	0. 53150	0. 53189	0. 53228	0. 53268	0. 53307	0. 53346	0. 53386	0. 53425	0. 53464	0. 53504
13. 60	. 53543	. 53583	. 53622	. 53661	. 53701	. 53740	. 53779	. 53819	. 53858	. 53898
13. 70	. 53937	. 53976	. 54016	. 54055	. 54094	. 54134	. 54173	. 54212	. 54252	. 54291
13. 80	. 54331	. 54370	. 54409	. 54449	. 54488	. 54527	. 54567	. 54606	. 54646	. 54685
13. 90	. 54724	. 54764	. 54803	. 54842	. 54882	. 54921	. 54961	. 55000	. 55039	. 55079
14.00	0. 55118	0. 55157	0. 55197	0. 55236	0. 55275	0. 55315	0. 55354	0. 55394	0. 55433	0. 55472
14. 10	. 55512	. 55551	. 55590	. 55630	. 55669	. 55709	. 55748	. 55787	. 55827	. 55866
14. 20	. 55905	. 55945	. 55984	. 56024	. 56063	. 56102	. 56142	. 56181	. 56220	. 56260
14. 30	. 56299	. 56338	. 56378	. 56417	. 56457	. 56496	. 56535	. 56575	. 56614	. 56653
14. 40	. 56693	. 56732	. 56772	. 56811	. 56850	. 56890	. 56929	. 56968	. 57008	. 57047

14.50	0. 57086	0. 57126	0. 57165	0. 57205	0. 57244	0. 57283	0. 57323	0. 57362	0. 57401	. 57441
14. 60	. 57480	. 57520	. 57559	. 57598	. 57638	. 57677	. 57716	. 57756	. 57795	. 57835
14. 70	. 57874	. 57913	. 57953	. 57992	. 58031	. 58071	. 58110	. 58149	. 58189	. 58228
14. 80	. 58268	. 58307	. 58346	. 58386	. 58425	. 58464	. 58504	. 58543	. 58583	. 58622
14. 90	. 58661	. 58701	. 58740	. 58779	. 58819	. 58858	. 58898	. 58937	. 58976	. 59016
15.00	0. 59055	0. 59094	0. 59134	0. 59173	0. 59212	0. 59252	0. 59291	0. 59331	0. 59370	0. 59409
15. 10	. 59449	. 59488	. 59527	. 59567	. 59606	. 59646	. 59685	. 59724	. 59764	. 59803
15. 20	. 59842	. 59882	. 59921	. 59961	. 60000	. 60039	. 60079	. 60118	. 60157	. 60197
15. 30	. 60236	. 60275	. 60315	. 60354	. 60394	. 60433	. 60472	. 60512	. 60551	. 60590
15. 40	. 60630	. 60669	. 60709	. 60748	. 60787	. 60827	. 60866	. 60905	. 60945	. 60984
15.50	0. 61024	0. 61063	0. 61102	0. 61142	0. 61181	0. 61220	0. 61260	0. 61299	0. 61338	0. 61378
15. 60	. 61417	. 61457	. 61496	. 61535	. 61575	. 61614	. 61653	. 61693	. 61732	. 61772
15. 70	. 61811	. 61850	. 61890	. 61929	. 61968	. 62008	. 62047	. 62086	. 62126	. 62165
15. 80	. 62205	. 62244	. 62283	. 62323	. 62362	. 62401	. 62441	. 62480	. 62520	. 62559
15. 90	. 62598	. 62638	. 62677	. 62716	. 62756	. 62795	. 62835	. 62874	. 62913	. 62953
16.00	0. 62992	0. 63031	0. 63071	0. 63110	0. 63149	0. 63189	0. 63228	0. 63268	0. 63307	0. 63346
16. 10	. 63386	. 63425	. 63464	. 63504	. 63543	. 63583	. 63622	. 63661	. 63701	. 63740
16. 20	. 63779	. 63819	. 63858	. 63898	. 63937	. 63976	. 64016	. 64055	. 64094	. 64134
16. 30	. 64173	. 64212	. 64252	. 64291	. 64331	. 64370	. 64409	. 64449	. 64488	. 64527
16. 40	. 64567	. 64606	. 64646	. 64685	. 64724	. 64764	. 64803	. 64842	. 64882	. 64921
16.50	0. 64960	0. 65000	0. 65039	0. 65079	0. 65118	0. 65157	0. 65197	0. 65236	0. 65275	0. 65315
16. 60	. 65354	. 65394	. 65433	. 65472	. 65512	. 65551	. 65590	. 65630	. 65669	. 65709
16. 70	. 65748	. 65787	. 65827	. 65866	. 65905	. 65945	. 65984	. 66023	. 66063	. 66102
16. 80	. 66142	. 66181	. 66220	. 66260	. 66299	. 66338	. 66378	. 66417	. 66457	. 66496
16. 90	. 66535	. 66575	. 66614	. 66653	. 66693	. 66732	. 66772	. 66811	. 66850	. 66890
17.00	0. 66929	0. 66968	0. 67008	0. 67047	0. 67086	0. 67126	0. 67165	0. 67205	0. 67244	0. 67283
17. 10	. 67323	. 67362	. 67401	. 67441	. 67480	. 67520	. 67559	. 67598	. 67638	. 67677
17. 20	. 67716	. 67756	. 67795	. 67835	. 67874	. 67913	. 67953	. 67992	. 68031	. 68071
17. 30	. 68110	. 68149	. 68189	. 68228	. 68268	. 68307	. 68346	. 68386	. 68425	. 68464
17. 40	. 68504	. 68543	. 68583	. 68622	. 68661	. 68701	. 68740	. 68779	. 68819	. 68858
17.50	0. 68898	0. 68937	0. 68976	0. 69016	0. 69055	0. 69094	0. 69134	0. 69173	0. 69212	0. 69252
17. 60	. 69291	. 69331	. 69370	. 69409	. 69449	. 69488	. 69527	. 69567	. 69606	. 69646
17. 70	. 69685	. 69724	. 69764	. 69803	. 69842	. 69882	. 69921	. 69960	. 70000	. 70039
17. 80	. 70079	. 70118	. 70157	. 70197	. 70236	. 70275	. 70315	. 70354	. 70394	. 70433
17. 90	. 70472	. 70512	. 70551	. 70590	. 70630	. 70669	. 70709	. 70748	. 70787	. 70827

LENGTH—MILLIMETERS TO INCHES—Continued

[From 0.00 to 25.40 millimeters by 0.01 millimeter. 1 millimeter=0.03937 inch.]

Millimeters	Hundredths of millimeters									
	0.00	0.01	0.02	0.03	0.04	0.05	0.06	0.07	0.08	0.09
	Inches	Inches	Inches	Inches	Inches	Inches	Inches	Inches	Inches	Inches
18.00	0. 70866	0. 70905	0. 70945	0. 70984	0. 71023	0. 71063	0. 71102	0. 71142	0. 71181	0. 71220
18. 10	. 71260	. 71299	. 71338	. 71378	. 71417	. 71457	. 71496	. 71535	. 71575	. 71614
18. 20	. 71653	. 71693	. 71732	. 71772	. 71811	. 71850	. 71890	. 71929	. 71968	. 72008
18. 30	. 72047	. 72086	. 72126	. 72165	. 72205	. 72244	. 72283	. 72323	. 72362	. 72401
18. 40	. 72441	. 72480	. 72520	. 72559	. 72598	. 72638	. 72677	. 72716	. 72756	. 72795
18.50	0. 72834	0. 72874	0. 72913	0. 72953	0. 72992	0. 73031	0. 73071	0. 73110	0. 73149	0. 73189
18. 60	. 73228	. 73268	. 73307	. 73346	. 73386	. 73425	. 73464	. 73504	. 73543	. 73583
18. 70	. 73622	. 73661	. 73701	. 73740	. 73779	. 73819	. 73858	. 73897	. 73937	. 73976
18. 80	. 74016	. 74055	. 74094	. 74134	. 74173	. 74212	. 74252	. 74291	. 74331	. 74370
18. 90	. 74409	. 74449	. 74488	. 74527	. 74567	. 74606	. 74646	. 74685	. 74724	. 74764
19.00	0. 74803	0. 74842	0. 74882	0. 74921	0. 74960	0. 75000	0. 75039	0. 75079	0. 75118	0. 75157
19. 10	. 75197	. 75236	. 75275	. 75315	. 75354	. 75394	. 75433	. 75472	. 75512	. 75551
19. 20	. 75590	. 75630	. 75669	. 75709	. 75748	. 75787	. 75827	. 75866	. 75905	. 75945
19. 30	. 75984	. 76023	. 76063	. 76102	. 76142	. 76181	. 76220	. 76260	. 76299	. 76338
19. 40	. 76378	. 76417	. 76457	. 76496	. 76535	. 76575	. 76614	. 76653	. 76693	. 76732
19.50	0. 76772	0. 76811	0. 76850	0. 76890	0. 76929	0. 76968	0. 77008	0. 77047	0. 77086	0. 77126
19. 60	. 77165	. 77205	. 77244	. 77283	. 77323	. 77362	. 77401	. 77441	. 77480	. 77520
19. 70	. 77559	. 77598	. 77638	. 77677	. 77716	. 77756	. 77795	. 77834	. 77874	. 77913
19. 80	. 77953	. 77992	. 78031	. 78071	. 78110	. 78149	. 78189	. 78228	. 78268	. 78307
19. 90	. 78346	. 78386	. 78425	. 78464	. 78504	. 78543	. 78583	. 78622	. 78661	. 78701
20.00	0. 78740	0. 78779	0. 78819	0. 78858	0. 78897	0. 78937	0. 78976	0. 79016	0. 79055	0. 79094
20. 10	. 79134	. 79173	. 79212	. 79252	. 79291	. 79331	. 79370	. 79409	. 79449	. 79488
20. 20	. 79527	. 79567	. 79606	. 79646	. 79685	. 79724	. 79764	. 79803	. 79842	. 79882
20. 30	. 79921	. 79960	. 80000	. 80039	. 80079	. 80118	. 80157	. 80197	. 80236	. 30275
20. 40	. 80315	. 80354	. 80394	. 80433	. 80472	. 80512	. 80551	. 80590	. 80630	. 80669

20.50	0.80708	0.80748	0.80787	0.80827	0.80866	0.80905	0.80945	0.80984	0.81023	0.81063
20.60	.81102	.81142	.81181	.81220	.81260	.81299	.81338	.81378	.81417	.81457
20.70	.81496	.81535	.81575	.81614	.81653	.81693	.81732	.81771	.81811	.81850
20.80	.81890	.81929	.81968	.82008	.82047	.82086	.82126	.82165	.82205	.82244
20.90	.82283	.82323	.82362	.82401	.82441	.82480	.82520	.82559	.82598	.82638
21.00	0.82677	0.82716	0.82756	0.82795	0.82834	0.82874	0.82913	0.82953	0.82992	0.83031
21.10	.83071	.83110	.83149	.83189	.83228	.83268	.83307	.83346	.83386	.83425
21.20	.83464	.83504	.83543	.83583	.83622	.83661	.83701	.83740	.83779	.83819
21.30	.83858	.83897	.83937	.83976	.84016	.84055	.84094	.84134	.84173	.84212
21.40	.84252	.84291	.84331	.84370	.84409	.84449	.84488	.84527	.84567	.84606
21.50	0.84646	0.84685	0.84724	0.84764	0.84803	0.84842	0.84882	0.84921	0.84960	0.85000
21.60	.85039	.85079	.85118	.85157	.85197	.85236	.85275	.85315	.85354	.85394
21.70	.85433	.85472	.85512	.85551	.85590	.85630	.85669	.85708	.85748	.85787
21.80	.85827	.85866	.85905	.85945	.85984	.86023	.86063	.86102	.86142	.86181
21.90	.86220	.86260	.86299	.86338	.86378	.86417	.86457	.86496	.86535	.86575
22.00	0.86614	0.86653	0.86693	0.86732	0.86771	0.86811	0.86850	0.86890	0.86929	0.86968
22.10	.87008	.87047	.87086	.87126	.87165	.87205	.87244	.87283	.87323	.87362
22.20	.87401	.87441	.87480	.87520	.87559	.87598	.87638	.87677	.87716	.87756
22.30	.87795	.87834	.87874	.87913	.87953	.87992	.88031	.88071	.88110	.88149
22.40	.88189	.88228	.88268	.88307	.88346	.88386	.88425	.88464	.88504	.88543
22.50	0.88582	0.88622	0.88661	0.88701	0.88740	0.88779	0.88819	0.88858	0.88897	0.88937
22.60	.88976	.89016	.89055	.89094	.89134	.89173	.89212	.89252	.89291	.89331
22.70	.89370	.89409	.89449	.89488	.89527	.89567	.89606	.89645	.89685	.89724
22.80	.89764	.89803	.89842	.89882	.89921	.89960	.90000	.90039	.90079	.90118
22.90	.90157	.90197	.90236	.90275	.90315	.90354	.90394	.90433	.90472	.90512
23.00	0.90551	0.90590	0.90630	0.90669	0.90708	0.90748	0.90787	0.90827	0.90866	0.90905
23.10	.90945	.90984	.91023	.91063	.91102	.91142	.91181	.91220	.91260	.91299
23.20	.91338	.91378	.91417	.91457	.91496	.91535	.91575	.91614	.91653	.91693
23.30	.91732	.91771	.91811	.91850	.91890	.91929	.91968	.92008	.92047	.92086
23.40	.92126	.92165	.92205	.92244	.92283	.92323	.92362	.92401	.92441	.92480
23.50	0.92520	0.92559	0.92598	0.92638	0.92677	0.92716	0.92756	0.92795	0.92834	0.92874
23.60	.92913	.92953	.92992	.93031	.93071	.93110	.93149	.93189	.93228	.93268
23.70	.93307	.93346	.93386	.93425	.93464	.93504	.93543	.93582	.93622	.93661
23.80	.93701	.93740	.93779	.93819	.93858	.93897	.93937	.93976	.94016	.94055
23.90	.94094	.94134	.94173	.94212	.94252	.94291	.94331	.94370	.94409	.94449

LENGTH—MILLIMETERS TO INCHES—Continued

[From 0.00 to 25.40 millimeters by 0.01 millimeter 1 millimeter=0.03937 inch.]

Millimeters	Hundredths of millimeters									
	0.00	0.01	0.02	0.03	0.04	0.05	0.06	0.07	0.08	0.09
	Inches	Inches	Inches	Inches	Inches	Inches	Inches	Inches	Inches	Inches
24.00	0.94488	0.94527	0.94567	0.94606	0.94645	0.94685	0.94724	0.94764	0.94803	0.94842
24.10	.94882	.94921	.94960	.95000	.95039	.95079	.95118	.95157	.95197	.95236
24.20	.95275	.95315	.95354	.95394	.95433	.95472	.95512	.95551	.95590	.95630
24.30	.95669	.95708	.95748	.95787	.95827	.95866	.95905	.95945	.95984	.96023
24.40	.96063	.96102	.96142	.96181	.96220	.96260	.96299	.96338	.96378	.96417
24.50	0.96456	0.96496	0.96535	0.96575	0.96614	0.96653	0.96693	0.96732	0.96771	0.96811
24.60	.96850	.96890	.96929	.96968	.97008	.97047	.97086	.97126	.97165	.97205
24.70	.97244	.97283	.97323	.97362	.97401	.97441	.97480	.97519	.97559	.97598
24.80	.97638	.97677	.97716	.97756	.97795	.97834	.97874	.97913	.97953	.97992
24.90	.98031	.98071	.98110	.98149	.98189	.98228	.98268	.98307	.98346	.98386
25.00	0.98425	0.98464	0.98504	0.98543	0.98582	0.98622	0.98661	0.98701	0.98740	0.98779
25.10	.98819	.98858	.98897	.98937	.98976	.99016	.99055	.99094	.99134	.99173
25.20	.99212	.99252	.99291	.99331	.99370	.99409	.99449	.99488	.99527	.99567
25.30	.99606	.99645	.99685	.99724	.99764	.99803	.99842	.99882	.99921	.99960
25.40	1.00000	....	....	....	....	....	....	....	....	....

"TWO-WAY" CONVERSION TABLES

TABLES OF EQUIVALENTS FROM 1 TO 999 UNITS

LENGTH—FEET TO METERS

[Reduction factor: 1 foot = 0.3048006096 meter]

Feet	Meters	Feet	Meters	Feet	Meters	Feet	Meters	Feet	Meters	Feet	Meters	Feet	Meters	Feet	Meters	Feet	Meters	Feet	Meters
0		**100**	30.48006	**200**	60.96012	**300**	91.44018	**400**	121.92024	**500**	152.40030	**600**	182.88037	**700**	213.36043	**800**	243.84049	**900**	274.32055
1	0.30480	1	30.78486	1	61.26492	1	91.74498	1	122.22504	1	152.70511	1	183.18517	1	213.66523	1	244.14529	1	274.62535
2	.60960	2	31.08966	2	61.56972	2	92.04978	2	122.52985	2	153.00991	2	183.48997	2	213.97003	2	244.45009	2	274.93015
3	.91440	3	31.39446	3	61.87452	3	92.35458	3	122.83465	3	153.31471	3	183.79477	3	214.27483	3	244.75489	3	275.23495
4	1.21920	4	31.69926	4	62.17932	4	92.65939	4	123.13945	4	153.61951	4	184.09957	4	214.57963	4	245.05969	4	275.53975
5	1.52400	5	32.00406	5	62.48412	5	92.96419	5	123.44425	5	153.92431	5	184.40437	5	214.88443	5	245.36449	5	275.84455
6	1.82880	6	32.30886	6	62.78893	6	93.26899	6	123.74905	6	154.22911	6	184.70917	6	215.18923	6	245.66929	6	276.14935
7	2.13360	7	32.61367	7	63.09373	7	93.57379	7	124.05385	7	154.53391	7	185.01397	7	215.49403	7	245.97409	7	276.45415
8	2.43840	8	32.91847	8	63.39853	8	93.87859	8	124.35865	8	154.83871	8	185.31877	8	215.79883	8	246.27889	8	276.75895
9	2.74321	9	33.22327	9	63.70333	9	94.18339	9	124.66345	9	155.14351	9	185.62357	9	216.10363	9	246.58369	9	277.06375
10	3.04801	**110**	33.52807	**210**	64.00813	**310**	94.48819	**410**	124.96825	**510**	155.44831	**610**	185.92837	**710**	216.40843	**810**	246.88849	**910**	277.36855
1	3.35281	1	33.83287	1	64.31293	1	94.79299	1	125.27305	1	155.75311	1	186.23317	1	216.71323	1	247.19329	1	277.67336
2	3.65761	2	34.13767	2	64.61773	2	95.09779	2	125.57785	2	156.05791	2	186.53797	2	217.01803	2	247.49809	2	277.97816
3	3.96241	3	34.44247	3	64.92253	3	95.40259	3	125.88265	3	156.36271	3	186.84277	3	217.32283	3	247.80290	3	278.28296
4	4.26721	4	34.74727	4	65.22733	4	95.70739	4	126.18745	4	156.66751	4	187.14757	4	217.62764	4	248.10770	4	278.58776
5	4.57201	5	35.05207	5	65.53213	5	96.01219	5	126.49225	5	156.97231	5	187.45237	5	217.93244	5	248.41250	5	278.89256
6	4.87681	6	35.35687	6	65.83693	6	96.31699	6	126.79705	6	157.27711	6	187.75718	6	218.23724	6	248.71730	6	279.19736
7	5.18161	7	35.66167	7	66.14173	7	96.62179	7	127.10185	7	157.58192	7	188.06198	7	218.54204	7	249.02210	7	279.50216
8	5.48641	8	35.96647	8	66.44653	8	96.92659	8	127.40665	8	157.88672	8	188.36678	8	218.84684	8	249.32690	8	279.80696
9	5.79121	9	36.27127	9	66.75133	9	97.23139	9	127.71146	9	158.19152	9	188.67158	9	219.15164	9	249.63170	9	280.11176
20	6.09601	**120**	36.57607	**220**	67.05613	**320**	97.53620	**420**	128.01626	**520**	158.49632	**620**	188.97638	**720**	219.45644	**820**	249.93650	**920**	280.41656
1	6.40081	1	36.88087	1	67.36093	1	97.84100	1	128.32106	1	158.80112	1	189.28118	1	219.76124	1	250.24130	1	280.72136
2	6.70561	2	37.18567	2	67.66574	2	98.14580	2	128.62586	2	159.10592	2	189.58598	2	220.06604	2	250.54610	2	281.02616
3	7.01041	3	37.49047	3	67.97054	3	98.45060	3	128.93066	3	159.41072	3	189.89078	3	220.37084	3	250.85090	3	281.33096
4	7.31521	4	37.79528	4	68.27534	4	98.75540	4	129.23546	4	159.71552	4	190.19558	4	220.67564	4	251.15570	4	281.63576
5	7.62002	5	38.10008	5	68.58014	5	99.06020	5	129.54026	5	160.02032	5	190.50038	5	220.98044	5	251.46050	5	281.94056
6	7.92482	6	38.40488	6	68.88494	6	99.36500	6	129.84506	6	160.32512	6	190.80518	6	221.28524	6	251.76530	6	282.24536
7	8.22962	7	38.70968	7	69.18974	7	99.66980	7	130.14986	7	160.62992	7	191.10998	7	221.59004	7	252.07010	7	282.55017
8	8.53442	8	39.01448	8	69.49454	8	99.97460	8	130.45466	8	160.93472	8	191.41478	8	221.89484	8	252.37490	8	282.85497
9	8.83922	9	39.31928	9	69.79934	9	100.27940	9	130.75946	9	161.23952	9	191.71958	9	222.19964	9	252.67971	9	283.15977
30	9.14402	**130**	39.62408	**230**	70.10414	**330**	100.58420	**430**	131.06426	**530**	161.54432	**630**	192.02438	**730**	222.50445	**830**	252.98451	**930**	283.46457
1	9.44882	1	39.92888	1	70.40894	1	100.88900	1	131.36906	1	161.84912	1	192.32918	1	222.80925	1	253.28931	1	283.76937
2	9.75362	2	40.23368	2	70.71374	2	101.19380	2	131.67386	2	162.15392	2	192.63399	2	223.11405	2	253.59411	2	284.07417
3	10.05842	3	40.53848	3	71.01854	3	101.49860	3	131.97866	3	162.45872	3	192.93879	3	223.41885	3	253.89891	3	284.37897
4	10.36322	4	40.84328	4	71.32334	4	101.80340	4	132.28346	4	162.76353	4	193.24359	4	223.72365	4	254.20371	4	284.68377
5	10.66802	5	41.14808	5	71.62814	5	102.10820	5	132.58827	5	163.06833	5	193.54839	5	224.02845	5	254.50851	5	284.98857
6	10.97282	6	41.45288	6	71.93294	6	102.41300	6	132.89307	6	163.37313	6	193.85319	6	224.33325	6	254.81331	6	285.29337
7	11.27762	7	41.75768	7	72.23774	7	102.71781	7	133.19787	7	163.67793	7	194.15799	7	224.63805	7	255.11811	7	285.59817
8	11.58242	8	42.06248	8	72.54255	8	103.02261	8	133.50267	8	163.98273	8	194.46279	8	224.94285	8	255.42291	8	285.90297
9	11.88722	9	42.36728	9	72.84735	9	103.32741	9	133.80747	9	164.28753	9	194.76759	9	225.24765	9	255.72771	9	286.20777
40	12.19202	**140**	42.67209	**240**	73.15215	**340**	103.63221	**440**	134.11227	**540**	164.59233	**640**	195.07239	**740**	225.55245	**840**	256.03251	**940**	286.51257
1	12.49682	1	42.97689	1	73.45695	1	103.93701	1	134.41707	1	164.89713	1	195.37719	1	225.85725	1	256.33731	1	286.81737
2	12.80163	2	43.28169	2	73.76175	2	104.24181	2	134.72187	2	165.20193	2	195.68199	2	226.16205	2	256.64211	2	287.12217
3	13.10643	3	43.58649	3	74.06655	3	104.54661	3	135.02667	3	165.50673	3	195.98679	3	226.46685	3	256.94691	3	287.42697
4	13.41123	4	43.89129	4	74.37135	4	104.85141	4	135.33147	4	165.81153	4	196.29159	4	226.77165	4	257.25171	4	287.73178
5	13.71603	5	44.19609	5	74.67615	5	105.15621	5	135.63627	5	166.11633	5	196.59639	5	227.07645	5	257.55652	5	288.03658
6	14.02083	6	44.50089	6	74.98095	6	105.46101	6	135.94107	6	166.42113	6	196.90119	6	227.38125	6	257.86132	6	288.34138
7	14.32563	7	44.80569	7	75.28575	7	105.76581	7	136.24587	7	166.72593	7	197.20599	7	227.68606	7	258.16612	7	288.64618
8	14.63043	8	45.11049	8	75.59055	8	106.07061	8	136.55067	8	167.03073	8	197.51080	8	227.99086	8	258.47092	8	288.95098
9	14.93523	9	45.41529	9	75.89535	9	106.37541	9	136.85547	9	167.33553	9	197.81560	9	228.29566	9	258.77572	9	289.25578

50	15. 24003	**150**	45. 72009	**250**	76. 20015	**350**	106. 68021	**450**	137. 16027	**550**	167. 64034	**650**	198. 12040	**750**	228. 60046	**850**	259. 08052	**950**	289. 56058
1	15. 54483	1	46. 02489	1	76. 50495	1	106. 98501	1	137. 46507	1	167. 94514	1	198. 42520	1	228. 90526	1	259. 38532	1	289. 86538
2	15. 84963	2	46. 32969	2	76. 80975	2	107. 28981	2	137. 76988	2	168. 24994	2	198. 73000	2	229. 21006	2	259. 69012	2	290. 17018
3	16. 15443	3	46. 63449	3	77. 11455	3	107. 59462	3	138. 07468	3	168. 55474	3	199. 03480	3	229. 51486	3	259. 99492	3	290. 47498
4	16. 45923	4	46. 93929	4	77. 41935	4	107. 89942	4	138. 37948	4	168. 85954	4	199. 33960	4	229. 81966	4	260. 29972	4	290. 77978
5	16. 76403	5	47. 24409	5	77. 72416	5	108. 20422	5	138. 68423	5	169. 16434	5	199. 64440	5	230. 12446	5	260. 60452	5	291. 08458
6	17. 06883	6	47. 54890	6	78. 02896	6	108. 50902	6	138. 98903	6	169. 46914	6	199. 94920	6	230. 42926	6	260. 90932	6	291. 33938
7	17. 37363	7	47. 85370	7	78. 33376	7	108. 81382	7	139. 29388	7	169. 77394	7	200. 25400	7	230. 73406	7	261. 21412	7	291. 69418
8	17. 67844	8	48. 15850	8	78. 63856	8	109. 11862	8	139. 59868	8	170. 07874	8	200. 55880	8	231. 03886	8	261. 51892	8	291. 99898
9	17. 98324	9	48. 46330	9	78. 94336	9	109. 42342	9	139. 90348	9	170. 38354	9	200. 86360	9	231. 34366	9	261. 82372	9	292. 30378
60	18. 28804	**160**	48. 76810	**260**	79. 24816	**360**	109. 72822	**460**	140. 20828	**560**	170. 68834	**660**	201. 16840	**760**	231. 64846	**860**	262. 12852	**960**	292. 60859
1	18. 59284	1	49. 07290	1	79. 55296	1	110. 03302	1	140. 51308	1	170. 99314	1	201. 47320	1	231. 95326	1	262. 43332	1	292. 91339
2	18. 89764	2	49. 37770	2	79. 85776	2	110. 33782	2	140. 81788	2	171. 29794	2	201. 77800	2	232. 25806	2	262. 73813	2	293. 21819
3	19. 20244	3	49. 68250	3	80. 16256	3	110. 64262	3	141. 12268	3	171. 60274	3	202. 08280	3	232. 56287	3	263. 04293	3	293. 52299
4	19. 50724	4	49. 98730	4	80. 46736	4	110. 94742	4	141. 42748	4	171. 90754	4	202. 38760	4	232. 86767	4	263. 34773	4	293. 82779
5	19. 81204	5	50. 29210	5	80. 77216	5	111. 25222	5	141. 73228	5	172. 21234	5	202. 69241	5	233. 17247	5	263. 65253	5	294. 13259
6	20. 11684	6	50. 59690	6	81. 07696	6	111. 55702	6	142. 03708	6	172. 51715	6	202. 99721	6	233. 47727	6	263. 95733	6	294. 43739
7	20. 42164	7	50. 90170	7	81. 38176	7	111. 86182	7	142. 34188	7	172. 82195	7	203. 30201	7	233. 78207	7	264. 26213	7	294. 74219
8	20. 72644	8	51. 20650	8	81. 68656	8	112. 16662	8	142. 64669	8	173. 12675	8	203. 60681	8	234. 08687	8	264. 56693	8	295. 04699
9	21. 03124	9	51. 51130	9	81. 99136	9	112. 47142	9	142. 95149	9	173. 43155	9	203. 91161	9	234. 39167	9	264. 87173	9	295. 35179
70	21. 33604	**170**	51. 81610	**270**	82. 29616	**370**	112. 77623	**470**	143. 25629	**570**	173. 73635	**670**	204. 21641	**770**	234. 69647	**870**	265. 17653	**970**	295. 65659
1	21. 64084	1	52. 12090	1	82. 60097	1	113. 08103	1	143. 56109	1	174. 04115	1	204. 52121	1	235. 00127	1	265. 48133	1	295. 96139
2	21. 94564	2	52. 42570	2	82. 90577	2	113. 38583	2	143. 86589	2	174. 34595	2	204. 82601	2	235. 30607	2	265. 78613	2	296. 26619
3	22. 25044	3	52. 73051	3	83. 21057	3	113. 69063	3	144. 17069	3	174. 65075	3	205. 13081	3	235. 61087	3	266. 09093	3	296. 57099
4	22. 55525	4	53. 03531	4	83. 51537	4	113. 99543	4	144. 47549	4	174. 95555	4	205. 43561	4	235. 91567	4	266. 39573	4	296. 87579
5	22. 86005	5	53. 34011	5	83. 82017	5	114. 30023	5	144. 78029	5	175. 26035	5	205. 74041	5	236. 22047	5	266. 70053	5	297. 18059
6	23. 16485	6	53. 64491	6	84. 12497	6	114. 60503	6	145. 08509	6	175. 56515	6	206. 04521	6	236. 52527	6	267. 00533	6	297. 48539
7	23. 46965	7	53. 94971	7	84. 42977	7	114. 90983	7	145. 38989	7	175. 86995	7	206. 35001	7	236. 83007	7	267. 31013	7	297. 79020
8	23. 77445	8	54. 25451	8	84. 73457	8	115. 21463	8	145. 69469	8	176. 17475	8	206. 65481	8	237. 13487	8	267. 61494	8	298. 09500
9	24. 07925	9	54. 55931	9	85. 03937	9	115. 51943	9	145. 99949	9	176. 47955	9	206. 95961	9	237. 43967	9	267. 91974	9	298. 39980
80	24. 38405	**180**	54. 86411	**280**	85. 34417	**380**	115. 82423	**480**	146. 30429	**580**	176. 78435	**680**	207. 26441	**780**	237. 74448	**880**	268. 22454	**980**	298. 70460
1	24. 68885	1	55. 16891	1	85. 64897	1	116. 12903	1	146. 60909	1	177. 08915	1	207. 56922	1	238. 04928	1	268. 52934	1	299. 00940
2	24. 99365	2	55. 47371	2	85. 95377	2	116. 43383	2	146. 91389	2	177. 39395	2	207. 87402	2	238. 35408	2	268. 83414	2	299. 31420
3	25. 29845	3	55. 77851	3	86. 25857	3	116. 73863	3	147. 21869	3	177. 69876	3	208. 17882	3	238. 65888	3	269. 13894	3	299. 61900
4	25. 60325	4	56. 08331	4	86. 56337	4	117. 04343	4	147. 52350	4	178. 00356	4	208. 48362	4	238. 96368	4	269. 44374	4	299. 92380
5	25. 90805	5	56. 38811	5	86. 86817	5	117. 34823	5	147. 82830	5	178. 30836	5	208. 78842	5	239. 26848	5	269. 74854	5	300. 22860
6	26. 21285	6	56. 69291	6	87. 17297	6	117. 65304	6	148. 13310	6	178. 61316	6	209. 09322	6	239. 57328	6	270. 05334	6	300. 53340
7	26. 51765	7	56. 99771	7	87. 47777	7	117. 95784	7	148. 43790	7	178. 91796	7	209. 39802	7	239. 87808	7	270. 35814	7	300. 83820
8	26. 82245	8	57. 30251	8	87. 78258	8	118. 26264	8	148. 74270	8	179. 22276	8	209. 70282	8	240. 18288	8	270. 66294	8	301. 14300
9	27. 12725	9	57. 60732	9	88. 08738	9	118. 56744	9	149. 04750	9	179. 52756	9	210. 00762	9	240. 48768	9	270. 96774	9	301. 44780
90	27. 43205	**190**	57. 91212	**290**	88. 39218	**390**	118. 87224	**490**	149. 35230	**590**	179. 83236	**690**	210. 31242	**790**	240. 79248	**890**	271. 27254	**990**	301 75260
1	27. 73686	1	58. 21692	1	88. 69698	1	119. 17704	1	149. 65710	1	180. 13716	1	210. 61722	1	241. 09728	1	271. 57734	1	302. 05740
2	28. 04166	2	58. 52172	2	89. 00178	2	119. 48184	2	149. 96190	2	180. 44196	2	210. 92202	2	241. 40208	2	271. 88214	2	302. 36220
3	28. 34646	3	58. 82652	3	89. 30658	3	119. 78664	3	150. 26670	3	180. 74676	3	211. 22682	3	241. 70688	3	272. 18694	3	302. 66701
4	28. 65126	4	59. 13132	4	89. 61138	4	120. 09144	4	150. 57150	4	181. 05156	4	211. 53162	4	242. 01168	4	272. 49174	4	302. 97181
5	28. 95606	5	59. 43612	5	89. 91618	5	120. 39624	5	150. 87630	5	181. 35636	5	211. 83642	5	242. 31648	5	272. 79655	5	303. 27661
6	29. 26086	6	59. 74092	6	90. 22098	6	120. 70104	6	151. 18110	6	181. 66116	6	212. 14122	6	242. 62129	6	273. 10135	6	303. 58141
7	29. 56566	7	60. 04572	7	90. 52578	7	121. 00584	7	151. 48590	7	181. 96596	7	212. 44602	7	242. 92609	7	273. 40615	7	303. 88621
8	29. 87046	8	60. 35052	8	90. 83058	8	121. 31064	8	151. 79070	8	182. 27076	8	212. 75083	8	243. 23089	8	273. 71095	8	304. 19101
9	30. 17526	9	60. 65532	9	91. 13538	9	121. 61544	9	152. 09550	9	182. 57557	9	213. 05563	9	243. 53569	9	274. 01575	9	304. 49581

1 inch = 0.02540 meter	**4 inches = 0.10160 meter**	**7 inches = 0.17780 meter**	**10 inches = 0.25400 meter**
2 inches = .05080 meter	**5 inches = .12700 meter**	**8 inches = .20320 meter**	**11 inches = .27940 meter**
3 inches = .07620 meter	**6 inches = .15240 meter**	**9 inches = .22860 meter**	**12 inches = .30480 meter**

LENGTH—METERS TO FEET

[Reduction factor: 1 meter = 3.280833333 feet]

Meters	Feet	Meters	Feet	Meters	Feet	Meters	Feet	Meters	Feet	Meters	Feet	Meters	Feet	Meters	Feet	Meters	Feet	Meters	Feet
0		**100**	328.08333	**200**	656.16667	**300**	984.25000	**400**	1,312.33333	**500**	1,640.41667	**600**	1,968.50000	**700**	2,296.58333	**800**	2,624.66667	**900**	2,952.75000
1	3.28083	1	331.36417	1	659.44750	1	987.53083	1	1,315.61417	1	1,643.69750	1	1,971.78083	1	2,299.86417	1	2,627.94750	1	2,956.03083
2	6.56167	2	334.64500	2	662.72833	2	990.81167	2	1,318.89500	2	1,646.97833	2	1,975.06167	2	2,303.14500	2	2,631.22833	2	2,959.31167
3	9.84250	3	337.92583	3	666.00917	3	994.09250	3	1,322.17583	3	1,650.25917	3	1,978.34250	3	2,306.42583	3	2,634.50917	3	2,962.59250
4	13.12333	4	341.20667	4	669.29000	4	997.37333	4	1,325.45667	4	1,653.54000	4	1,981.62333	4	2,309.70667	4	2,637.79000	4	2,965.87333
5	16.40417	5	344 48750	5	672.57083	5	1,000.65417	5	1,328.73750	5	1,656.82083	5	1,984.90417	5	2,312.98750	5	2,641.07083	5	2,969.15417
6	19.68500	6	347.76833	6	675.85167	6	1,003.93500	6	1,332.01833	6	1,660.10167	6	1,988.18500	6	2,316.26833	6	2,644.35167	6	2,972.43500
7	22.96583	7	351.04917	7	679.13250	7	1,007.21583	7	1,335.29917	7	1,663.38250	7	1,991.46583	7	2,319.54917	7	2,647.63250	7	2,975.71583
8	26.24667	8	354.33000	8	682.41333	8	1,010.49667	8	1,338.58000	8	1,666.66333	8	1,994.74667	8	2,322.83000	8	2,650.91333	8	2,978.99667
9	29.52750	9	357.61083	9	685.69417	9	1,013.77750	9	1,341.86083	9	1,669.94417	9	1,998.02750	9	2,326.11083	9	2,654.19417	9	2,982.27750
10	32.80833	**110**	360.89167	**210**	688.97500	**310**	1,017.05833	**410**	1,345.14167	**510**	1,673.22500	**610**	2,001.30833	**710**	2,329.39167	**810**	2,657.47500	**910**	2,985.55833
1	36.08917	1	364.17250	1	692.25583	1	1,020 33917	1	1,348.42250	1	1,676.50583	1	2,004.58917	1	2,332.67250	1	2,660.75583	1	2,988.83917
2	39.37000	2	367.45333	2	695.53667	2	1,023.62000	2	1,351.70333	2	1,679.78667	2	2,307.87000	2	2,335.95333	2	2,664.03667	2	2,992.12000
3	42.65083	3	370.73417	3	698.81750	3	1,026.90083	3	1,354.98417	3	1,683.06750	3	2,011.15083	3	2,339.23417	3	2,667.31750	3	2,995.40083
4	45.93167	4	374.01500	4	702.09833	4	1,030.18167	4	1,358.26500	4	1,686.34833	4	2,014.43167	4	2,342.51500	4	2,670.59833	4	2,998.68167
5	49.21250	5	377.29583	5	705.37917	5	1,033.46250	5	1,361.54583	5	1,689.62917	5	2,017.71250	5	2,345.79583	5	2,673.87917	5	3,001.96250
6	52.49333	6	380.57667	6	708.66000	6	1,036.74333	6	1,364.82667	6	1,692.91000	6	2,020.99333	6	2,349.07667	6	2,677.16000	6	3,005.24333
7	55.77417	7	383.85750	7	711.94083	7	1,040.02417	7	1,368.10750	7	1,696.19083	7	2,024.27417	7	2,352.35750	7	2,680.44083	7	3,008.52417
8	59.05500	8	387.13833	8	715.22167	8	1,043.30500	8	1,371.38833	8	1,699.47167	8	2,027.55500	8	2,355.63833	8	2,683.72167	8	3,011.80500
9	62.33583	9	390.41917	9	718.50250	9	1,046.58583	9	1,374.66917	9	1,702.75250	9	2,030.83583	9	2,358.91917	9	2,687.00250	9	3,015.08583
20	65.61667	**120**	393.70000	**220**	721.78333	**320**	1,049.86667	**420**	1,377.95000	**520**	1,706.03333	**620**	2,034.11667	**720**	2,362.20000	**820**	2,690.28333	**920**	3,018.36667
1	68.89750	1	396.98083	1	725.06417	1	1,053.14750	1	1,381.23083	1	1,709.31417	1	2,037.39750	1	2,365.48083	1	2,693.56417	1	3,021.64750
2	72.17833	2	400.26167	2	728.34500	2	1,056.42833	2	1,384.51167	2	1,712.59500	2	2,040.67833	2	2,368.76167	2	2,696.84500	2	3,024.92833
3	75.45917	3	403.54250	3	731.62583	3	1,059.70917	3	1,387.79250	3	1,715.87583	3	2,043.95917	3	2,372.04250	3	2,700.12583	3	3,028.20917
4	78.74000	4	406.82333	4	734.90667	4	1,062.99000	4	1,391.07333	4	1,719.15667	4	2,047.24000	4	2,375.32333	4	2,703.40667	4	3,031.49000
5	82.02083	5	410.10417	5	738.18750	5	1,066.27083	5	1,394.35417	5	1,722.43750	5	2,050.52083	5	2,378.60417	5	2,706.68750	5	3,034.77083
6	85.30167	6	413.38500	6	741.46833	6	1,069.55167	6	1,397.63500	6	1,725.71833	6	2,053.80167	6	2,381.88500	6	2,709.96833	6	3,038.05167
7	88.58250	7	416.66583	7	744.74917	7	1,072.83250	7	1,400.91583	7	1,728.99917	7	2,057.08250	7	2,385.16583	7	2,713.24917	7	3,041.33250
8	91.86333	8	419.94667	8	748.03000	8	1,076.11333	8	1,404.19667	8	1,732.28000	8	2,060.36333	8	2,388.44667	8	2,716.53000	8	3,044.61333
9	95.14417	9	423.22750	9	751.31083	9	1,079.39417	9	1,407.47750	9	1,735.56083	9	2,063.64417	9	2,391.72750	9	2,719.81083	9	3,047.89417
30	98.42500	**130**	426.50833	**230**	754.59167	**330**	1,082.67500	**430**	1,410.75833	**530**	1,738.84167	**630**	2,066.92500	**730**	2,395.00833	**830**	2,723.09167	**930**	3,051.17500
1	101.70583	1	429.78917	1	757.87250	1	1,085.95583	1	1,414.03917	1	1,742.12250	1	2,070.20583	1	2,398.28917	1	2,726.37250	1	3,054.45583
2	104.98667	2	433.07000	2	761.15333	2	1,089.23667	2	1,417.32000	2	1,745.40333	2	2,073.48667	2	2,401.57000	2	2,729.65333	2	3,057.73667
3	108.26750	3	436.35083	3	764.43417	3	1,092.51750	3	1,420.60083	3	1,748.68417	3	2,076.76750	3	2,404.85083	3	2,732.93417	3	3,061.01750
4	111.54833	4	439.63167	4	767.71500	4	1,095.79833	4	1,423.88167	4	1,751.96500	4	2,080.04833	4	2,408.13167	4	2,736.21500	4	3,064.29833
5	114.82917	5	442.91250	5	770.99583	5	1,099.07917	5	1,427.16250	5	1,755.24583	5	2,083.32917	5	2,411.41250	5	2,739.49583	5	3,067.57917
6	118.11000	6	446.19333	6	774.27667	6	1,102.36000	6	1,430.44333	6	1,758.52667	6	2,086.61000	6	2,414.69333	6	2,742.77667	6	3,070.86000
7	121.39083	7	449.47417	7	777.55750	7	1,105.64083	7	1,433.72417	7	1,761.80750	7	2,089.89083	7	2,417.97417	7	2,746.05750	7	3,074.14083
8	124.67167	8	452.75500	8	780.83833	8	1,108.92167	8	1,437.00500	8	1,765.08833	8	2,093.17167	8	2,421.25500	8	2,749.33833	8	3,077.42167
9	127.95250	9	456.03583	9	784.11917	9	1,112.20250	9	1,440.28583	9	1,768.36917	9	2,096.45250	9	2,424.53583	9	2,752.61917	9	3,080.70250
40	131.23333	**140**	459.31667	**240**	787.40000	**340**	1,115.48333	**440**	1,443.56667	**540**	1,771.65000	**640**	2,099.73333	**740**	2,427.81667	**840**	2,755.90000	**940**	3,083.98333
1	134.51417	1	462.59750	1	790.68083	1	1,118.76417	1	1,446.84750	1	1,774.93083	1	2,103.01417	1	2,431.09750	1	2,759.18083	1	3,087.26417
2	137.79500	2	465.87833	2	793.96167	2	1,122.04500	2	1,450.12833	2	1,778.21167	2	2,106.29500	2	2,434.37833	2	2,762.46167	2	3,090.54500
3	141.07583	3	469.15917	3	797.24250	3	1,125.32583	3	1,453.40917	3	1,781.49250	3	2,109.57583	3	2,437.65917	3	2,765.74250	3	3,093.82583
4	144.35667	4	472.44000	4	800.52333	4	1,128.60667	4	1,456.69000	4	1,784.77333	4	2,112.85667	4	2,440.94000	4	2,769.02333	4	3,097.10667
5	147.63750	5	475.72083	5	803.80417	5	1,131.88750	5	1,459.97083	5	1,788.05417	5	2,116.13750	5	2,444.22083	5	2,772.30417	5	3,100.38750
6	150.91833	6	479.00167	6	807.08500	6	1,135.16833	6	1,463.25167	6	1,791.33500	6	2,119.41833	6	2,447.50167	6	2,775.58500	6	3,103.66833
7	154.19917	7	482.28250	7	810.36583	7	1,138.44917	7	1,466.53250	7	1,794.61583	7	2,122.69917	7	2,450.78250	7	2,778.86583	7	3,106.94917
8	157.48000	8	485.56333	8	813.64667	8	1,141.73000	8	1,469.81333	8	1,797.89667	8	2,125.98000	8	2,454.06333	8	2,782.14667	8	3,110.23000
9	160.76083	9	488.84417	9	816.92750	9	1,145.01083	9	1,473.09417	9	1,801.17750	9	2,129.26083	9	2,457.34417	9	2,785.42750	9	3,113.51083

50	164.04167	150	492.12500	250	820.20833	350	1,148.29167	450	1,476.37500	550	1,804.45833	650	2,132.54167	750	2,460.62500	850	2,788.70833	950	3,116.79167
1	167.32250	1	495.40583	1	823.48917	1	1,151.57250	1	1,479.65583	1	1,807.73917	1	2,135.82250	1	2,463.90583	1	2,791.98917	1	3,120.07250
2	170.60333	2	498.68667	2	826.77000	2	1,154.85333	2	1,482.93667	2	1,811.02000	2	2,139.10333	2	2,467.18667	2	2,795.27000	2	3,123.35333
3	173.88417	3	501.96750	3	830.05083	3	1,158.13417	3	1,486.21750	3	1,814.30083	3	2,142.38417	3	2,470.46750	3	2,798.55083	3	3,126.63417
4	177.16500	4	505.24833	4	833.33167	4	1,161.41500	4	1,489.49833	4	1,817.58167	4	2,145.66500	4	2,473.74833	4	2,801.83167	4	3,129.91500
5	180.44583	5	508.52917	5	836.61250	5	1,164.69583	5	1,492.77917	5	1,820.86250	5	2,148.94583	5	2,477.02917	5	2,805.11250	5	3,133.19583
6	183.72667	6	511.81000	6	839.89333	6	1,167.97667	6	1,496.06000	6	1,824.14333	6	2,152.22667	6	2,480.31000	6	2,808.39333	6	3,136.47667
7	187.00750	7	515.09083	7	843.17417	7	1,171.25750	7	1,499.34083	7	1,827.42417	7	2,155.50750	7	2,483.59083	7	2,811.67417	7	3,139.75750
8	190.28833	8	518.37167	8	846.45500	8	1,174.53833	8	1,502.62167	8	1,830.70500	8	2,158.78833	8	2,486.87167	8	2,814.95500	8	3,143.03833
9	193.56917	9	521.65250	9	849.73583	9	1,177.81917	9	1,505.90250	9	1,833.98583	9	2,162.06917	9	2,490.15250	9	2,818.23583	9	3,146.31917
60	196.85000	160	524.93333	260	853.01667	360	1,181.10000	460	1,509.18333	560	1,837.26667	660	2,165.35000	760	2,493.43333	860	2,821.51667	960	3,149.60000
1	200.13083	1	528.21417	1	856.29750	1	1,184.38083	1	1,512.46417	1	1,840.54750	1	2,168.63083	1	2,496.71417	1	2,824.79750	1	3,152.88083
2	203.41167	2	531.49500	2	859.57833	2	1,187.66167	2	1,515.74500	2	1,843.82833	2	2,171.91167	2	2,499.99500	2	2,828.07833	2	3,156.16167
3	206.69250	3	534.77583	3	862.85917	3	1,190.94250	3	1,519.02583	3	1,847.10917	3	2,175.19250	3	2,503.27583	3	2,831.35917	3	3,159.44250
4	209.97333	4	538.05667	4	866.14000	4	1,194.22333	4	1,522.30667	4	1,850.39000	4	2,178.47333	4	2,506.55667	4	2,834.64000	4	3,162.72333
5	213.25417	5	541.33750	5	869.42083	5	1,197.50417	5	1,525.58750	5	1,853.67083	5	2,181.75417	5	2,509.83750	5	2,837.92083	5	3,166.00417
6	216.53500	6	544.61833	6	872.70167	6	1,200.78500	6	1,528.86833	6	1,856.95167	6	2,185.03500	6	2,513.11833	6	2,841.20167	6	3,169.28500
7	219.81583	7	547.89917	7	875.98250	7	1,204.06583	7	1,532.14917	7	1,860.23250	7	2,188.31583	7	2,516.39917	7	2,844.48250	7	3,172.56583
8	223.09667	8	551.18000	8	879.26333	8	1,207.34667	8	1,535.43000	8	1,863.51333	8	2,191.59667	8	2,519.68000	8	2,847.76333	8	3,175.84667
9	226.37750	9	554.46083	9	882.54417	9	1,210.62750	9	1,538.71083	9	1,866.79417	9	2,194.87750	9	2,522.96083	9	2,851.04417	9	3,179.12750
70	229.65833	170	557.74167	270	885.82500	370	1,213.90833	470	1,541.99167	570	1,870.07500	670	2,198.15833	770	2,526.24167	870	2,854.32500	970	3,182.40833
1	232.93917	1	561.02250	1	889.10583	1	1,217.18917	1	1,545.27250	1	1,873.35583	1	2,201.43917	1	2,529.52250	1	2,857.60583	1	3,185.68917
2	236.22000	2	564.30333	2	892.38667	2	1,220.47000	2	1,548.55333	2	1,876.63667	2	2,204.72000	2	2,532.80333	2	2,860.88667	2	3,188.97000
3	239.50083	3	567.58417	3	895.66750	3	1,223.75083	3	1,551.83417	3	1,879.91750	3	2,208.00083	3	2,536.08417	3	2,864.16750	3	3,192.25083
4	242.78167	4	570.86500	4	898.94833	4	1,227.03167	4	1,555.11500	4	1,883.19833	4	2,211.28167	4	2,539.36500	4	2,867.44833	4	3,195.53167
5	246.06250	5	574.14583	5	902.22917	5	1,230.31250	5	1,558.39583	5	1,886.47917	5	2,214.56250	5	2,542.64583	5	2,870.72917	5	3,198.81250
6	249.34333	6	577.42667	6	905.51000	6	1,233.59333	6	1,561.67667	6	1,889.76000	6	2,217.84333	6	2,545.92667	6	2,874.01000	6	3,202.09333
7	252.62417	7	580.70750	7	908.79083	7	1,236.87417	7	1,564.95750	7	1,893.04083	7	2,221.12417	7	2,549.20750	7	2,877.29083	7	3,205.37417
8	255.90500	8	583.98833	8	912.07167	8	1,240.15500	8	1,568.23833	8	1,896.32167	8	2,224.40500	8	2,552.48833	8	2,880.57167	8	3,208.65500
9	259.18583	9	587.26917	9	915.35250	9	1,243.43583	9	1,571.51917	9	1,899.60250	9	2,227.68583	9	2,555.76917	9	2,883.85250	9	3,211.93583
80	262.46667	180	590.55000	280	918.63333	380	1,246.71667	480	1,574.80000	580	1,902.88333	680	2,230.96667	780	2,559.05000	880	2,887.13333	980	3,215.21667
1	265.74750	1	593.83083	1	921.91417	1	1,249.99750	1	1,578.08083	1	1,906.16417	1	2,234.24750	1	2,562.33083	1	2,890.41417	1	3,218.49750
2	269.02833	2	597.11167	2	925.19500	2	1,253.27833	2	1,581.36167	2	1,909.44500	2	2,237.52833	2	2,565.61167	2	2,893.69500	2	3,221.77833
3	272.30917	3	600.39250	3	928.47583	3	1,256.55917	3	1,584.64250	3	1,912.72583	3	2,240.80917	3	2,568.89250	3	2,896.97583	3	3,225.05917
4	275.59000	4	603.67333	4	931.75667	4	1,259.84000	4	1,587.92333	4	1,916.00667	4	2,244.09000	4	2,572.17333	4	2,900.25667	4	3,228.34000
5	278.87083	5	606.95417	5	935.03750	5	1,263.12083	5	1,591.20417	5	1,919.28750	5	2,247.37083	5	2,575.45417	5	2,903.53750	5	3,231.62083
6	282.15167	6	610.23500	6	938.31833	6	1,266.40167	6	1,594.48500	6	1,922.56833	6	2,250.65167	6	2,578.73500	6	2,906.81833	6	3,234.90167
7	285.43250	7	613.51583	7	941.59917	7	1,269.68250	7	1,597.76583	7	1,925.84917	7	2,253.93250	7	2,582.01583	7	2,910.09917	7	3,238.18250
8	288.71333	8	616.79667	8	944.88000	8	1,272.96333	8	1,601.04667	8	1,929.13000	8	2,257.21333	8	2,585.29667	8	2,913.38000	8	3,241.46333
9	291.99417	9	620.07750	9	948.16083	9	1,276.24417	9	1,604.32750	9	1,932.41083	9	2,260.49417	9	2,588.57750	9	2,916.66083	9	3,244.74417
90	295.27500	190	623.35833	290	951.44167	390	1,279.52500	490	1,607.60833	590	1,935.69167	690	2,263.77500	790	2,591.85833	890	2,919.94167	990	3,248.02500
1	298.55583	1	626.63917	1	954.72250	1	1,282.80583	1	1,610.88917	1	1,938.97250	1	2,267.05583	1	2,595.13917	1	2,923.22250	1	3,251.30583
2	301.83667	2	629.92000	2	958.00333	2	1,286.08667	2	1,614.17000	2	1,942.25333	2	2,270.33667	2	2,598.42000	2	2,926.50333	2	3,254.58667
3	305.11750	3	633.20083	3	961.28417	3	1,289.36750	3	1,617.45083	3	1,945.53417	3	2,273.61750	3	2,601.70083	3	2,929.78417	3	3,257.86750
4	308.39833	4	636.48167	4	964.56500	4	1,292.64833	4	1,620.73167	4	1,948.81500	4	2,276.89833	4	2,604.98167	4	2,933.06500	4	3,261.14833
5	311.67917	5	639.76250	5	967.84583	5	1,295.92917	5	1,624.01250	5	1,952.09583	5	2,280.17917	5	2,608.26250	5	2,936.34583	5	3,264.42917
6	314.96000	6	643.04333	6	971.12667	6	1,299.21000	6	1,627.29333	6	1,955.37667	6	2,283.46000	6	2,611.54333	6	2,939.62667	6	3,267.71000
7	318.24083	7	646.32417	7	974.40750	7	1,302.49083	7	1,630.57417	7	1,958.65750	7	2,286.74083	7	2,614.82417	7	2,942.90750	7	3,270.99083
8	321.52167	8	649.60500	8	977.68833	8	1,305.77167	8	1,633.85500	8	1,961.93833	8	2,290.02167	8	2,618.10500	8	2,946.18833	8	3,274.27167
9	324.80250	9	652.88583	9	980.96917	9	1,309.05250	9	1,637.13583	9	1,965.21917	9	2,293.30250	9	2,621.38583	9	2,949.46917	9	3,277.55250

LENGTH—MILES TO KILOMETERS

[Reduction factor: 1 mile=1.609347219 kilometers]

Miles	Kilo-meters	Miles	Kilo-meters	Miles	Kilo-meters	Miles	Kilo-meters	Miles	Kilo-meters	Miles	Kilo-meters	Miles	Kilo-meters	Miles	Kilo-meters	Miles	Kilo-meters	Miles	Kilo-meters
0		**100**	160.9347	**200**	321.8694	**300**	482.8042	**400**	643.7389	**500**	804.6736	**600**	965.6083	**700**	1,126.5431	**800**	1,287.4778	**900**	1,448.4125
1	1.6093	1	162.5441	1	323.4788	1	484.4135	1	645.3482	1	806.2830	1	967.2177	1	1,128.1524	1	1,289.0871	1	1,450.0218
2	3.2187	2	164.1534	2	325.0881	2	486.0229	2	646.9576	2	807.8923	2	968.8270	2	1,129.7617	2	1,290.6965	2	1,451.6312
3	4.8280	3	165.7628	3	326.6975	3	487.6322	3	648.5669	3	809.5017	3	970.4364	3	1,131.3711	3	1,292.3058	3	1,453.2405
4	6.4374	4	167.3721	4	328.3068	4	489.2416	4	650.1763	4	811.1110	4	972.0457	4	1,132.9804	4	1,293.9152	4	1,454.8499
5	8.0467	5	168.9815	5	329.9162	5	490.8509	5	651.7856	5	812.7203	5	973.6551	5	1,134.5898	5	1,295.5245	5	1,456.4592
6	9.6561	6	170.5908	6	331.5255	6	492.4602	6	653.3950	6	814.3297	6	975.2644	6	1,136.1991	6	1,297.1339	6	1,458.0686
7	11.2654	7	172.2002	7	333.1349	7	494.0696	7	655.0043	7	815.9390	7	976.8738	7	1,137.8085	7	1,298.7432	7	1,459.6779
8	12.8748	8	173.8095	8	334.7442	8	495.6789	8	656.6137	8	817.5484	8	978.4831	8	1,139.4178	8	1,300.3526	8	1,461.2873
9	14.4841	9	175.4188	9	336.3536	9	497.2883	9	658.2230	9	819.1577	9	980.0925	9	1,141.0272	9	1,301.9619	9	1,462.8966
10	16.0935	**110**	177.0282	**210**	337.9629	**310**	498.8976	**410**	659.8324	**510**	820.7671	**610**	981.7018	**710**	1,142.6365	**810**	1,303.5712	**910**	1,464.5060
1	17.7028	1	178.6375	1	339.5723	1	500.5070	1	661.4417	1	822.3764	1	983.3112	1	1,144.2459	1	1,305.1806	1	1,466.1153
2	19.3122	2	180.2469	2	341.1816	2	502.1163	2	663.0511	2	823.9858	2	984.9205	2	1,145.8552	2	1,306.7899	2	1,467.7247
3	20.9215	3	181.8562	3	342.7910	3	503.7257	3	664.6604	3	825.5951	3	986.5298	3	1,147.4646	3	1,308.3993	3	1,469.3340
4	22.5309	4	183.4656	4	344.4003	4	505.3350	4	666.2697	4	827.2045	4	988.1392	4	1,149.0739	4	1,310.0086	4	1,470.9434
5	24.1402	5	185.0749	5	346.0097	5	506.9444	5	667.8791	5	828.8138	5	989.7485	5	1,150.6833	5	1,311.6180	5	1,472.5527
6	25.7496	6	186.6843	6	347.6190	6	508.5537	6	669.4884	6	830.4232	6	991.3579	6	1,152.2926	6	1,313.2273	6	1,474.1621
7	27.3589	7	188.2936	7	349.2283	7	510.1631	7	671.0978	7	832.0325	7	992.9672	7	1,153.9020	7	1,314.8367	7	1,475.7714
8	28.9682	8	189.9030	8	350.8377	8	511.7724	8	672.7071	8	833.6419	8	994.5766	8	1,155.5113	8	1,316.4460	8	1,477.3807
9	30.5776	9	191.5123	9	352.4470	9	513.3818	9	674.3165	9	835.2512	9	996.1859	9	1,157.1207	9	1,318.0554	9	1,478.9901
20	32.1869	**120**	193.1217	**220**	354.0564	**320**	514.9911	**420**	675.9258	**520**	836.8606	**620**	997.7953	**720**	1,158.7300	**820**	1,319.6647	**920**	1,480.5994
1	33.7963	1	194.7310	1	355.6657	1	516.6005	1	677.5352	1	838.4699	1	999.4046	1	1,160.3393	1	1,321.2741	1	1,482.2088
2	35.4056	2	196.3404	2	357.2751	2	518.2098	2	679.1445	2	840.0792	2	1,001.0140	2	1,161.9487	2	1,322.8834	2	1,483.8181
3	37.0150	3	197.9497	3	358.8844	3	519.8192	3	680.7539	3	841.6886	3	1,002.6233	3	1,163.5580	3	1,324.4928	3	1,485.4275
4	38.6243	4	199.5591	4	360.4938	4	521.4285	4	682.3632	4	843.2979	4	1,004.2327	4	1,165.1674	4	1,326.1021	4	1,487.0368
5	40.2337	5	201.1684	5	362.1031	5	523.0378	5	683.9726	5	844.9073	5	1,005.8420	5	1,166.7767	5	1,327.7115	5	1,488.6462
6	41.8430	6	202.7777	6	363.7125	6	524.6472	6	685.5819	6	846.5166	6	1,007.4514	6	1,168.3861	6	1,329.3208	6	1,490.2555
7	43.4524	7	204.3871	7	365.3218	7	526.2565	7	687.1913	7	848.1260	7	1,009.0607	7	1,169.9954	7	1,330.9301	7	1,491.8649
8	45.0617	8	205.9954	8	366.9312	8	527.8659	8	688.8006	8	849.7353	8	1,010.6701	8	1,171.6048	8	1,332.5395	8	1,493.4742
9	46.6711	9	207.6058	9	368.5405	9	529.4752	9	690.4100	9	851.3447	9	1,012.2794	9	1,173.2141	9	1,334.1488	9	1,495.0836
30	48.2804	**130**	209.2151	**230**	370.1499	**330**	531.0846	**430**	692.0193	**530**	852.9540	**630**	1,013.8887	**730**	1,174.8235	**830**	1,335.7582	**930**	1,496.6929
1	49.8898	1	210.8245	1	371.7592	1	532.6939	1	693.6287	1	854.5634	1	1,015.4981	1	1,176.4328	1	1,337.3675	1	1,498.3023
2	51.4991	2	212.4338	2	373.3686	2	534.3033	2	695.2380	2	856.1727	2	1,017.1074	2	1,178.0422	2	1,338.9769	2	1,499.9116
3	53.1085	3	214.0432	3	374.9779	3	535.9126	3	696.8473	3	857.7821	3	1,018.7168	3	1,179.6515	3	1,340.5862	3	1,501.5210
4	54.7178	4	215.6525	4	376.5872	4	537.5220	4	698.4567	4	859.3914	4	1,020.3261	4	1,181.2609	4	1,342.1956	4	1,503.1303
5	56.3272	5	217.2619	5	378.1966	5	539.1313	5	700.0660	5	861.0008	5	1,021.9355	5	1,182.8702	5	1,343.8049	5	1,504.7396
6	57.9365	6	218.8712	6	379.8059	6	540.7407	6	701.6754	6	862.6101	6	1,023.5448	6	1,184.4796	6	1,345.4143	6	1,506.3490
7	59.5458	7	220.4806	7	381.4153	7	542.3500	7	703.2847	7	864.2195	7	1,025.1542	7	1,186.0889	7	1,347.0236	7	1,507.9583
8	61.1552	8	222.0899	8	383.0246	8	543.9594	8	704.8941	8	865.8288	8	1,026.7635	8	1,187.6982	8	1,348.6330	8	1,509.5677
9	62.7645	9	223.6993	9	384.6340	9	545.5687	9	706.5034	9	867.4382	9	1,028.3729	9	1,189.3076	9	1,350.2423	9	1,511.1770
40	64.3739	**140**	225.3086	**240**	386.2433	**340**	547.1781	**440**	708.1128	**540**	869.0475	**640**	1,029.9822	**740**	1,190.9169	**840**	1,351.8517	**940**	1,512.7864
1	65.9832	1	226.9180	1	387.8527	1	548.7874	1	709.7221	1	870.6568	1	1,031.5916	1	1,192.5263	1	1,353.4610	1	1,514.3957
2	67.5926	2	228.5273	2	389.4620	2	550.3967	2	711.3315	2	872.2662	2	1,033.2009	2	1,194.1356	2	1,355.0704	2	1,516.0051
3	69.2019	3	230.1366	3	391.0714	3	552.0061	3	712.9408	3	873.8755	3	1,034.8103	3	1,195.7450	3	1,356.6797	3	1,517.6144
4	70.8113	4	231.7460	4	392.6807	4	553.6154	4	714.5502	4	875.4849	4	1,036.4196	4	1,197.3543	4	1,358.2891	4	1,519.2238
5	72.4206	5	233.3553	5	394.2901	5	555.2248	5	716.1595	5	877.0942	5	1,038.0290	5	1,198.9637	5	1,359.8984	5	1,520.8331
6	74.0300	6	234.9647	6	395.8994	6	556.8341	6	717.7689	6	878.7036	6	1,039.6383	6	1,200.5730	6	1,361.5077	6	1,522.4425
7	75.6393	7	236.5740	7	397.5088	7	558.4435	7	719.3782	7	880.3129	7	1,041.2477	7	1,202.1824	7	1,363.1171	7	1,524.0518
8	77.2487	8	238.1834	8	399.1181	8	560.0528	8	720.9876	8	881.9223	8	1,042.8570	8	1,203.7917	8	1,364.7264	8	1,525.6612
9	78.8580	9	239.7927	9	400.7275	9	561.6622	9	722.5969	9	883.5316	9	1,044.4663	9	1,205.4011	9	1,366.3358	9	1,527.2705

50	80.4674	**150**	241.4021	**250**	402.3368	**350**	563.2715	**450**	724.2062	**550**	885.1410	**650**	1,046.0757	**750**	1,207.0104	**850**	1,367.9451	**950**	1,528.8799
1	82.0767	1	243.0114	1	403.9461	1	564.8809	1	725.8156	1	886.7503	1	1,047.6850	1	1,208.6198	1	1,369.5545	1	1,530.4892
2	83.6861	2	244.6208	2	405.5555	2	566.4902	2	727.4249	2	888.3597	2	1,049.2944	2	1,210.2291	2	1,371.1638	2	1,532.0986
3	85.2954	3	246.2301	3	407.1648	3	568.0996	3	729.0343	3	889.9690	3	1,050.9037	3	1,211.8385	3	1,372.7732	3	1,533.7079
4	86.9047	4	247.8395	4	408.7742	4	569.7089	4	730.6436	4	891.5784	4	1,052.5131	4	1,213.4478	4	1,374.3825	4	1,535.3172
5	88.5141	5	249.4488	5	410.3835	5	571.3183	5	732.2530	5	893.1877	5	1,054.1224	5	1,215.0572	5	1,375.9919	5	1,536.9266
6	90.1234	6	251.0582	6	411.9929	6	572.9276	6	733.8623	6	894.7971	6	1,055.7318	6	1,216.6665	6	1,377.6012	6	1,538.5359
7	91.7328	7	252.6675	7	413.6022	7	574.5370	7	735.4717	7	896.4064	7	1,057.3411	7	1,218.2758	7	1,379.2106	7	1,540.1453
8	93.3421	8	254.2769	8	415.2116	8	576.1463	8	737.0810	8	898.0157	8	1,058.9505	8	1,219.8852	8	1,380.8199	8	1,541.7546
9	94.9515	9	255.8862	9	416.8209	9	577.7557	9	738.6904	9	899.6251	9	1,060.5598	9	1,221.4945	9	1,382.4293	9	1,543.3640
60	96.5608	**160**	257.4956	**260**	418.4303	**360**	579.3650	**460**	740.2997	**560**	901.2344	**660**	1,062.1692	**760**	1,223.1039	**860**	1,384.0386	**960**	1,544.9733
1	98.1702	1	259.1049	1	420.0396	1	580.9743	1	741.9091	1	902.8438	1	1,063.7785	1	1,224.7132	1	1,385.6480	1	1,546.5827
2	99.7795	2	260.7142	2	421.6490	2	582.5837	2	743.5184	2	904.4531	2	1,065.3879	2	1,226.3226	2	1,387.2573	2	1,548.1920
3	101.3889	3	262.3236	3	423.2583	3	584.1930	3	745.1278	3	906.0625	3	1,066.9972	3	1,227.9319	3	1,388.8666	3	1,549.8014
4	102.9982	4	263.9329	4	424.8677	4	585.8024	4	746.7371	4	907.6718	4	1,068.6066	4	1,229.5413	4	1,390.4760	4	1,551.4107
5	104.6076	5	265.5423	5	426.4770	5	587.4117	5	748.3465	5	909.2812	5	1,070.2159	5	1,231.1506	5	1,392.0853	5	1,553.0201
6	106.2169	6	267.1516	6	428.0864	6	589.0211	6	749.9558	6	910.8905	6	1,071.8252	6	1,232.7600	6	1,393.6947	6	1,554.6294
7	107.8263	7	268.7610	7	429.6957	7	590.6304	7	751.5652	7	912.4999	7	1,073.4346	7	1,234.3693	7	1,395.3040	7	1,556.2388
8	109.4356	8	270.3703	8	431.3051	8	592.2398	8	753.1745	8	914.1092	8	1,075.0439	8	1,235.9787	8	1,396.9134	8	1,557.8481
9	111.0450	9	271.9797	9	432.9144	9	593.8491	9	754.7838	9	915.7186	9	1,076.6533	9	1,237.5880	9	1,398.5227	9	1,559.4575
70	112.6543	**170**	273.5890	**270**	434.5237	**370**	595.4585	**470**	756.3932	**570**	917.3279	**670**	1,078.2626	**770**	1,239.1974	**870**	1,400.1321	**970**	1,561.0668
1	114.2637	1	275.1984	1	436.1331	1	597.0678	1	758.0025	1	918.9373	1	1,079.8720	1	1,240.8067	1	1,401.7414	1	1,562.6761
2	115.8730	2	276.8077	2	437.7424	2	598.6772	2	759.6119	2	920.5466	2	1,081.4813	2	1,242.4161	2	1,403.3508	2	1,564.2855
3	117.4823	3	278.4171	3	439.3518	3	600.2865	3	761.2212	3	922.1560	3	1,083.0907	3	1,244.0254	3	1,404.9601	3	1,565.8948
4	119.0917	4	280.0264	4	440.9611	4	601.8959	4	762.8306	4	923.7653	4	1,084.7000	4	1,245.6347	4	1,406.5695	4	1,567.5042
5	120.7010	5	281.6358	5	442.5705	5	603.5052	5	764.4399	5	925.3747	5	1,086.3094	5	1,247.2441	5	1,408.1788	5	1,569.1135
6	122.3104	6	283.2451	6	444.1798	6	605.1145	6	766.0493	6	926.9840	6	1,087.9187	6	1,248.8534	6	1,409.7882	6	1,570.7229
7	123.9197	7	284.8545	7	445.7892	7	606.7239	7	767.6586	7	928.5933	7	1,089.5281	7	1,250.4628	7	1,411.3975	7	1,572.3322
8	125.5291	8	286.4638	8	447.3985	8	608.3332	8	769.2680	8	930.2027	8	1,091.1374	8	1,252.0721	8	1,413.0069	8	1,573.9416
9	127.1384	9	288.0732	9	449.0079	9	609.9426	9	770.8773	9	931.8120	9	1,092.7468	9	1,253.6815	9	1,414.6162	9	1,575.5509
80	128.7478	**180**	289.6825	**280**	450.6172	**380**	611.5519	**480**	772.4867	**580**	933.4214	**680**	1,094.3561	**780**	1,255.2908	**880**	1,416.2256	**980**	1,577.1603
1	130.3571	1	291.2918	1	452.2266	1	613.1613	1	774.0960	1	935.0307	1	1,095.9655	1	1,256.9002	1	1,417.8349	1	1,578.7696
2	131.9665	2	292.9012	2	453.8359	2	614.7706	2	775.7054	2	936.6401	2	1,097.5748	2	1,258.5095	2	1,419.4442	2	1,580.3790
3	133.5758	3	294.5105	3	455.4453	3	616.3800	3	777.3147	3	938.2494	3	1,099.1842	3	1,260.1189	3	1,421.0536	3	1,581.9883
4	135.1852	4	296.1199	4	457.0546	4	617.9893	4	778.9241	4	939.8588	4	1,100.7935	4	1,261.7282	4	1,422.6629	4	1,583.5977
5	136.7945	5	297.7292	5	458.6640	5	619.5987	5	780.5334	5	941.4681	5	1,102.4028	5	1,263.3376	5	1,424.2723	5	1,585.2070
6	138.4039	6	299.3386	6	460.2733	6	621.2080	6	782.1427	6	943.0775	6	1,104.0122	6	1,264.9469	6	1,425.8816	6	1,586.8164
7	140.0132	7	300.9479	7	461.8827	7	622.8174	7	783.7521	7	944.6868	7	1,105.6215	7	1,266.5563	7	1,427.4910	7	1,588.4257
8	141.6226	8	302.5573	8	463.4920	8	624.4267	8	785.3614	8	946.2962	8	1,107.2309	8	1,268.1656	8	1,429.1003	8	1,590.0351
9	143.2319	9	304.1666	9	465.1013	9	626.0361	9	786.9708	9	947.9055	9	1,108.8402	9	1,269.7750	9	1,430.7097	9	1,591.6444
90	144.8412	**190**	305.7760	**290**	466.7107	**390**	627.6454	**490**	788.5801	**590**	949.5149	**690**	1,110.4496	**790**	1,271.3843	**890**	1,432.3190	**990**	1,593.2537
1	146.4506	1	307.3853	1	468.3200	1	629.2548	1	790.1895	1	951.1242	1	1,112.0589	1	1,272.9936	1	1,433.9284	1	1,594.8631
2	148.0599	2	308.9947	2	469.9294	2	630.8641	2	791.7988	2	952.7336	2	1,113.6683	2	1,274.6030	2	1,435.5377	2	1,596.4724
3	149.6693	3	310.6040	3	471.5387	3	632.4735	3	793.4082	3	954.3429	3	1,115.2776	3	1,276.2123	3	1,437.1471	3	1 598.0818
4	151.2786	4	312.2134	4	473.1481	4	634.0828	4	795.0175	4	955.9522	4	1,116.8870	4	1,277.8217	4	1,438.7564	4	1,599.6911
5	152.8880	5	313.8227	5	474.7574	5	635.6922	5	796.6269	5	957.5616	5	1,118.4963	5	1,279.4310	5	1,440.3658	5	1,601.3005
6	154.4973	6	315.4321	6	476.3668	6	637.3015	6	798.2362	6	959.1709	6	1,120.1057	6	1,281.0404	6	1,441.9751	6	1,602.9098
7	156.1067	7	317.0414	7	477.9761	7	638.9108	7	799.8456	7	960.7803	7	1,121.7150	7	1,282.6497	7	1,443.5845	7	1,604.5192
8	157.7160	8	318.6507	8	479.5855	8	640.5202	8	801.4549	8	962.3896	8	1,123.3244	8	1,284.2591	8	1,445.1938	8	1,606.1285
9	159.3254	9	320.2601	9	481.1948	9	642.1295	9	803.0643	9	963.9990	9	1,124.9337	9	1,285.8684	9	1,446.8031	9	1,607.7379

LENGTH—KILOMETERS TO MILES

[Reduction factor: 1 kilometer=0.6213699495 mile]

Kilometers	Miles	Kilometers	Miles	Kilometers	Miles	Kilometers	Miles	Kilometers	Miles	Kilometers	Miles	Kilometers	Miles	Kilometers	Miles	Kilometers	Miles	Kilometers	Miles
0		100	62.13699	200	124.27399	300	186.41098	400	248.54798	500	310.68497	600	372.82197	700	434.95896	800	497.09596	900	559.23295
1	0.62137	1	62.75836	1	124.89536	1	187.03235	1	249.16935	1	311.30634	1	373.44334	1	435.58033	1	497.71733	1	559.85432
2	1.24274	2	63.37973	2	125.51673	2	187.65372	2	249.79072	2	311.92771	2	374.06471	2	436.20170	2	498.33870	2	560.47569
3	1.86411	3	64.00110	3	126.13810	3	188.27509	3	250.41209	3	312.54908	3	374.68608	3	436.82307	3	498.96007	3	561.09706
4	2.48548	4	64.62247	4	126.75947	4	188.89646	4	251.03346	4	313.17045	4	375.30745	4	437.44444	4	499.58144	4	561.71843
5	3.10685	5	65.24384	5	127.38084	5	189.51783	5	251.65483	5	313.79182	5	375.92882	5	438.06581	5	500.20281	5	562.33980
6	3.72822	6	65.86521	6	128.00221	6	190.13920	6	252.27620	6	314.41319	6	376.55019	6	438.68718	6	500.82418	6	562.96117
7	4.34959	7	66.48658	7	128.62358	7	190.76057	7	252.89757	7	315.03456	7	377.17156	7	439.30855	7	501.44555	7	563.58254
8	4.97096	8	67.10795	8	129.24495	8	191.38194	8	253.51894	8	315.65593	8	377.79293	8	439.92992	8	502.06692	8	564.20391
9	5.59233	9	67.72932	9	129.86632	9	192.00331	9	254.14031	9	316.27730	9	378.41430	9	440.55129	9	502.68829	9	564.82528
10	6.21370	110	68.35069	210	130.48769	310	192.62468	410	254.76168	510	316.89867	610	379.03567	.710	441.17266	810	503.30966	910	565.44665
1	6.83507	1	68.97206	1	131.10906	1	193.24605	1	255.38305	1	317.52004	1	379.65704	1	441.79403	1	503.93103	1	566.06802
2	7.45644	2	69.59343	2	131.73043	2	193.86742	2	256.00442	2	318.14141	2	380.27841	2	442.41540	2	504.55240	2	566.68939
3	8.07781	3	70.21480	3	132.35180	3	194.48879	3	256.62579	3	318.76278	3	380.89978	3	443.03677	3	505.17377	3	567.31076
4	8.69918	4	70.83617	4	132.97317	4	195.11016	4	257.24716	4	319.38415	4	381.52115	4	443.65814	4	505.79514	4	567.93213
5	9.32055	5	71.45754	5	133.59454	5	195.73153	5	257.86853	5	320.00552	5	382.14252	5	444.27951	5	506.41651	5	568.55350
6	9.94192	6	72.07891	6	134.21591	6	196.35290	6	258.48990	6	320.62689	6	382.76389	6	444.90088	6	507.03788	6	569.17487
7	10.56329	7	72.70028	7	134.83728	7	196.97427	7	259.11127	7	321.24826	7	383.38526	7	445.52225	7	507.65925	7	569.79624
8	11.18466	8	73.32165	8	135.45865	8	197.59564	8	259.73264	8	321.86963	8	384.00663	8	446.14362	8	508.28062	8	570.41761
9	11.80603	9	73.94302	9	136.08002	9	198.21701	9	260.35401	9	322.49100	9	384.62800	9	446.76499	9	508.90199	9	571.03898
20	12.42740	120	74.56439	220	136.70139	320	198.83838	420	260.97538	520	323.11237	620	385.24937	720	447.38636	820	509.52336	920	571.66035
1	13.04877	1	75.18576	1	137.32276	1	199.45975	1	261.59675	1	323.73374	1	335.87074	1	448.00773	1	510.14473	1	572.28172
2	13.67014	2	75.80713	2	137.94413	2	200.08112	2	262.21812	2	324.35511	2	386.49211	2	448.62910	2	510.76610	2	572.90309
3	14.29151	3	76.42850	3	138.56550	3	200.70249	3	262.83949	3	324.97648	3	387.11348	3	449.25047	3	511.38747	3	573.52446
4	14.91288	4	77.04987	4	139.18687	4	201.32386	4	263.46086	4	325.59785	4	387.73485	4	449.87184	4	512.00884	4	574.14583
5	15.53425	5	77.67124	5	139.80824	5	201.94523	5	264.08223	5	326.21922	5	388.35622	5	450.49321	5	512.63021	5	574.76720
6	16.15562	6	78.29261	6	140.42961	6	202.56660	6	264.70360	6	326.84059	6	388.97759	6	451.11458	6	513.25158	6	575.38857
7	16.77699	7	78.91398	7	141.05098	7	203.18797	7	265.32497	7	327.46196	7	389.59896	7	451.73595	7	513.87295	7	576.00994
8	17.39836	8	79.53535	8	141.67235	8	203.80934	8	265.94634	8	328.08333	8	390.22033	8	452.35732	8	514.49432	8	576.63131
9	18.01973	9	80.15672	9	142.29372	9	204.43071	9	266.56771	9	328.70470	9	390.84170	9	452.97869	9	515.11569	9	577.25268
30	18.64110	130	80.77809	230	142.91509	330	205.05208	430	267.18908	530	329.32607	630	391.46307	730	453.60006	830	515.73706	930	577.87405
1	19.26247	1	81.39946	1	143.53646	1	205.67345	1	267.81045	1	329.94744	1	392.08444	1	454.22143	1	516.35843	1	578.49542
2	19.88384	2	82.02083	2	144.15783	2	206.29482	2	268.43182	2	330.56881	2	392.70581	2	454.84280	2	516.97980	2	579.11679
3	20.50521	3	82.64220	3	144.77920	3	206.91619	3	269.05319	3	331.19018	3	393.32718	3	455.46417	3	517.60117	3	579.73816
4	21.12658	4	83.26357	4	145.40057	4	207.53756	4	269.67456	4	331.81155	4	393.94855	4	456.08554	4	518.22254	4	580.35953
5	21.74795	5	83.88494	5	146.02194	5	208.15893	5	270.29593	5	332.43292	5	394.56992	5	456.70691	5	518.84391	5	580.98090
6	22.36932	6	84.50631	6	146.64331	6	208.78030	6	270.91730	6	333.05429	6	395.19129	6	457.32828	6	519.46528	6	581.60227
7	22.99069	7	85.12768	7	147.26468	7	209.40167	7	271.53867	7	333.67566	7	395.81266	7	457.94965	7	520.08665	7	582.22364
8	23.61206	8	85.74905	8	147.88605	8	210.02304	8	272.16004	8	334.29703	8	396.43403	8	458.57102	8	520.70802	8	582.84501
9	24.23343	9	86.37042	9	148.50742	9	210.64441	9	272.78141	9	334.91840	9	397.05540	9	459.19239	9	521.32939	9	583.46638
40	24.85480	140	86.99179	240	149.12879	340	211.26578	440	273.40278	540	335.53977	640	397.67677	740	459.81376	840	521.95076	940	584.08775
1	25.47617	1	87.61316	1	149.75016	1	211.88715	1	274.02415	1	336.16114	1	398.29814	1	460.43513	1	522.57213	1	584.70912
2	26.09754	2	88.23453	2	150.37153	2	212.50852	2	274.64552	2	336.78251	2	398.91951	2	461.05650	2	523.19350	2	585.33049
3	26.71891	3	88.85590	3	150.99290	3	213.12989	3	275.26689	3	337.40388	3	399.54088	3	461.67787	3	523.81487	3	585.95186
4	27.34028	4	89.47727	4	151.61427	4	213.75126	4	275.88826	4	338.02525	4	400.16225	4	462.29924	4	524.43624	4	586.57323
5	27.96165	5	90.09864	5	152.23564	5	214.37263	5	276.50963	5	338.64662	5	400.78362	5	462.92061	5	525.05761	5	587.19460
6	28.58302	6	90.72001	6	152.85701	6	214.99400	6	277.13100	6	339.26799	6	401.40499	6	463.54198	6	525.67898	6	587.81597
7	29.20439	7	91.34138	7	153.47838	7	215.61537	7	277.75237	7	339.88936	7	402.02636	7	464.16335	7	526.30035	7	588.43734
8	29.82576	8	91.96275	8	154.09975	8	216.23674	8	278.37374	8	340.51073	8	402.64773	8	464.78472	8	526.92172	8	589.05871
9	30.44713	9	92.58412	9	154.72112	9	216.85811	9	278.99511	9	341.13210	9	403.26910	9	465.40609	9	527.54309	9	589.68008

50	31.06850	**150**	93.20549	**250**	155.34249	**350**	217.47948	**450**	279.61648	**550**	341.75347	**650**	403.89047	**750**	466.02746	**850**	528.16446	**950**	590.30145
1	31.68987	1	93.82686	1	155.96386	1	218.10085	1	280.23785	1	342.37484	1	404.51184	1	466.64883	1	528.78583	1	590.92282
2	32.31124	2	94.44823	2	156.58523	2	218.72222	2	280.85922	2	342.99621	2	405.13321	2	467.27020	2	529.40720	2	591.54419
3	32.93261	3	95.06960	3	157.20660	3	219.34359	3	281.48059	3	343.61758	3	405.75458	3	467.89157	3	530.02857	3	592.16556
4	33.55398	4	95.69097	4	157.82797	4	219.96496	4	282.10196	4	344.23895	4	406.37595	4	468.51294	4	530.64994	4	592.78693
5	34.17535	5	96.31234	5	158.44934	5	220.58633	5	282.72333	5	344.86032	5	406.99732	5	469.13431	5	531.27131	5	593.40830
6	34.79672	6	96.93371	6	159.07071	6	221.20770	6	283.34470	6	345.48169	6	407.61869	6	469.75568	6	531.89268	6	594.02967
7	35.41809	7	97.55508	7	159.69208	7	221.82907	7	283.96607	7	346.10306	7	408.24006	7	470.37705	7	532.51405	7	594.65104
8	36.03946	8	98.17645	8	160.31345	8	222.45044	8	284.58744	8	346.72443	8	408.86143	8	470.99842	8	533.13542	8	595.27241
9	36.66083	9	98.79782	9	160.93482	9	223.07181	9	285.20881	9	347.34580	9	409.48280	9	471.61979	9	533.75679	9	595.89378
60	37.28220	**160**	99.41919	**260**	161.55619	**360**	223.69318	**460**	285.83018	**560**	347.96717	**660**	410.10417	**760**	472.24116	**860**	534.37816	**960**	596.51515
1	37.90357	1	100.04056	1	162.17756	1	224.31455	1	286.45155	1	348.58854	1	410.72554	1	472.86253	1	534.99953	1	597.13652
2	38.52494	2	100.66193	2	162.79893	2	224.93592	2	287.07292	2	349.20991	2	411.34691	2	473.48390	2	535.62090	2	597.75789
3	39.14631	3	101.28330	3	163.42030	3	225.55729	3	287.69429	3	349.83128	3	411.96828	3	474.10527	3	536.24227	3	598.37926
4	39.76768	4	101.90467	4	164.04167	4	226.17866	4	288.31566	4	350.45265	4	412.58965	4	474.72664	4	536.86364	4	599.00063
5	40.38905	5	102.52604	5	164.66304	5	226.80003	5	288.93703	5	351.07402	5	413.21102	5	475.34801	5	537.48501	5	599.62200
6	41.01042	6	103.14741	6	165.28441	6	227.42140	6	289.55840	6	351.69539	6	413.83239	6	475.96938	6	538.10638	6	600.24337
7	41.63179	7	103.76878	7	165.90578	7	228.04277	7	290.17977	7	352.31676	7	414.45376	7	476.59075	7	538.72775	7	600.86474
8	42.25316	8	104.39015	8	166.52715	8	228.66414	8	290.80114	8	352.93813	8	415.07513	8	477.21212	8	539.34912	8	601.48611
9	42.87453	9	105.01152	9	167.14852	9	229.28551	9	291.42251	9	353.55950	9	415.69650	9	477.83349	9	539.97049	9	602.10748
70	43.49590	**170**	105.63289	**270**	167.76989	**370**	229.90688	**470**	292.04388	**570**	354.18087	**670**	416.31787	**770**	478.45486	**870**	540.59186	**970**	602.72885
1	44.11727	1	106.25426	1	168.39126	1	230.52825	1	292.66525	1	354.80224	1	416.93924	1	479.07623	1	541.21323	1	603.35022
2	44.73864	2	106.87563	2	169.01263	2	231.14962	2	293.28662	2	355.42361	2	417.56061	2	479.69760	2	541.83460	2	603.97159
3	45.36001	3	107.49700	3	169.63400	3	231.77099	3	293.90799	3	356.04498	3	418.18198	3	480.31897	3	542.45597	3	604.59296
4	45.98138	4	108.11837	4	170.25537	4	232.39236	4	294.52936	4	356.66635	4	418.80335	4	480.94034	4	543.07734	4	605.21433
5	46.60275	5	108.73974	5	170.87674	5	233.01373	5	295.15073	5	357.28772	5	419.42472	5	481.56171	5	543.69871	5	605.83570
6	47.22412	6	109.36111	6	171.49811	6	233.63510	6	295.77210	6	357.90909	6	420.04609	6	482.18308	6	544.32008	6	606.45707
7	47.84549	7	109.98248	7	172.11948	7	234.25647	7	296.39347	7	358.53046	7	420.66746	7	482.80445	7	544.94145	7	607.07844
8	48.46686	8	110.60385	8	172.74085	8	234.87784	8	297.01484	8	359.15183	8	421.28883	8	483.42582	8	545.56282	8	607.69981
9	49.08823	9	111.22522	9	173.36222	9	235.49921	9	297.63621	9	359.77320	9	421.91020	9	484.04719	9	546.18419	9	608.32118
80	49.70960	**180**	111.84659	**280**	173.98359	**380**	236.12058	**480**	298.25758	**580**	360.39457	**680**	422.53157	**780**	484.66856	**880**	546.80556	**980**	608.94255
1	50.33097	1	112.46796	1	174.60496	1	236.74195	1	298.87895	1	361.01594	1	423.15294	1	485.28993	1	547.42693	1	609.56392
2	50.95234	2	113.08933	2	175.22633	2	237.36332	2	299.50032	2	361.63731	2	423.77431	2	485.91130	2	548.04830	2	610.18529
3	51.57371	3	113.71070	3	175.84770	3	237.98469	3	300.12169	3	362.25868	3	424.39568	3	486.53267	3	548.66967	3	610.80666
4	52.19508	4	114.33207	4	176.46907	4	238.60606	4	300.74306	4	362.88005	4	425.01705	4	487.15404	4	549.29104	4	611.42803
5	52.81645	5	114.95344	5	177.09044	5	239.22743	5	301.36443	5	363.50142	5	425.63842	5	487.77541	5	549.91241	5	612.04940
6	53.43782	6	115.57481	6	177.71181	6	239.84880	6	301.98580	6	364.12279	6	426.25979	6	488.39678	6	550.53378	6	612.67077
7	54.05919	7	116.19618	7	178.33318	7	240.47017	7	302.60717	7	364.74416	7	426.88116	7	489.01815	7	551.15515	7	613.29214
8	54.68056	8	116.81755	8	178.95455	8	241.09154	8	303.22854	8	365.36553	8	427.50253	8	489.63952	8	551.77652	8	613.91351
9	55.30193	9	117.43892	9	179.57592	9	241.71291	9	303.84991	9	365.98690	9	428.12390	9	490.26089	9	552.39789	9	614.53488
90	55.92330	**190**	118.06029	**290**	180.19729	**390**	242.33428	**490**	304.47128	**590**	366.60827	**690**	428.74527	**790**	490.88226	**890**	553.01926	**990**	615.15625
1	56.54467	1	118.68166	1	180.81866	1	242.95565	1	305.09265	1	367.22964	1	429.36664	1	491.50363	1	553.64062	1	615.77762
2	57.16604	2	119.30303	2	181.44003	2	243.57702	2	305.71402	2	367.85101	2	429.98801	2	492.12500	2	554.26199	2	616.39899
3	57.78741	3	119.92440	3	182.06140	3	244.19839	3	306.33539	3	368.47238	3	430.60937	3	492.74637	3	554.88336	3	617.02036
4	58.40878	4	120.54577	4	182.68277	4	244.81976	4	306.95676	4	369.09375	4	431.23074	4	493.36774	4	555.50473	4	617.64173
5	59.03015	5	121.16714	5	183.30414	5	245.44113	5	307.57812	5	369.71512	5	431.85211	5	493.98911	5	556.12610	5	618.26310
6	59.65152	6	121.78851	6	183.92551	6	246.06250	6	308.19949	6	370.33649	6	432.47348	6	494.61048	6	556.74747	6	618.88447
7	60.27289	7	122.40988	7	184.54687	7	246.68387	7	308.82086	7	370.95786	7	433.09485	7	495.23185	7	557.36884	7	619.50584
8	60.89426	8	123.03125	8	185.16824	8	247.30524	8	309.44223	8	371.57923	8	433.71622	8	495.85322	8	557.99021	8	620.12721
9	61.51562	9	123.65262	9	185.78961	9	247.92661	9	310.06360	9	372.20060	9	434.33759	9	496.47459	9	558.61158	9	620.74858

AREA—ACRES TO HECTARES

[Reduction factor: 1 acre=0.4046872610 hectare]

Acres	Hectares	Acres	Hectares	Acres	Hectares	Acres	Hectares	Acres	Hectares	Acres	Hectares	Acres	Hectares	Acres	Hectares	Acres	Hectares	Acres	Hectares
0		**100**	40.46873	**200**	80.93745	**300**	121.40618	**400**	161.87490	**500**	202.34363	**600**	242.81236	**700**	283.28108	**800**	323.74981	**900**	364.21853
1	0.40469	1	40.87341	1	81.34214	1	121.81087	1	162.27959	1	202.74832	1	243.21704	1	283.68577	1	324.15450	1	364.62322
2	0.80937	2	41.27810	2	81.74683	2	122.21555	2	162.68428	2	203.15301	2	243.62173	2	284.09046	2	324.55918	2	365.02791
3	1.21406	3	41.68279	3	82.15151	3	122.62024	3	163.08897	3	203.55769	3	244.02642	3	284.49514	3	324.96387	3	365.43260
4	1.61875	4	42.08748	4	82.55620	4	123.02493	4	163.49365	4	203.96238	4	244.43111	4	284.89983	4	325.36856	4	365.83728
5	2.02344	5	42.49216	5	82.96089	5	123.42961	5	163.89834	5	204.36707	5	244.83579	5	285.30452	5	325.77325	5	366.24197
6	2.42812	6	42.89685	6	83.36558	6	123.83430	6	164.30303	6	204.77175	6	245.24048	6	285.70921	6	326.17793	6	366.64666
7	2.83281	7	43.30154	7	83.77026	7	124.23899	7	164.70772	7	205.17644	7	245.64517	7	286.11389	7	326.58262	7	367.05135
8	3.23750	8	43.70622	8	84.17495	8	124.64368	8	165.11240	8	205.58113	8	246.04985	8	286.51858	8	326.98731	8	367.45603
9	3.64219	9	44.11091	9	84.57964	9	125.04836	9	165.51709	9	205.98582	9	246.45454	9	286.92327	9	327.39199	9	367.86072
10	4.04687	**110**	44.51560	**210**	84.98432	**310**	125.45305	**410**	165.92178	**510**	206.39050	**610**	246.85923	**710**	287.32796	**810**	327.79668	**910**	368.26541
1	4.45156	1	44.92029	1	85.38901	1	125.85774	1	166.32646	1	206.79519	1	247.26392	1	287.73264	1	328.20137	1	368.67009
2	4.85625	2	45.32497	2	85.79370	2	126.26243	2	166.73115	2	207.19988	2	247.66860	2	288.13733	2	328.60606	2	369.07478
3	5.26093	3	45.72966	3	86.19839	3	126.66711	3	167.13584	3	207.60456	3	248.07329	3	288.54202	3	329.01074	3	369.47947
4	5.66562	4	46.13435	4	86.60307	4	127.07180	4	167.54053	4	208.00925	4	248.47798	4	288.94670	4	329.41543	4	369.88416
5	6.07031	5	46.53904	5	87.00776	5	127.47649	5	167.94521	5	208.41394	5	248.88267	5	289.35139	5	329.82012	5	370.28884
6	6.47300	6	46.94372	6	87.41245	6	127.88117	6	168.34990	6	208.81863	6	249.28735	6	289.75608	6	330.22480	6	370.69353
7	6.87968	7	47.34841	7	87.81714	7	128.28586	7	168.75459	7	209.22331	7	249.69204	7	290.16077	7	330.62949	7	371.09822
8	7.28437	8	47.75310	8	88.22182	8	128.69055	8	169.15928	8	209.62800	8	250.09673	8	290.56545	8	331.03418	8	371.50291
9	7.68906	9	48.15778	9	88.62651	9	129.09524	9	169.56396	9	210.03269	9	250.50141	9	290.97014	9	331.43887	9	371.90759
20	8.09375	**120**	48.56247	**220**	89.03120	**320**	129.49992	**420**	169.96865	**520**	210.43738	**620**	250.90610	**720**	291.37483	**820**	331.84355	**920**	372.31228
1	8.49843	1	48.96716	1	89.43588	1	129.90461	1	170.37334	1	210.84206	1	251.31079	1	291.77952	1	332.24824	1	372.71697
2	8.90312	2	49.37185	2	89.84057	2	130.30930	2	170.77802	2	211.24675	2	251.71548	2	292.18420	2	332.65293	2	373.12165
3	9.30781	3	49.77653	3	90.24526	3	130.71399	3	171.18271	3	211.65144	3	252.12016	3	292.58889	3	333.05762	3	373.52634
4	9.71249	4	50.18122	4	90.64995	4	131.11867	4	171.58740	4	212.05612	4	252.52485	4	292.99358	4	333.46230	4	373.93103
5	10.11718	5	50.58591	5	91.05463	5	131.52336	5	171.99209	5	212.46081	5	252.92954	5	293.39826	5	333.86699	5	374.33572
6	10.52187	6	50.99059	6	91.45932	6	131.92805	6	172.39677	6	212.86550	6	253.33423	6	293.80295	6	334.27168	6	374.74040
7	10.92656	7	51.39528	7	91.86401	7	132.33273	7	172.80146	7	213.27019	7	253.73891	7	294.20764	7	334.67636	7	375.14509
8	11.33124	8	51.79997	8	92.26870	8	132.73742	8	173.20615	8	213.67487	8	254.14360	8	294.61233	8	335.08105	8	375.54978
9	11.73593	9	52.20466	9	92.67338	9	133.14211	9	173.61083	9	214.07956	9	254.54829	9	295.01701	9	335.48574	9	375.95447
30	12.14062	**130**	52.60934	**230**	93.07807	**330**	133.54680	**430**	174.01552	**530**	214.48425	**630**	254.95297	**730**	295.42170	**830**	335.89043	**930**	376.35915
1	12.54531	1	53.01403	1	93.48276	1	133.95148	1	174.42021	1	214.88894	1	255.35766	1	295.82639	1	336.29511	1	376.76384
2	12.94999	2	53.41872	2	93.88744	2	134.35617	2	174.82490	2	215.29362	2	255.76235	2	296.23108	2	336.69980	2	377.16853
3	13.35468	3	53.82341	3	94.29213	3	134.76086	3	175.22958	3	215.69831	3	256.16704	3	296.63576	3	337.10449	3	377.57321
4	13.75937	4	54.22809	4	94.69682	4	135.16555	4	175.63427	4	216.10300	4	256.57172	4	297.04045	4	337.50918	4	377.97790
5	14.16405	5	54.63278	5	95.10151	5	135.57023	5	176.03896	5	216.50768	5	256.97641	5	297.44514	5	337.91386	5	378.38259
6	14.56874	6	55.03747	6	95.50619	6	135.97492	6	176.44365	6	216.91237	6	257.38110	6	297.84982	6	338.31855	6	378.78728
7	14.97343	7	55.44215	7	95.91088	7	136.37961	7	176.84833	7	217.31706	7	257.78579	7	298.25451	7	338.72324	7	379.19196
8	15.37812	8	55.84684	8	96.31557	8	136.78429	8	177.25302	8	217.72175	8	258.19047	8	298.65920	8	339.12792	8	379.59665
9	15.78280	9	56.25153	9	96.72026	9	137.18898	9	177.65771	9	218.12643	9	258.59516	9	299.06389	9	339.53261	9	380.00134
40	16.18749	**140**	56.65622	**240**	97.12494	**340**	137.59367	**440**	178.06239	**540**	218.53112	**640**	258.99985	**740**	299.46857	**840**	339.93730	**940**	380.40603
1	16.59218	1	57.06090	1	97.52963	1	137.99836	1	178.46708	1	218.93581	1	259.40453	1	299.87326	1	340.34199	1	380.81071
2	16.99686	2	57.46559	2	97.93432	2	138.40304	2	178.87177	2	219.34050	2	259.80922	2	300.27795	2	340.74667	2	381.21540
3	17.40155	3	57.87028	3	98 33900	3	138.80773	3	179.27646	3	219.74518	3	260.21391	3	300.68263	3	341.15136	3	381.62009
4	17.80624	4	58.27497	4	98.74369	4	139.21242	4	179.68114	4	220.14987	4	260.61860	4	301.08732	4	341.55605	4	382.02477
5	18.21093	5	58.67965	5	99.14838	5	139.61711	5	180.08583	5	220.55456	5	261.02328	5	301.49201	5	341.96074	5	382.42946
6	18.61561	6	59.08434	6	99.55307	6	140.02179	6	180.49052	6	220.95924	6	261.42797	6	301.89670	6	342.36542	6	382.83415
7	19.02030	7	59.48903	7	99.95775	7	140.42648	7	180.89521	7	221.36393	7	261.83266	7	302.30138	7	342.77011	7	383.23884
8	19.42499	8	59.89371	8	100.36244	8	140.83117	8	181.29989	8	221.76862	8	262.23735	8	302.70607	8	343.17480	8	383.64352
9	19.82968	9	60.29840	9	100.76713	9	141.23585	9	181.70458	9	222.17331	9	262.64203	9	303.11076	9	343.57948	9	384.04821

50	20.23436	**150**	60.70309	**250**	101.17182	**350**	141.64054	**450**	182.10927	**550**	222.57799	**650**	263.04672	**750**	303.51545	**850**	343.98417	**950**	384.45290
1	20.63905	1	61.10778	1	101.57650	1	142.04523	1	182.51395	1	222.98268	1	263.45141	1	303.92013	1	344.38886	1	384.85759
2	21.04374	2	61.51246	2	101.98119	2	142.44992	2	182.91864	2	223.38737	2	263.85609	2	304.32482	2	344.79355	2	385.26227
3	21.44842	3	61.91715	3	102.38588	3	142.85460	3	183.32333	3	223.79206	3	264.26078	3	304.72951	3	345.19823	3	385.66696
4	21.85311	4	62.32184	4	102.79056	4	143.25929	4	183.72802	4	224.19674	4	264.66547	4	305.13419	4	345.60292	4	386.07165
5	22.25780	5	62.72653	5	103.19525	5	143.66398	5	184.13270	5	224.60143	5	265.07016	5	305.53888	5	346.00761	5	386.47633
6	22.66249	6	63.13121	6	103.59994	6	144.06866	6	184.53739	6	225.00612	6	265.47484	6	305.94357	6	346.41230	6	386.88102
7	23.06717	7	63.53590	7	104.00463	7	144.47335	7	184.94208	7	225.41080	7	265.87953	7	306.34826	7	346.81698	7	387.28571
8	23.47186	8	63.94059	8	104.40931	8	104.87804	8	185.34677	8	225.81549	8	266.28422	8	306.75294	8	347.22167	8	387.69040
9	23.87655	9	64.34527	9	104.81400	9	145.28273	9	185.75145	9	226.22018	9	266.68890	9	307.15763	9	347.62636	9	388.09508
60	24.28124	**160**	64.74996	**260**	105.21869	**360**	145.68741	**460**	186.15614	**560**	226.62487	**660**	267.09359	**760**	307.56232	**860**	348.03104	**960**	388.49977
1	24.68592	1	65.15465	1	105.62338	1	146.09210	1	186.56083	1	227.02955	1	267.49828	1	307.96701	1	348.43573	1	388.90446
2	25.09061	2	65.55934	2	106.02806	2	146.49679	2	186.96551	2	227.43424	2	267.90297	2	308.37169	2	348.84042	2	389.30915
3	25.49530	3	65.96402	3	106.43275	3	146.90148	3	187.37020	3	227.83893	3	268.30765	3	308.77638	3	349.24511	3	389.71383
4	25.89998	4	66.36871	4	106.83744	4	147.30616	4	187.77489	4	228.24362	4	268.71234	4	309.18107	4	349.64979	4	390.11852
5	26.30467	5	66.77340	5	107.24212	5	147.71085	5	188.17958	5	228.64830	5	269.11703	5	309.58575	5	350.05448	5	390.52321
6	26.70936	6	67.17809	6	107.64681	6	148.11554	6	188.58426	6	229.05299	6	269.52172	6	309.99044	6	350.45917	6	390.92789
7	27.11405	7	67.58277	7	108.05150	7	148.52022	7	188.98895	7	229.45768	7	269.92640	7	310.39513	7	350.86386	7	391.33258
8	27.51873	8	67.98746	8	108.45619	8	148.92491	8	189.39364	8	229.86236	8	270.33109	8	310.79982	8	351.26854	8	391.73727
9	27.92342	9	68.39215	9	108.86087	9	149.32960	9	189.79833	9	230.26705	9	270.73578	9	311.20450	9	351.67323	9	392.14196
70	28.32811	**170**	68.79683	**270**	109.26556	**370**	149.73429	**470**	190.20301	**570**	230.67174	**670**	271.14046	**770**	311.60919	**870**	352.07792	**970**	392.54664
1	28.73280	1	69.20152	1	109.67025	1	150.13897	1	190.60770	1	231.07643	1	271.54515	1	312.01388	1	352.48260	1	392.95133
2	29.13748	2	69.60621	2	110.07493	2	150.54366	2	191.01239	2	231.48111	2	271.94984	2	312.41857	2	352.88729	2	393.35602
3	29.54217	3	70.01090	3	110.47962	3	150.94835	3	191.41707	3	231.88580	3	272.35453	3	312.82325	3	353.29198	3	393.76070
4	29.94686	4	70.41558	4	110.88431	4	151.35304	4	191.82176	4	232.29049	4	272.75921	4	313.22794	4	353.69667	4	394.16539
5	30.35154	5	70.82027	5	111.28900	5	151.75772	5	192.22645	5	232.69518	5	273.16390	5	313.63263	5	354.10135	5	394.57008
6	30.75623	6	71.22496	6	111.69368	6	152.16241	6	192.63114	6	233.09986	6	273.56859	6	314.03731	6	354.50604	6	394.97477
7	31.16092	7	71.62965	7	112.09837	7	152.56710	7	193.03582	7	233.50455	7	273.97328	7	314.44200	7	354.91073	7	395.37945
8	31.56561	8	72.03433	8	112.50306	8	152.97178	8	193.44051	8	233.90924	8	274.37796	8	314.84669	8	355.31542	8	395.78414
9	31.97029	9	72.43902	9	112.90775	9	153.37647	9	193.84520	9	234.31392	9	274.78265	9	315.25138	9	355.72010	9	396.18883
80	32.37498	**180**	72.84371	**280**	113.31243	**380**	153.78116	**480**	194.24989	**580**	234.71861	**680**	275.18734	**780**	315.65606	**880**	356.12479	**980**	396.59352
1	32.77967	1	73.24839	1	113.71712	1	154.18585	1	194.65457	1	235.12330	1	275.59202	1	316.06075	1	356.52948	1	396.99820
2	33.18436	2	73.65308	2	114.12181	2	154.59053	2	195.05926	2	235.52799	2	275.99671	2	316.46544	2	356.93416	2	397.40289
3	33.58904	3	74.05777	3	114.52649	3	154.99522	3	195.46395	3	235.93267	3	276.40140	3	316.87013	3	357.33885	3	397.80758
4	33.99373	4	74.46246	4	114.93118	4	155.39991	4	195.86863	4	236.33736	4	276.80609	4	317.27481	4	357.74354	4	398.21226
5	34.39842	5	74.86714	5	115.33587	5	155.80460	5	196.27332	5	236.74205	5	277.21077	5	317.67950	5	358.14823	5	398.61695
6	34.80310	6	75.27183	6	115.74056	6	156.20928	6	196.67801	6	237.14673	6	277.61546	6	318.08419	6	358.55291	6	399.02164
7	35.20779	7	75.67652	7	116.14524	7	156.61397	7	197.08270	7	237.55142	7	278.02015	7	318.48887	7	358.95760	7	399.42633
8	35.61248	8	76.08121	8	116.54993	8	157.01866	8	197.48738	8	237.95611	8	278.42484	8	318.89356	8	359.36229	8	399.83101
9	36.01717	9	76.48589	9	116.95462	9	157.42334	9	197.89207	9	238.36080	9	278.82952	9	319.29825	9	359.76698	9	400.23570
90	36.42185	**190**	76.89058	**290**	117.35931	**390**	157.82803	**490**	198.29676	**590**	238.76548	**690**	279.23421	**790**	319.70294	**890**	360.17166	**990**	400.64039
1	36.82654	1	77.29527	1	117.76399	1	158.23272	1	198.70145	1	239.17017	1	279.63890	1	320.10762	1	360.57635	1	401.04508
2	37.23123	2	77.69995	2	118.16868	2	158.63741	2	199.10613	2	239.57486	2	280.04358	2	320.51231	2	360.98104	2	401.44976
3	37.63592	3	78.10464	3	118.57337	3	159.04209	3	199.51082	3	239.97955	3	280.44827	3	320.91700	3	361.38572	3	401.85445
4	38.04060	4	78.50933	4	118.97805	4	159.44678	4	199.91551	4	240.38423	4	280.85296	4	321.32169	4	361.79041	4	402.25914
5	38.44529	5	78.91402	5	119.38274	5	159.85147	5	200.32019	5	240.78892	5	281.25765	5	321.72637	5	362.19510	5	402.66382
6	38.84998	6	79.31870	6	119.78743	6	160.25616	6	200.72488	6	241.19361	6	281.66233	6	322.13106	6	362.59979	6	403.06851
7	39.25466	7	79.72339	7	120.19212	7	160.66084	7	201.12957	7	241.59829	7	282.06702	7	322.53575	7	363.00447	7	403.47320
8	39.65935	8	80.12808	8	120.59680	8	161.06553	8	201.53426	8	242.00298	8	282.47171	8	322.94043	8	363.40916	8	403.87789
9	40.06404	9	80.53276	9	121.00149	9	161.47022	9	201.93894	9	242.40767	9	282.87640	9	323.34512	9	363.81385	9	404.28257

AREA—HECTARES TO ACRES

[Reduction factor: 1 hectare=2.471043930 acres]

Hectares	Acres	Hectares	Acres	Hectares	Acres	Hectares	Acres	Hectares	Acres	Hectares	Acres	Hectares	Acres	Hectares	Acres	Hectares	Acres	Hectares	Acres
0		100	247.10439	200	494.20879	300	741.31318	400	988.41757	500	1,235.52197	600	1,482.62636	700	1,729.73075	800	1,976.83514	900	2,223.93954
1	2.47104	1	249.57544	1	496.67983	1	743.78422	1	990.88862	1	1,237.99301	1	1,485.09740	1	1,732.20180	1	1,979.30619	1	2,226.41058
2	4.94209	2	252.04648	2	499.15087	2	746.25527	2	993.35966	2	1,240.46405	2	1,487.56845	2	1,734.67284	2	1,981.77723	2	2,228.88163
3	7 41313	3	254.51752	3	501.62192	3	748.72631	3	995.83070	3	1,242.93510	3	1,490.03949	3	1,737.14388	3	1,984.24828	3	2,231.35267
4	9.88418	4	256.98857	4	504.09296	4	751.19735	4	998.30175	4	1,245.40614	4	1,492.51053	4	1,739.61493	4	1,986.71932	4	2,233.82371
5	12.35522	5	259.45961	5	506.56401	5	753.66840	5	1,000.77279	5	1,247.87718	5	1,494.98158	5	1,742.08597	5	1,989.19036	5	2,236.29476
6	14.82626	6	261.93066	6	509.03505	6	756.13944	6	1,003.24384	6	1,250.34823	6	1,497.45262	6	1,744.55701	6	1,991.66141	6	2,238.76580
7	17.29731	7	264.40170	7	511.50609	7	758.61049	7	1,005.71488	7	1,252.81927	7	1,499.92367	7	1,747.02806	7	1,994.13245	7	2,241.23684
8	19.76835	8	266.87274	8	513.97714	8	761.08153	8	1,008.18592	8	1,255.29032	8	1,502.39471	8	1,749.49910	8	1,996.60350	8	2,243.70789
9	22.23940	9	269.34379	9	516.44818	9	763.55257	9	1,010.65697	9	1,257.76136	9	1,504.86575	9	1,751.97015	9	1,999.07454	9	2,246.17893
10	24.71044	110	271.81483	210	518.91923	310	766.02362	410	1,013.12801	510	1,260.23240	610	1,507.33680	710	1,754.44119	810	2,001.54558	910	2,248.64998
1	27.18148	1	274.28588	1	521.39027	1	768.49466	1	1,015.59906	1	1,262.70345	1	1,509.80784	1	1,756.91223	1	2,004.01663	1	2,251.12102
2	29.65253	2	276.75692	2	523.86131	2	770.96571	2	1,018.07010	2	1,265.17449	2	1,512.27889	2	1,759.38328	2	2,006.48767	2	2,253.59206
3	32.12357	3	279.22796	3	526.33236	3	773.43675	3	1,020.54114	3	1,267.64554	3	1,514.74993	3	1,761.85432	3	2,008.95872	3	2,256.06311
4	34.59462	4	281.69901	4	528.80340	4	775.90779	4	1,023.01219	4	1,270.11658	4	1,517.22097	4	1,764.32537	4	2,011.42976	4	2,258.53415
5	37.06566	5	284.17005	5	531.27444	5	778.37884	5	1,025.48323	5	1,272.58762	5	1,519.69202	5	1,766.79641	5	2,013.90080	5	2,261.00520
6	39.53670	6	286.64110	6	533.74549	6	780.84988	6	1,027.95427	6	1,275.05867	6	1,522.16306	6	1,769.26745	6	2,016.37185	6	2,263.47624
7	42.00775	7	289.11214	7	536.21653	7	783.32093	7	1,030.42532	7	1,277.52971	7	1,524.63411	7	1,771.73850	7	2,018.84289	7	2,265.94728
8	44.47879	8	291.58318	8	538.68758	8	785.79197	8	1,032.89636	8	1,280.00076	8	1,527.10515	8	1,774.20954	8	2,021.31394	8	2,268.41833
9	46.94983	9	294.05423	9	541.15862	9	788.26301	9	1,035.36741	9	1,282.47180	9	1,529.57619	9	1,776.68059	9	2,023.78498	9	2,270.88937
20	49.42088	120	296.52527	220	543.62966	320	790.73406	420	1,037.83845	520	1,284.94284	620	1,532.04724	720	1,779.15163	820	2,026.25602	920	2,273.36042
1	51.89192	1	298.99632	1	546.10071	1	793.20510	1	1,040.30949	1	1,287.41389	1	1,534.51828	1	1,781.62267	1	2,028.72707	1	2,275.83146
2	54.36297	2	301.46736	2	548.57175	2	795.67615	2	1,042.78054	2	1,289.88493	2	1,536.98932	2	1,784.09372	2	2,031.19811	2	2,278.30250
3	56.83401	3	303.93840	3	551.04280	3	798.14719	3	1,045.25158	3	1,292.35598	3	1,539.46037	3	1,786.56476	3	2,033.66915	3	2,280.77355
4	59.30505	4	306.40945	4	553.51384	4	800.61823	4	1,047.72263	4	1,294.82702	4	1,541.93141	4	1,789.03581	4	2,036.14020	4	2,283.24459
5	61.77610	5	308.88049	5	555.98488	5	803.08928	5	1,050.19367	5	1,297.29806	5	1,544.40246	5	1,791.50685	5	2,038.61124	5	2,285.71564
6	64.24714	6	311.35154	6	558.45593	6	805.56032	6	1,052.66471	6	1,299.76911	6	1,546.87350	6	1,793.97789	6	2,041.08229	6	2,288.18668
7	66.71819	7	313.82258	7	560.92697	7	808.03137	7	1,055.13576	7	1,302.24015	7	1,549.34454	7	1,796.44894	7	2,043.55333	7	2,290.65772
8	69.18923	8	316.29362	8	563.39802	8	810.50241	8	1,057.60680	8	1,304.71120	8	1,551.81559	8	1,798.91998	8	2,046.02437	8	2,293.12877
9	71.66027	9	318 76467	9	565.86906	9	812.97345	9	1,060.07785	9	1,307.18224	9	1,554.28663	9	1,801.39103	9	2,048.49542	9	2,295.59981
30	74.13132	130	321 23571	230	568.34010	330	815.44450	430	1,062.54889	530	1,309.65328	630	1,556.75768	730	1,803.86207	830	2,050.96646	930	2,298.07086
1	76.60236	1	323 70675	1	570.81115	1	817.91554	1	1,065.01993	1	1,312.12433	1	1,559.22872	1	1,806.33311	1	2,053.43751	1	2,300.54190
2	79.07341	2	326 17780	2	573.28219	2	820.38658	2	1,067.49098	2	1,314.59537	2	1,561.69976	2	1,808.80416	2	2,055.90855	2	2,303.01294
3	81.54445	3	328 64884	3	575.75324	3	822.85763	3	1,069.96202	3	1,317.06641	3	1,564.17081	3	1,811.27520	3	2,058.37959	3	2,305.48399
4	84.01549	4	331.11989	4	578.22428	4	825.32867	4	1,072.43307	4	1,319.53746	4	1,566.64185	4	1,813.74624	4	2,060.85064	4	2,307.95503
5	86.48654	5	333 59093	5	580.69532	5	827.79972	5	1,074.90411	5	1,322.00850	5	1,569.11290	5	1,816.21729	5	2,063.32168	5	2,310.42607
6	88.95758	6	336.06197	6	583.16637	6	830.27076	6	1,077.37515	6	1,324.47555	6	1,571.58394	6	1,818.68833	6	2,065.79273	6	2,312.89712
7	91.42863	7	338.53302	7	585.63741	7	832.74180	7	1,079.84620	7	1,326.95059	7	1,574.05498	7	1,821.15938	7	2,068.26377	7	2,315.36816
8	93.89967	8	341.00406	8	588.10846	8	835.21285	8	1,082.31724	8	1,329.42163	8	1,576.52603	8	1,823.63042	8	2,070.73481	8	2,317.83921
9	96.37071	9	343 47511	9	590.57950	9	837.68389	9	1,084.78829	9	1,331.89268	9	1,578.99707	9	1,826.10146	9	2,073.20586	9	2,320.31025
40	98.84176	140	345.94615	240	593.05054	340	840.15494	440	1,087.25933	540	1,334.36372	640	1,581.46812	740	1,828.57251	840	2,075.67690	940	2,322.78129
1	101.31280	1	348.41719	1	595.52159	1	842.62598	1	1,089.73037	1	1,336.83477	1	1,583.93916	1	1,831.04355	1	2,078.14795	1	2,325.25234
2	103 78385	2	350.88824	2	597.99263	2	845.09702	2	1,092.20142	2	1,339.30581	2	1,586.41020	2	1,833.51460	2	2,080.61899	2	2,327.72338
3	106.25489	3	353.35928	3	600.46367	3	847.56807	3	1,094.67246	3	1,341.77685	3	1,588.88125	3	1,835.98564	3	2,083.09003	3	2,330.19443
4	108.72593	4	355.83033	4	602.93472	4	850.03911	4	1,097 14350	4	1,344.24790	4	1,591.35229	4	1,838.45668	4	2,085.56108	4	2,332.66547
5	111.19698	5	358.30137	5	605.40576	5	852.51016	5	1,099.61455	5	1,346.71894	5	1,593.82334	5	1,840.92773	5	2,088.03212	5	2,335.13651
6	113.66802	6	360.77241	6	607.87681	6	854.98120	6	1,102.08559	6	1,349.18999	6	1,596.29438	6	1,843.39877	6	2,090.50317	6	2,337.60756
7	116.13906	7	363.24346	7	610.34785	7	857.45224	7	1,104.55664	7	1,351.66103	7	1,598.76542	7	1,845.86982	7	2,092.97421	7	2,340.07860
8	118.61011	8	365.71450	8	612.81889	8	859.92329	8	1,107.02768	8	1,354.13207	8	1,601.23647	8	1,848.34086	8	2,095.44525	8	2,342.54965
9	121.08115	9	368.18555	9	615.28994	9	862.39433	9	1,109.49872	9	1,356.60312	9	1,603.70751	9	1,850.81190	9	2,097.91630	9	2,345.02069

50	123.55220	**150**	370.65659	**250**	617.76098	**350**	864.86538	**450**	1,111.96977	**550**	1,359.07416	**650**	1,606.17855	**750**	1,853.28295	**850**	2,100.38734	**950**	2,347.49173
1	126.02324	1	373.12763	1	620.23203	1	867.33642	1	1,114.44081	1	1,361.54521	1	1,608.64960	1	1,855.75399	1	2,102.85838	1	2,349.96278
2	128.49428	2	375.59868	2	622.70307	2	869.80746	2	1,116.91186	2	1,364.01625	2	1,611.12064	2	1,858.22504	2	2,105.32943	2	2,352.43382
3	130.96533	3	378.06972	3	625.17411	3	872.27851	3	1,119.38290	3	1,366.48729	3	1,613.59169	3	1,860.69608	3	2,107.80047	3	2,354.90487
4	133.43637	4	380.54077	4	627.64516	4	874.74955	4	1,121.85394	4	1,368.95834	4	1,616.06273	4	1,863.16712	4	2,110.27152	4	2,357.37591
5	135.90742	5	383.01181	5	630.11620	5	877.22060	5	1,124.32499	5	1,371.42938	5	1,618.53377	5	1,865.63817	5	2,112.74256	5	2,359.84695
6	138.37846	6	385.48285	6	632.58725	6	879.69164	6	1,126.79603	6	1,373.90043	6	1,621.00482	6	1,868.10921	6	2,115.21360	6	2,362.31800
7	140.84950	7	387.95390	7	635.05829	7	882.16268	7	1,129.26708	7	1,376.37147	7	1,623.47586	7	1,870.58026	7	2,117.68465	7	2,364.78904
8	143.32055	8	390.42494	8	637.52933	8	884.63373	8	1,131.73812	8	1,378.84251	8	1,625.94691	8	1,873.05130	8	2,120.15569	8	2,367.26009
9	145.79159	9	392.89598	9	640.00038	9	887.10477	9	1,134.20916	9	1,381.31356	9	1,628.41795	9	1,875.52234	9	2,122.62674	9	2,369.73113
60	148.26264	**160**	395.36703	**260**	642.47142	**360**	889.57581	**460**	1,136.68021	**560**	1,383.78460	**660**	1,630.88899	**760**	1,877.99339	**860**	2,125.09778	**960**	2,372.20217
1	150.73368	1	397.83807	1	644.94247	1	892.04686	1	1,139.15125	1	1,386.25564	1	1,633.36004	1	1,880.46443	1	2,127.56882	1	2,374.67322
2	153.20472	2	400.30912	2	647.41351	2	894.51790	2	1,141.62230	2	1,388.72669	2	1,635.83108	2	1,882.93547	2	2,130.03987	2	2,377.14426
3	155.67577	3	402.78016	3	649.88455	3	896.98895	3	1,144.09334	3	1,391.19773	3	1,638.30213	3	1,885.40652	3	2,132.51091	3	2,379.61530
4	158.14681	4	405.25120	4	652.35560	4	899.45999	4	1,146.56438	4	1,393.66878	4	1,640.77317	4	1,887.87756	4	2,134.98196	4	2,382.08635
5	160.61786	5	407.72225	5	654.82664	5	901.93103	5	1,149.03543	5	1,396.13982	5	1,643.24421	5	1,890.34861	5	2,137 45300	5	2,384.55739
6	163.08890	6	410.19329	6	657.29769	6	904.40208	6	1,151.50647	6	1,398.61086	6	1,645.71526	6	1,892.81965	6	2,139.92404	6	2,387.02844
7	165.55994	7	412.66434	7	659.76873	7	906.87312	7	1,153.97752	7	1,401.08191	7	1,648.18630	7	1,895.29069	7	2,142.39509	7	2,389.49948
8	168.03099	8	415.13538	8	662.23977	8	909.34417	8	1,156.44856	8	1,403.55295	8	1,650.65735	8	1,897.76174	8	2,1ı4.86613	8	2,391.97052
9	170.50203	9	417.60642	9	664.71082	9	911.81521	9	1,158.91960	9	1,406.02400	9	1,653.12839	9	1,900.23278	9	2,1ı7.33718	9	2,394.44157
70	172.97308	**170**	420.07747	**270**	667.18186	**370**	914.28625	**470**	1,161.39065	**570**	1,408.49504	**670**	1,655.59943	**770**	1,902.70383	**870**	2,149.80822	**970**	2,396.91261
1	175.44412	1	422.54851	1	669.65291	1	916.75730	1	1,163.86169	1	1,410.96608	1	1,658.07048	1	1,905.17487	1	2,152.27926	1	2,399.38366
2	177.91516	2	425.01956	2	672.12395	2	919.22834	2	1,166.33273	2	1,413.43713	2	1,660.54152	2	1,907.64591	2	2,154.75031	2	2,401.85470
3	180.38621	3	427.49060	3	674.59499	3	921.69939	3	1,168.80378	3	1,415.90817	3	1,663.01257	3	1,910.11696	3	2,157.22135	3	2,404.32574
4	182.85725	4	429.96164	4	677.06604	4	924.17043	4	1,171.27482	4	1,418.37922	4	1,665.48361	4	1,912.58800	4	2,159.69240	4	2,406.79679
5	185.32829	5	432.43269	5	679.53708	5	926.64147	5	1,173.74587	5	1,420.85026	5	1,667.95465	5	1,915.05905	5	2,162.16344	5	2,409.26783
6	187.79934	6	434.90373	6	682.00812	6	929.11252	6	1,176.21691	6	1,423.32130	6	1,670.42570	6	1,917.53009	6	2,164.63448	6	2,411.73888
7	190.27038	7	437.37478	7	684.47917	7	931.58356	7	1,178.68795	7	1,425.79235	7	1,672.89674	7	1,920.00113	7	2,167.10553	7	2,414.20992
8	192.74143	8	439.84582	8	686.95021	8	934.05461	8	1,181.15900	8	1,428.26339	8	1,675.36778	8	1,922.47218	8	2,169.57657	8	2,416.68096
9	195.21247	9	442.31686	9	689.42126	9	936.52565	9	1,183.63004	9	1,430.73444	9	1,677.83883	9	1,924.94322	9	2,172.04761	9	2,419.15201
80	197.68351	**180**	444.78791	**280**	691.89230	**380**	938.99669	**480**	1,186.10109	**580**	1,433.20548	**680**	1,680.30987	**780**	1,927.41427	**880**	2,174.51866	**980**	2,421.62305
1	200.15456	1	447.25895	1	694.36334	1	941.46774	1	1,188.57213	1	1,435.67652	1	1,682.78092	1	1,929.88531	1	2,176.98970	1	2,424.09410
2	202.62560	2	449.73000	2	696.83439	2	943.93878	2	1,191.04317	2	1,438.14757	2	1,685.25196	2	1,932.35635	2	2,179.46075	2	2,426.56514
3	205.09665	3	452.20104	3	699.30543	3	946.40983	3	1,193.51422	3	1,440.61861	3	1,687.72300	3	1,934.82740	3	2,181.93179	3	2,429.03618
4	207.56769	4	454.67208	4	701.77648	4	948.88087	4	1,195.98526	4	1,443.08966	4	1,690.19405	4	1,937.29844	4	2,184.40283	4	2,431.50723
5	210.03873	5	457.14313	5	704.24752	5	951.35191	5	1,198.45631	5	1,445.56070	5	1,692.66509	5	1,939.76949	5	2,186.87388	5	2,433.97827
6	212.50978	6	459.61417	6	706.71856	6	953.82296	6	1,200.92735	6	1,448.03174	6	1,695.13614	6	1,942.24053	6	2,189.34492	6	2,436.44932
7	214.98082	7	462.08521	7	709.18961	7	956.29400	7	1,203.39839	7	1,450.50279	7	1,697.60718	7	1,944.71157	7	2,191.81597	7	2,438.92036
8	217.45187	8	464.55626	8	711.66065	8	958.76504	8	1,205.86944	8	1,452.97383	8	1,700.07822	8	1,947.18262	8	2,194.28701	8	2,441.39140
9	219.92291	9	467.02730	9	714.13170	9	961.23609	9	1,208.34048	9	1,455.44487	9	1,702.54927	9	1,949.65366	9	2,196.75805	9	2,443.86245
90	222.39395	**190**	469.49835	**290**	716.60274	**390**	963.70713	**490**	1,210.81153	**590**	1,457.91592	**690**	1,705.02031	**790**	1,952.12471	**890**	2,199.22910	**990**	2,446.33349
1	224.86500	1	471.96939	1	719.07378	1	966.17818	1	1,213.28257	1	1,460.38696	1	1,707.49136	1	1,954.59575	1	2,201.70014	1	2,448.80454
2	227.33604	2	474.44043	2	721.54483	2	968.64922	2	1,215.75361	2	1,462.85801	2	1,709.96240	2	1,957.06679	2	2,204.17119	2	2,451.27558
3	229.80709	3	476.91148	3	724.01587	3	971.12026	3	1,218.22466	3	1,465.32905	3	1,712.43344	3	1,959.53784	3	2,206.64223	3	2,453.74662
4	232.27813	4	479.38252	4	726.48692	4	973.59131	4	1,220.69570	4	1,467.80009	4	1,714.90449	4	1,962.00888	4	2,209.11327	4	2,456.21767
5	234.74917	5	481.85357	5	728.95796	5	976.06235	5	1,223.16675	5	1,470.27114	5	1,717.37553	5	1,964.47992	5	2,211.58432	5	2,458.68871
6	237.22022	6	484.32461	6	731.42900	6	978.53340	6	1,225.63779	6	1,472.74218	6	1,719.84658	6	1,966.95097	6	2,214.05536	6	2,461.15975
7	239.69126	7	486.79565	7	733.90005	7	981.00444	7	1,228.10883	7	1,475.21323	7	1,722.31762	7	1,969.42201	7	2,216.52641	7	2,463.63080
8	242.16231	8	489.26670	8	736.37109	8	983.47548	8	1,230.57988	8	1,477.68427	8	1,724.78866	8	1,971.89306	8	2,218.99745	8	2,466.10184
9	244.63335	9	491.73774	9	738.84214	9	985.94653	9	1,233.05092	9	1,480.15531	9	1,727.25971	9	1,974.36410	9	2,221.46849	9	2,468.57289

VOLUME—CUBIC YARDS TO CUBIC METERS

[Reduction factor: 1 cubic yard=0.7645594453 cubic meter]

Cubic yards	Cubic meters	Cubic yards	Cubic meters	Cubic yards	Cubic meters	Cubic yards	Cubic meters	Cubic yards	Cubic meters	Cubic yards	Cubic meters	Cubic yards	Cubic meters	Cubic yards	Cubic meters	Cubic yards	Cubic meters	Cubic yards	Cubic meters
0		**100**	76.45594	**200**	152.91189	**300**	229.36783	**400**	305.82378	**500**	382.27972	**600**	458.73567	**700**	535.19161	**800**	611.64756	**900**	688.10350
1	0.76456	1	77.22050	1	153.67645	1	230.13239	1	306.58834	1	333.04428	1	459.50023	1	535.95617	1	612.41212	1	688.86806
2	1.52912	2	77.98506	2	154.44101	2	230.89695	2	307.35290	2	383.80884	2	460.26479	2	536.72073	2	613.17668	2	689.63262
3	2.29368	3	78.74962	3	155.20557	3	231.66151	3	308.11746	3	384.57340	3	461.02935	3	537.48529	3	613.94123	3	690.39718
4	3.05824	4	79.51418	4	155.97013	4	232.42607	4	308.88202	4	385.33796	4	461.79390	4	538.24985	4	614.70579	4	691.16174
5	3.82280	5	80.27874	5	156.73469	5	233.19063	5	309.64658	5	386.10252	5	462.55846	5	539.01441	5	615.47035	5	691.92630
6	4.58736	6	81.04330	6	157.49925	6	233.95519	6	310.41113	6	386.86708	6	463.32302	6	539.77897	6	616.23491	6	692.69086
7	5.35192	7	81.80786	7	158.26381	7	234.71975	7	311.17569	7	387.63164	7	464.08758	7	540.54353	7	616.99947	7	693.45542
8	6.11648	8	82.57242	8	159.02836	8	235.48431	8	311.94025	8	388.39620	8	464.85214	8	541.30809	8	617.76403	8	694.21998
9	6.88104	9	83.33698	9	159.79292	9	236.24887	9	312.70481	9	389.16076	9	465.61670	9	542.07265	9	618.52859	9	694.98454
10	7.64559	**110**	84.10154	**210**	160.55748	**310**	237.01343	**410**	313.46937	**510**	389.92532	**610**	466.38126	**710**	542.83721	**810**	619.29315	**910**	695.74910
1	8.41015	1	84.86610	1	161.32204	1	237.77799	1	314.23393	1	390.68988	1	467.14582	1	543.60177	1	620.05771	1	696.51365
2	9.17471	2	85.63066	2	162.08660	2	238.54255	2	314.99849	2	391.45444	2	467.91038	2	544.36633	2	620.82227	2	697.27821
3	9.93927	3	86.39522	3	162.85116	3	239.30711	3	315.76305	3	392.21900	3	468.67494	3	545.13088	3	621.58683	3	698.04277
4	10.70383	4	87.15978	4	163.61572	4	240.07167	4	316.52761	4	392.98355	4	469.43950	4	545.89544	4	622.35139	4	698.80733
5	11.46839	5	87.92434	5	164.38028	5	240.83623	5	317.29217	5	393.74811	5	470.20406	5	546.66000	5	623.11595	5	699.57189
6	12.23295	6	88.68890	6	165.14484	6	241.60078	6	318.05673	6	394.51267	6	470.96862	6	547.42456	6	623.88051	6	700.33645
7	12.99751	7	89.45346	7	165.90940	7	242.36534	7	318.82129	7	395.27723	7	471.73318	7	548.18912	7	624.64507	7	701.10101
8	13.76207	8	90.21801	8	166.67396	8	243.12990	8	319.58585	8	396.04179	8	472.49774	8	548.95368	8	625.40963	8	701.86557
9	14.52663	9	90.98257	9	167.43852	9	243.89446	9	320.35041	9	396.80635	9	473.26230	9	549.71824	9	626.17419	9	702.63013
20	15.29119	**120**	91.74713	**220**	168.20308	**320**	244.65902	**420**	321.11497	**520**	397.57091	**620**	474.02686	**720**	550.48280	**820**	626.93875	**920**	703.39469
1	16.05575	1	92.51169	1	168.96764	1	245.42358	1	321.87953	1	398.33547	1	474.79142	1	551.24736	1	627.70330	1	704.15925
2	16.82031	2	93.27625	2	169.73220	2	246.18814	2	322.64409	2	399.10003	2	475.55597	2	552.01192	2	628.46786	2	704.92381
3	17.58487	3	94.04081	3	170.49676	3	246.95270	3	323.40865	3	399.86459	3	476.32053	3	552.77648	3	629.23242	3	705.68837
4	18.34943	4	94.80537	4	171.26132	4	247.71726	4	324.17320	4	400.62915	4	477.08509	4	553.54104	4	629.99698	4	706.45293
5	19.11399	5	95.56993	5	172.02588	5	248.48182	5	324.93776	5	401.39371	5	477.84965	5	554.30560	5	630.76154	5	707.21749
6	19.87855	6	96.33449	6	172.79043	6	249.24638	6	325.70232	6	402.15827	6	478.61421	6	555.07016	6	631.52610	6	707.98205
7	20.64311	7	97.09905	7	173.55499	7	250.01094	7	326.46688	7	402.92283	7	479.37877	7	555.83472	7	632.29066	7	708.74661
8	21.40766	8	97.86361	8	174.31955	8	250.77550	8	327.23144	8	403.68739	8	480.14333	8	556.59928	8	633.05522	8	709.51117
9	22.17222	9	98.62817	9	175.08411	9	251.54006	9	327.99600	9	404.45195	9	480.90789	9	557.36384	9	633.81978	9	710.27572
30	22.93678	**130**	99.39273	**230**	175.84867	**330**	252.30462	**430**	328.76056	**530**	405.21651	**630**	481.67245	**730**	558.12840	**830**	634.58434	**930**	711.04028
1	23.70134	1	100.15729	1	176.61323	1	253.06918	1	329.52512	1	405.98107	1	482.43701	1	558.89295	1	635.34890	1	711.80484
2	24.46590	2	100.92185	2	177.37779	2	253.83374	2	330.28968	2	406.74562	2	483.20157	2	559.65751	2	636.11346	2	712.56940
3	25.23046	3	101.68641	3	178.14235	3	254.59830	3	331.05424	3	407.51018	3	483.96613	3	560.42207	3	636.87802	3	713.33396
4	25.99502	4	102.45097	4	178.90691	4	255.36285	4	331.81880	4	408.27474	4	484.73069	4	561.18663	4	637.64258	4	714.09852
5	26.75958	5	103.21553	5	179.67147	5	256.12741	5	332.58336	5	409.03930	5	485.49525	5	561.95119	5	638.40714	5	714.86308
6	27.52414	6	103.98008	6	180.43603	6	256.89197	6	333.34792	6	409.80386	6	486.25981	6	562.71575	6	639.17170	6	715.62764
7	28.28870	7	104.74464	7	181.20059	7	257.65653	7	334.11248	7	410.56842	7	487.02437	7	563.48031	7	639.93626	7	716.39220
8	29.05326	8	105.50920	8	181.96515	8	258.42109	8	334.87704	8	411.33298	8	487.78893	8	564.24487	8	640.70082	8	717.15676
9	29.81782	9	106.27376	9	182.72971	9	259.18565	9	335.64160	9	412.09754	9	488.55349	9	565.00943	9	641.46537	9	717.92132
40	30.58238	**140**	107.03832	**240**	183.49427	**340**	259.95021	**440**	336.40616	**540**	412.86210	**640**	489.31804	**740**	565.77399	**840**	642.22993	**940**	718.68588
1	31.34694	1	107.80288	1	184.25883	1	260.71477	1	337.17072	1	413.62666	1	490.08260	1	566.53855	1	642.99449	1	719.45044
2	32.11150	2	108.56744	2	185.02339	2	261.47933	2	337.93527	2	414.39122	2	490.84716	2	567.30311	2	643.75905	2	720.21500
3	32.87606	3	109.33200	3	185.78795	3	262.24389	3	338.69983	3	415.15578	3	491.61172	3	568.06767	3	644.52361	3	720.97956
4	33.64062	4	110.09656	4	186.55250	4	263.00845	4	339.46439	4	415.92034	4	492.37628	4	568.83223	4	645.28817	4	721.74412
5	34.40518	5	110.86112	5	187.31706	5	263.77301	5	340.22895	5	416.68490	5	493.14084	5	569.59679	5	646.05273	5	722.50868
6	35.16973	6	111.62568	6	188.08162	6	264.53757	6	340.99351	6	417.44946	6	493.90540	6	570.36135	6	646.81729	6	723.27324
7	35.93429	7	112.39024	7	188.84618	7	265.30213	7	341.75807	7	418.21402	7	494.66996	7	571.12591	7	647.58185	7	724.03779
8	36.69885	8	113.15480	8	189.61074	8	266.06669	8	342.52263	8	418.97858	8	495.43452	8	571.89047	8	648.34641	8	724.80235
9	37.46341	9	113.91936	9	190.37530	9	266.83125	9	343.28719	9	419.74314	9	496.19908	9	572.65502	9	649.11097	9	725.56691

50	38.22797	**150**	114.68392	**250**	191.13986	**350**	267.59581	**450**	344.05175	**550**	420.50769	**650**	496.96364	**750**	573.41958	**850**	649.87553	**950**	726.33147
1	38.99253	1	115.44848	1	191.90442	1	268.36037	1	344.81631	1	421.27225	1	497.72820	1	574.18414	1	650.64009	1	727.09603
2	39.75709	2	116.21304	2	192.66898	2	269.12492	2	345.58087	2	422.03681	2	498.49276	2	574.94870	2	651.40465	2	727.86059
3	40.52165	3	116.97760	3	193.43354	3	269.88948	3	346.34543	3	422.80137	3	499.25732	3	575.71326	3	652.16921	3	728.62515
4	41.28621	4	117.74215	4	194.19810	4	270.65404	4	347.10999	4	423.56593	4	500.02188	4	576.47782	4	652.93377	4	729.38971
5	42.05077	5	118.50671	5	194.96266	5	271.41860	5	347.87455	5	424.33049	5	500.78644	5	577.24238	5	653.69833	5	730.15427
6	42.81533	6	119.27127	6	195.72722	6	272.18316	6	348.63911	6	425.09505	6	501.55100	6	578.00694	6	654.46289	6	730.91883
7	43.57989	7	120.03583	7	196.49178	7	272.94772	7	349.40367	7	425.85961	7	502.31556	7	578.77150	7	655.22744	7	731.68339
8	44.34445	8	120.80039	8	197.25634	8	273.71228	8	350.16823	8	426.62417	8	503.08012	8	579.53606	8	655.99200	8	732.44795
9	45.10901	9	121.56495	9	198.02090	9	274.47684	9	350.93279	9	427.38873	9	503.84467	9	580.30062	9	656.75656	9	733.21251
60	45.87357	**160**	122.32951	**260**	198.78546	**360**	275.24140	**460**	351.69734	**560**	428.15329	**660**	504.60923	**760**	581.06518	**860**	657.52112	**960**	733.97707
1	46.63813	1	123.09407	1	199.55002	1	276.00596	1	352.46190	1	428.91785	1	505.37379	1	581.82974	1	658.28568	1	734.74163
2	47.40269	2	123.85863	2	200.31457	2	276.77052	2	353.22646	2	429.68241	2	506.13835	2	582.59430	2	659.05024	2	735.50619
3	48.16725	3	124.62319	3	201.07913	3	277.53508	3	353.99102	3	430.44697	3	506.90291	3	583.35886	3	659.81480	3	736.27075
4	48.93180	4	125.38775	4	201.84369	4	278.29964	4	354.75558	4	431.21153	4	507.66747	4	584.12342	4	660.57936	4	737.03531
5	49.69636	5	126.15231	5	202.60825	5	279.06420	5	355.52014	5	431.97609	5	508.43203	5	584.88798	5	661.34392	5	737.79986
6	50.46092	6	126.91687	6	203.37281	6	279.82876	6	356.28470	6	432.74065	6	509.19659	6	585.65254	6	662.10848	6	738.56442
7	51.22548	7	127.68143	7	204.13737	7	280.59332	7	357.04926	7	433.50521	7	509.96115	7	586.41709	7	662.87304	7	739.32898
8	51.99004	8	128.44599	8	204.90193	8	281.35788	8	357.81382	8	434.26976	8	510.72571	8	587.18165	8	663.63760	8	740.09354
9	52.75460	9	129.21055	9	205.66649	9	282.12244	9	358.57838	9	435.03432	9	511.49027	9	587.94621	9	664.40216	9	740.85810
70	53.51916	**170**	129.97511	**270**	206.43105	**370**	282.88699	**470**	359.34294	**570**	435.79888	**670**	512.25483	**770**	588.71077	**870**	665.16672	**970**	741.62266
1	54.28372	1	130.73967	1	207.19561	1	283.65155	1	360.10750	1	436.56344	1	513.01939	1	589.47533	1	665.93128	1	742.38722
2	55.04828	2	131.50422	2	207.96017	2	284.41611	2	360.87206	2	437.32800	2	513.78395	2	590.23989	2	666.69584	2	743.15178
3	55.81284	3	132.26878	3	208.72473	3	285.18067	3	361.63662	3	438.09256	3	514.54851	3	591.00445	3	667.46040	3	743.91634
4	56.57740	4	133.03334	4	209.48929	4	285.94523	4	362.40118	4	438.85712	4	515.31307	4	591.76901	4	668.22496	4	744.68090
5	57.34196	5	133.79790	5	210.25385	5	286.70979	5	363.16574	5	439.62168	5	516.07763	5	592.53357	5	668.98951	5	745.44546
6	58.10652	6	134.56246	6	211.01841	6	287.47435	6	363.93030	6	440.38624	6	516.84219	6	593.29813	6	669.75407	6	746.21002
7	58.87108	7	135.32702	7	211.78297	7	288.23891	7	364.69486	7	441.15080	7	517.60674	7	594.06269	7	670.51863	7	746.97458
8	59.63564	8	136.09158	8	212.54753	8	289.00347	8	365.45941	8	441.91536	8	518.37130	8	594.82725	8	671.28319	8	747.73914
9	60.40020	9	136.85614	9	213.31209	9	289.76803	9	366.22397	9	442.67992	9	519.13586	9	595.59181	9	672.04775	9	748.50370
80	61.16476	**180**	137.62070	**280**	214.07664	**380**	290.53259	**480**	366.98853	**580**	443.44448	**680**	519.90042	**780**	596.35637	**880**	672.81231	**980**	749.26826
1	61.92932	1	138.38526	1	214.84120	1	291.29715	1	367.75309	1	444.20904	1	520.66498	1	597.12093	1	673.57687	1	750.03282
2	62.69387	2	139.14982	2	215.60576	2	292.06171	2	368.51765	2	444.97360	2	521.42954	2	597.88549	2	674.34143	2	750.79738
3	63.45843	3	139.91438	3	216.37032	3	292.82627	3	369.28221	3	445.73816	3	522.19410	3	598.65005	3	675.10599	3	751.56193
4	64.22299	4	140.67894	4	217.13488	4	293.59083	4	370.04677	4	446.50272	4	522.95866	4	599.41461	4	675.87055	4	752.32649
5	64.98755	5	141.44350	5	217.89944	5	594.35539	5	370.81133	5	447.26728	5	523.72322	5	600.17916	5	676.63511	5	753.09105
6	65.75211	6	142.20806	6	218.66400	6	295.11995	6	371.57589	6	448.03183	6	524.48778	6	600.94372	6	677.39967	6	753.85561
7	66.51667	7	142.97262	7	219.42856	7	295.88451	7	372.34045	7	448.79639	7	525.25234	7	601.70828	7	678.16423	7	754.62017
8	67.28123	8	143.73718	8	220.19312	8	296.64906	8	373.10501	8	449.56095	8	526.01690	8	602.47284	8	678.92879	8	755.38473
9	68.04579	9	144.50174	9	220.95768	9	297.41362	9	373.86957	9	450.32551	9	526.78146	9	603.23740	9	679.69335	9	756.14929
90	68.81035	**190**	145.26629	**290**	221.72224	**390**	298.17818	**490**	374.63413	**590**	451.09007	**690**	527.54602	**790**	604.00196	**890**	680.45791	**990**	756.91385
1	69.57491	1	146.03085	1	222.48680	1	298.94274	1	375.39869	1	451.85463	1	528.31058	1	604.76652	1	681.22247	1	757.67841
2	70.33947	2	146.79541	2	223.25136	2	299.70730	2	376.16325	2	452.61919	2	529.07514	2	605.53108	2	681.98703	2	758.44297
3	71.10403	3	147.55997	3	224.01592	3	300.47186	3	376.92781	3	453.38375	3	529.83970	3	606.29564	3	682.75158	3	759.20753
4	71.86859	4	148.32453	4	224.78048	4	301.23642	4	377.69237	4	454.14831	4	530.60426	4	607.06020	4	683.51614	4	759.97209
5	72.63315	5	149.08909	5	225.54504	5	302.00098	5	378.45693	5	454.91287	5	531.36881	5	607.82476	5	684.28070	5	760.73665
6	73.39771	6	149.85365	6	226.30960	6	302.76554	6	379.22148	6	455.67743	6	532.13337	6	608.58932	6	685.04526	6	761.50121
7	74.16227	7	150.61821	7	227.07416	7	303.53010	7	379.98604	7	456.44199	7	532.89793	7	609.35388	7	685.80982	7	762.26577
8	74.92683	8	151.38277	8	227.83871	8	304.29466	8	380.75060	8	457.20655	8	533.66249	8	610.11844	8	686.57438	8	763.03033
9	75.69139	9	152.14733	9	228.60327	9	305.05922	9	381.51516	9	457.97111	9	534.42705	9	610.88300	9	687.33894	9	763.79489

VOLUME—CUBIC METERS TO CUBIC YARDS

[Reduction factor: 1 cubic meter=1.307942772 cubic yards]

Cubic meters	Cubic yards	Cubic meters	Cubic yards	Cubic meters	Cubic yards	Cubic meters	Cubic yards	Cubic meters	Cubic yards	Cubic meters	Cubic yards	Cubic meters	Cubic yards	Cubic meters	Cubic yards	Cubic meters	Cubic yards	Cubic meters	Cubic yards
0		**100**	130.79428	**200**	261.58855	**300**	392.38283	**400**	523.17711	**500**	653.97139	**600**	784.76566	**700**	915.55994	**800**	1,046.35422	**900**	1,177.14849
1	1.30794	1	132.10222	1	262.89650	1	393.69077	1	524.48505	1	655.27933	1	786.07361	1	916.86788	1	1,047.66216	1	1,178.45644
2	2.61589	2	133.41016	2	264.20444	2	394.99872	2	525.79299	2	656.58727	2	787.38155	2	918.17583	2	1,048.97010	2	1,179.76438
3	3.92383	3	134.71811	3	265.51238	3	396.30666	3	527.10094	3	657.89521	3	788.68949	3	919.48377	3	1,050.27805	3	1,181.07232
4	5.23177	4	136.02605	4	266.82033	4	397.61460	4	528.40888	4	659.20316	4	789.99743	4	920.79171	4	1,051.58599	4	1,182.38027
5	6.53971	5	137.33399	5	268.12827	5	398.92255	5	529.71682	5	660.51110	5	791.30538	5	922.09965	5	1,052.89393	5	1,183.68821
6	7.84766	6	138.64193	6	269.43621	6	400.23049	6	531.02477	6	661.81904	6	792.61332	6	923.40760	6	1,054.20187	6	1,184.99615
7	9.15560	7	139.94988	7	270.74415	7	401.53843	7	532.33271	7	663.12699	7	793.92126	7	924.71554	7	1,055.50982	7	1,186.30409
8	10.46354	8	141.25782	8	272.05210	8	402.84637	8	533.64065	8	664.43493	8	795.22921	8	926.02348	8	1,056.81776	8	1,187.61204
9	11.77148	9	142.56576	9	273.36004	9	404.15432	9	534.94859	9	665.74287	9	796.53715	9	927.33143	9	1,058.12570	9	1,188.91998
10	13.07943	**110**	143.87370	**210**	274.66798	**310**	405.46226	**410**	536.25654	**510**	667.05081	**610**	797.84509	**710**	928.63937	**810**	1,059.43365	**910**	1,190.22792
1	14.38737	1	145.18165	1	275.97592	1	406.77020	1	537.56448	1	668.35876	1	799.15303	1	929.94731	1	1,060.74159	1	1,191.53587
2	15.69531	2	146.48959	2	277.28387	2	408.07814	2	538.87242	2	669.66670	2	800.46098	2	931.25525	2	1,062.04953	2	1,192.84381
3	17.00326	3	147.79753	3	278.59181	3	409.38609	3	540.18036	3	670.97464	3	801.76892	3	932.56320	3	1,063.35747	3	1,194.15175
4	18.31120	4	149.10548	4	279.89975	4	410.69403	4	541.48831	4	672.28258	4	803.07686	4	933.87114	4	1,064.66542	4	1,195.45969
5	19.61914	5	150.41342	5	281.20770	5	412.00197	5	542.79625	5	673.59053	5	804.38480	5	935.17908	5	1,065.97336	5	1,196.76764
6	20.92708	6	151.72136	6	282.51564	6	413.30992	6	544.10419	6	674.89847	6	805.69275	6	936.48702	6	1,067.28130	6	1,198.07558
7	22.23503	7	153.02930	7	283.82358	7	414.61786	7	545.41214	7	676.20641	7	807.00069	7	937.79497	7	1,068.58924	7	1,199.38352
8	23.54297	8	154.33725	8	285.13152	8	415.92580	8	546.72008	8	677.51436	8	808.30863	8	939.10291	8	1,069.89719	8	1,200.69146
9	24.85091	9	155.64519	9	286.43947	9	417.23374	9	548.02802	9	678.82230	9	809.61658	9	940.41085	9	1,071.20513	9	1,201.99941
20	26.15886	**120**	156.95313	**220**	287.74741	**320**	418.54169	**420**	549.33596	**520**	680.13024	**620**	810.92452	**720**	941.71880	**820**	1,072.51307	**920**	1,203.30735
1	27.46680	1	158.26108	1	289.05535	1	419.84963	1	550.64391	1	681.43818	1	812.23246	1	943.02674	1	1,073.82102	1	1,204.61529
2	28.77474	2	159.56902	2	290.36330	2	421.15757	2	551.95185	2	682.74613	2	813.54040	2	944.33468	2	1,075.12896	2	1,205.92324
3	30.08268	3	160.87696	3	291.67124	3	422.46552	3	553.25979	3	684.05407	3	814.84835	3	945.64262	3	1,076.43690	3	1,207.23118
4	31.39063	4	162.18490	4	292.97918	4	423.77346	4	554.56774	4	685.36201	4	816.15629	4	946.95057	4	1,077.74484	4	1,208.53912
5	32.69857	5	163.49285	5	294.28712	5	425.08140	5	555.87568	5	686.66996	5	817.46423	5	948.25851	5	1,079.05279	5	1,209.84706
6	34.00651	6	164.80079	6	295.59507	6	426.38934	6	557.18362	6	687.97790	6	818.77218	6	949.56645	6	1,080.36073	6	1,211.15501
7	35.31445	7	166.10873	7	296.90301	7	427.69729	7	558.49156	7	689.28584	7	820.08012	7	950.87440	7	1,081.66867	7	1,212.46295
8	36.62240	8	167.41667	8	298.21095	8	429.00523	8	559.79951	8	690.59378	8	821.38806	8	952.18234	8	1,082.97662	8	1,213.77089
9	37.93034	9	168.72462	9	299.51889	9	430.31317	9	561.10745	9	691.90173	9	822.69600	9	953.49028	9	1,084.28456	9	1,215.07884
30	39.23828	**130**	170.03256	**230**	300.82684	**330**	431.62111	**430**	562.41539	**530**	693.20967	**630**	824.00395	**730**	954.79822	**830**	1,085.59250	**930**	1,216.38678
1	40.54623	1	171.34050	1	302.13478	1	432.92906	1	563.72333	1	694.51761	1	825.31189	1	956.10617	1	1,086.90044	1	1,217.69472
2	41.85417	2	172.64845	2	303.44272	2	434.23700	2	565.03128	2	695.82555	2	826.61983	2	957.41411	2	1,088.20839	2	1,219.00266
3	43.16211	3	173.95639	3	304.75067	3	435.54494	3	566.33922	3	697.13350	3	827.92777	3	958.72205	3	1,089.51633	3	1,220.31061
4	44.47005	4	175.26433	4	306.05861	4	436.85289	4	567.64716	4	698.44144	4	829.23572	4	960.02999	4	1,090.82427	4	1,221.61855
5	45.77800	5	176.57227	5	307.36655	5	438.16083	5	568.95511	5	699.74938	5	830.54366	5	961.33794	5	1,092.13221	5	1,222.92649
6	47.08594	6	177.88022	6	308.67449	6	439.46877	6	570.26305	6	701.05733	6	831.85160	6	962.64588	6	1,093.44016	6	1,224.23443
7	48.39388	7	179.18816	7	309.98244	7	440.77671	7	571.57099	7	702.36527	7	833.15955	7	963.95382	7	1,094.74810	7	1,225.54238
8	49.70183	8	180.49610	8	311.29038	8	442.08466	8	572.87893	8	703.67321	8	834.46749	8	965.26177	8	1,096.05604	8	1,226.85032
9	51.00977	9	181.80405	9	312.59832	9	443.39260	9	574.18688	9	704.98115	9	835.77543	9	966.56971	9	1,097.36399	9	1,228.15826
40	52.31771	**140**	183.11199	**240**	313.90627	**340**	444.70054	**440**	575.49482	**540**	706.28910	**640**	837.08337	**740**	967.87765	**840**	1,098.67193	**940**	1,229.46621
1	53.62565	1	184.41993	1	315.21421	1	446.00849	1	576.80276	1	707.59704	1	838.39132	1	969.18559	1	1,099.97987	1	1,230.77415
2	54.93360	2	185.72787	2	316.52215	2	447.31643	2	578.11071	2	708.90498	2	839.69926	2	970.49354	2	1,101.28781	2	1,232.08209
3	56.24154	3	187.03582	3	317.83009	3	448.62437	3	579.41865	3	710.21293	3	841.00720	3	971.80148	3	1,102.59576	3	1,233.39003
4	57.54948	4	188.34376	4	319.13804	4	449.93231	4	580.72659	4	711.52087	4	842.31515	4	973.10942	4	1,103.90370	4	1,234.69798
5	58.85742	5	189.65170	5	320.44598	5	451.24026	5	582.03453	5	712.82881	5	843.62309	5	974.41737	5	1,105.21164	5	1,236.00592
6	60.16537	6	190.95964	6	321.75392	6	452.54820	6	583.34248	6	714.13675	6	844.93103	6	975.72531	6	1,106.51959	6	1,237.31386
7	61.47331	7	192.26759	7	323.06186	7	453.85614	7	584.65042	7	715.44470	7	846.23897	7	977.03325	7	1,107.82753	7	1,238.62181
8	62.78125	8	193.57553	8	324.36981	8	455.16408	8	585.95836	8	716.75264	8	847.54692	8	978.34119	8	1,109.13547	8	1,239.92975
9	64.08920	9	194.88347	9	325.67775	9	456.47203	9	587.26630	9	718.06058	9	848.85486	9	979.64914	9	1,110.44341	9	1,241.23769

50	65.39714	**150**	196.19142	**250**	326.98569	**350**	457.77997	**450**	588.57425	**550**	719.36852	**650**	850.16280	**750**	980.95708	**850**	1,111.75136	**950**	1,242.54563
1	66.70508	1	197.49936	1	328.29364	1	459.08791	1	589.88219	1	720.67647	1	851.47074	1	982.26502	1	1,113.05930	1	1,243.85358
2	68.01302	2	198.80730	2	329.60158	2	460.39586	2	591.19013	2	721.98441	2	852.77869	2	983.57296	2	1,114.36724	2	1,245.16152
3	69.32097	3	200.11524	3	330.90952	3	461.70380	3	592.49808	3	723.29235	3	854.08663	3	984.88091	3	1,115.67518	3	1,246.46946
4	70.62891	4	201.42319	4	332.21746	4	463.01174	4	593.80602	4	724.60030	4	855.39457	4	986.18885	4	1,116.98313	4	1,247.77740
5	71.93685	5	202.73113	5	333.52541	5	464.31968	5	595.11396	5	725.90824	5	856.70252	5	987.49679	5	1,118.29107	5	1,249.08535
6	73.24480	6	204.03907	6	334.83335	6	465.62763	6	596.42190	6	727.21618	6	858.01046	6	988.80474	6	1,119.59901	6	1,250.39329
7	74.55274	7	205.34702	7	336.14129	7	466.93557	7	597.72985	7	728.52412	7	859.31840	7	990.11268	7	1,120.90696	7	1,251.70123
8	75.86068	8	206.65496	8	337.44924	8	468.24351	8	599.03779	8	729.83207	8	860.62634	8	991.42062	8	1,122.21490	8	1,253.00918
9	77.16862	9	207.96290	9	338.75718	9	469.55146	9	600.34573	9	731.14001	9	861.93429	9	992.72856	9	1,123.52284	9	1,254.31712
60	78.47657	**160**	209.27084	**260**	340.06512	**360**	470.85940	**460**	601.65368	**560**	732.44795	**660**	863.24223	**760**	994.03651	**860**	1,124.83078	**960**	1,255.62506
1	79.78451	1	210.57879	1	341.37306	1	472.16734	1	602.96162	1	733.75590	1	864.55017	1	995.34445	1	1,126.13873	1	1,256.93300
2	81.09245	2	211.88673	2	342.68101	2	473.47528	2	604.26956	2	735.06384	2	865.85812	2	996.65239	2	1,127.44667	2	1,258.24095
3	82.40039	3	213.19467	3	343.98895	3	474.78323	3	605.57750	3	736.37178	3	867.16606	3	997.96034	3	1,128.75461	3	1,259.54889
4	83.70834	4	214.50261	4	345.29689	4	476.09117	4	606.88545	4	737.67972	4	868.47400	4	999.26828	4	1,130.06256	4	1,260.85683
5	85.01628	5	215.81056	5	346.60483	5	477.39911	5	608.19339	5	738.98767	5	869.78194	5	1,000.57622	5	1,131.37050	5	1,262.16477
6	86.32422	6	217.11850	6	347.91278	6	478.70705	6	609.50133	6	740.29561	6	871.08989	6	1,001.88416	6	1,132.67844	6	1,263.47272
7	87.63217	7	218.42644	7	349.22072	7	480.01500	7	610.80927	7	741.60355	7	872.39783	7	1,003.19211	7	1,133.98638	7	1,264.78066
8	88.94011	8	219.73439	8	350.52866	8	481.32294	8	612.11722	8	742.91149	8	873.70577	8	1,004.50005	8	1,135.29433	8	1,266.08860
9	90.24805	9	221.04233	9	351.83661	9	482.63088	9	613.42516	9	744.21944	9	875.01371	9	1,005.80799	9	1,136.60227	9	1,267.39655
70	91.55599	**170**	222.35027	**270**	353.14455	**370**	483.93883	**470**	614.73310	**570**	745.52738	**670**	876.32166	**770**	1,007.11593	**870**	1,137.91021	**970**	1,268.70449
1	92.86394	1	223.65821	1	354.45249	1	485.24677	1	616.04105	1	746.83532	1	877.62960	1	1,008.42388	1	1,139.21815	1	1,270.01243
2	94.17188	2	224.96616	2	355.76043	2	486.55471	2	617.34899	2	748.14327	2	878.93754	2	1,009.73182	2	1,140.52610	2	1,271.32037
3	95.47982	3	226.27410	3	357.06838	3	487.86265	3	618.65693	3	749.45121	3	880.24549	3	1,011.03976	3	1,141.83404	3	1,272.62832
4	96.78777	4	227.58204	4	358.37632	4	489.17060	4	619.96487	4	750.75915	4	881.55343	4	1,012.34771	4	1,143.14198	4	1,273.93626
5	98.09571	5	228.88999	5	359.68426	5	490.47854	5	621.27282	5	752.06709	5	882.86137	5	1,013.65565	5	1,144.44993	5	1,275.24420
6	99.40365	6	230.19793	6	360.99221	6	491.78648	6	622.58076	6	753.37504	6	884.16931	6	1,014.96359	6	1,145.75787	6	1,276.55215
7	100.71159	7	231.50587	7	362.30015	7	493.09443	7	623.88870	7	754.68298	7	885.47726	7	1,016.27153	7	1,147.06581	7	1,277.86009
8	102.01954	8	232.81381	8	363.60809	8	494.40237	8	625.19665	8	755.99092	8	886.78520	8	1,017.57948	8	1,148.37375	8	1,279.16803
9	103.32748	9	234.12176	9	364.91603	9	495.71031	9	626.50459	9	757.29886	9	888.09314	9	1,018.88742	9	1,149.68170	9	1,280.47597
80	104.63542	**180**	235.42970	**280**	366.22398	**380**	497.01825	**480**	627.81253	**580**	758.60681	**680**	889.40108	**780**	1,020.19536	**880**	1,150.98964	**980**	1,281.78392
1	105.94336	1	236.73764	1	367.53192	1	498.32620	1	629.12047	1	759.91475	1	890.70903	1	1,021.50330	1	1,152.29758	1	1,283.09186
2	107.25131	2	238.04558	2	368.83986	2	499.63414	2	630.42842	2	761.22269	2	892.01697	2	1,022.81125	2	1,153.60552	2	1,284.39980
3	108.55925	3	239.35353	3	370.14780	3	500.94208	3	631.73636	3	762.53064	3	893.32491	3	1,024.11919	3	1,154.91347	3	1,285.70774
4	109.86719	4	240.66147	4	371.45575	4	502.25002	4	633.04430	4	763.83858	4	894.63286	4	1,025.42713	4	1,156.22141	4	1,287.01569
5	111.17514	5	241.96941	5	372.76369	5	503.55797	5	634.35224	5	765.14652	5	895.94080	5	1,026.73508	5	1,157.52935	5	1,288.32363
6	112.48308	6	243.27736	6	374.07163	6	504.86591	6	635.66019	6	766.45446	6	897.24874	6	1,028.04302	6	1,158.83730	6	1,289.63157
7	113.79102	7	244.58530	7	375.37958	7	506.17385	7	636.96813	7	767.76241	7	898.55668	7	1,029.35096	7	1,160.14524	7	1,290.93952
8	115.09896	8	245.89324	8	376.68752	8	507.48180	8	638.27607	8	769.07035	8	899.86463	8	1,030.65890	8	1,161.45318	8	1,292.24746
9	116.40691	9	247.20118	9	377.99546	9	508.78974	9	639.58402	9	770.37829	9	901.17257	9	1,031.96685	9	1,162.76112	9	1,293.55540
90	117.71485	**190**	248.50913	**290**	379.30340	**390**	510.09768	**490**	640.89196	**590**	771.68624	**690**	902.48051	**790**	1,033.27479	**890**	1,164.06907	**990**	1,294.86334
1	119.02279	1	249.81707	1	380.61135	1	511.40562	1	642.19990	1	772.99418	1	903.78846	1	1,034.58273	1	1,165.37701	1	1,296.17129
2	120.33074	2	251.12501	2	381.91929	2	512.71357	2	643.50784	2	774.30212	2	905.09640	2	1,035.89068	2	1,166.68495	2	1,297.47923
3	121.63868	3	252.43295	3	383.22723	3	514.02151	3	644.81579	3	775.61006	3	906.40434	3	1,037.19862	3	1,167.99290	3	1,298.78717
4	122.94662	4	253.74090	4	384.53517	4	515.32945	4	646.12373	4	776.91801	4	907.71228	4	1,038.50656	4	1,169.30084	4	1,300.09512
5	124.25456	5	255.04884	5	385.84312	5	516.63739	5	647.43167	5	778.22595	5	909.02023	5	1,039.81450	5	1,170.60878	5	1,301.40306
6	125.56251	6	256.35678	6	387.15106	6	517.94534	6	648.73961	6	779.53389	6	910.32817	6	1,041.12245	6	1,171.91672	6	1,302.71100
7	126.87045	7	257.66473	7	388.45900	7	519.25328	7	650.04756	7	780.84183	7	911.63611	7	1,042.43039	7	1,173.22467	7	1,304.01894
8	128.17839	8	258.97267	8	389.76695	8	520.56122	8	651.35550	8	782.14978	8	912.94405	8	1,043.73833	8	1,174.53261	8	1,305.32689
9	129.48633	9	260.28061	9	391.07489	9	521.86917	9	652.66344	9	783.45772	9	914.25200	9	1,045.04627	9	1,175.84055	9	1,306.63483

CAPACITY—LIQUID QUARTS TO LITERS

[Reduction factor: 1 liquid quart=0.94633307 liter]

Liquid quarts	Liters	Liquid quarts	Liters	Liquid quarts	Liters	Liquid quarts	Liters	Liquid quarts	Liters	Liquid quarts	Liters	Liquid quarts	Liters	Liquid quarts	Liters	Liquid quarts	Liters	Liquid quarts	Liters
0		**100**	94.633	**200**	189.267	**300**	283.900	**400**	378.533	**500**	473.167	**600**	567.800	**700**	662.433	**800**	757.066	**900**	851.700
1	0.9463	1	95.580	1	190.213	1	284.846	1	379.480	1	474.113	1	568.746	1	663.379	1	758.013	1	852.646
2	1.8927	2	96.526	2	191.159	2	285.793	2	380.426	2	475.059	2	569.693	2	664.326	2	758.959	2	853.592
3	2.8390	3	97.472	3	192.106	3	286.739	3	381.372	3	476.006	3	570.639	3	665.272	3	759.905	3	854.539
4	3.7853	4	98.419	4	193.052	4	287.685	4	382.319	4	476.952	4	571.585	4	666.218	4	760.852	4	855.485
5	4.7317	5	99.365	5	193.998	5	288.632	5	383.265	5	477.898	5	572.532	5	667.165	5	761.798	5	856.431
6	5.6780	6	100.311	6	194.945	6	289.578	6	384.211	6	478.845	6	573.478	6	668.111	6	762.744	6	857.378
7	6.6243	7	101.258	7	195.891	7	290.524	7	385.158	7	479.791	7	574.424	7	669.057	7	763.691	7	858.324
8	7.5707	8	102.204	8	196.837	8	291.471	8	386.104	8	480.737	8	575.371	8	670.004	8	764.637	8	859.270
9	8.5170	9	103.150	9	197.784	9	292.417	9	387.050	9	481.684	9	576.317	9	670.950	9	765.583	9	860.217
10	9.4633	**110**	104.097	**210**	198.730	**310**	293.363	**410**	387.997	**510**	482.630	**610**	577.263	**710**	671.896	**810**	766.530	**910**	861.163
1	10.4097	1	105.043	1	199.676	1	294.310	1	388.943	1	483.576	1	578.210	1	672.843	1	767.476	1	862.109
2	11.3560	2	105.989	2	200.623	2	295.256	2	389.889	2	484.523	2	579.156	2	673.789	2	768.422	2	863.056
3	12.3023	3	106.936	3	201.569	3	296.202	3	390.836	3	485.469	3	580.102	3	674.735	3	769.369	3	864.002
4	13.2487	4	107.882	4	202.515	4	297.149	4	391.782	4	486.415	4	581.049	4	675.682	4	770.315	4	864.948
5	14.1950	5	108.828	5	203.462	5	298.095	5	392.728	5	487.362	5	581.995	5	676.628	5	771.261	5	865.895
6	15.1413	6	109.775	6	204.408	6	299.041	6	393.675	6	488.308	6	582.941	6	677.574	6	772.208	6	866.841
7	16.0877	7	110.721	7	205.354	7	299.988	7	394.621	7	489.254	7	583.888	7	678.521	7	773.154	7	867.787
8	17.0340	8	111.667	8	206.301	8	300.934	8	395.567	8	490.201	8	584.834	8	679.467	8	774.100	8	868.734
9	17.9803	9	112.614	9	207.247	9	301.880	9	396.514	9	491.147	9	585.780	9	680.413	9	775.047	9	869.680
20	18.9267	**120**	113.560	**220**	208.193	**320**	302.827	**420**	397.460	**520**	492.093	**620**	586.727	**720**	681.360	**820**	775.993	**920**	870.626
1	19.8730	1	114.506	1	209.140	1	303.773	1	398.406	1	493.040	1	587.673	1	682.306	1	776.939	1	871.573
2	20.8193	2	115.453	2	210.086	2	304.719	2	399.353	2	493.986	2	588.619	2	683.252	2	777.886	2	872.519
3	21.7657	3	116.399	3	211.032	3	305.666	3	400.299	3	494.932	3	589.566	3	684.199	3	778.832	3	873.465
4	22.7120	4	117.345	4	211.979	4	306.612	4	401.245	4	495.879	4	590.512	4	685.145	4	779.778	4	874.412
5	23.6583	5	118.292	5	212.925	5	307.558	5	402.192	5	496.825	5	591.458	5	686.091	5	780.725	5	875.358
6	24.6047	6	119.238	6	213.871	6	308.505	6	403.138	6	497.771	6	592.405	6	687.038	6	781.671	6	876.304
7	25.5510	7	120.184	7	214.818	7	309 451	7	404.084	7	498.718	7	593.351	7	687.984	7	782.617	7	877.251
8	26.4973	8	121.131	8	215.764	8	310.397	8	405.031	8	499.664	8	594.297	8	688.930	8	783.564	8	878.197
9	27.4437	9	122.077	9	216.710	9	311.344	9	405.977	9	500.610	9	595.244	9	689.877	9	784.510	9	879.143
30	28.3900	**130**	123.023	**230**	217.657	**330**	312.290	**430**	406.923	**530**	501.557	**630**	596.190	**730**	690.823	**830**	785.456	**930**	880.090
1	29.3363	1	123.970	1	218.603	1	313.236	1	407.870	1	502.503	1	597.136	1	691.769	1	786.403	1	881.036
2	30.2827	2	124.916	2	219.549	2	314.183	2	408.816	2	503.449	2	598.083	2	692.716	2	787.349	2	881.982
3	31.2290	3	125.862	3	220.496	3	315.129	3	409.762	3	504.396	3	599.029	3	693.662	3	788.295	3	882.929
4	32.1753	4	126.809	4	221.442	4	316.075	4	410.709	4	505.342	4	599.975	4	694.608	4	789.242	4	883.875
5	33.1217	5	127.755	5	222.388	5	317.022	5	411.655	5	506.288	5	600.922	5	695.555	5	790.188	5	884.821
6	34.0680	6	128.701	6	223.335	6	317.968	6	412.601	6	507.235	6	601.868	6	696.501	6	791.134	6	885.768
7	35.0143	7	129.648	7	224.281	7	318.914	7	413.548	7	508.181	7	602.814	7	697.447	7	792.081	7	886.714
8	35.9607	8	130.594	8	225.227	8	319.861	8	414.494	8	509.127	8	603.761	8	698.394	8	793.027	8	887.660
9	36.9070	9	131.540	9	226.174	9	320.807	9	415.440	9	510.074	9	604.707	9	699.340	9	793.973	9	888.607
40	37.8533	**140**	132.487	**240**	227.120	**340**	321.753	**440**	416.387	**540**	511.020	**640**	605.653	**740**	700.286	**840**	794.920	**940**	889.553
1	38.7997	1	133.433	1	228.066	1	322.700	1	417.333	1	511.966	1	606.599	1	701.233	1	795.866	1	890.499
2	39.7460	2	134.379	2	229.013	2	323.646	2	418.279	2	512.913	2	607.546	2	702.179	2	796.812	2	891.446
3	40.6923	3	135.326	3	229.959	3	324.592	3	419.226	3	513.859	3	608.492	3	703.125	3	797.759	3	892.392
4	41.6387	4	136.272	4	230.905	4	325.539	4	420.172	4	514.805	4	609.438	4	704.072	4	798.705	4	893.338
5	42.5850	5	137.218	5	231.852	5	326.485	5	421.118	5	515.752	5	610.385	5	705.018	5	799.651	5	894.285
6	43.5313	6	138.165	6	232.798	6	327.431	6	422.065	6	516.698	6	611.331	6	705.964	6	800.598	6	895.231
7	44.4777	7	139.111	7	233.744	7	328.378	7	423.011	7	517.644	7	612.277	7	706.911	7	801.544	7	896.177
8	45.4240	8	140.057	8	234.691	8	329.324	8	423.957	8	518.591	8	613.224	8	707.857	8	802.490	8	897.124
9	46.3703	9	141.004	9	235.637	9	330.270	9	424.904	9	519.537	9	614.170	9	708.803	9	803.437	9	898.070

50	47.3167	150	141.950	250	236.583	350	331.217	450	425.850	550	520.483	650	615.116	750	709.750	850	804.383	950	899.016
1	48.2630	1	142.896	1	237.530	1	332.163	1	426.796	1	521.430	1	616.063	1	710.696	1	805.329	1	899.963
2	49.2093	2	143.843	2	238.476	2	333.109	2	427.743	2	522.376	2	617.009	2	711.642	2	806.276	2	900.909
3	50.1557	3	144.789	3	239.422	3	334.056	3	428.689	3	523.322	3	617.955	3	712.589	3	807.222	3	901.855
4	51.1020	4	145.735	4	240.369	4	335.002	4	429.635	4	524.269	4	618.902	4	713.535	4	808.168	4	902.802
5	52.0483	5	146.682	5	241.315	5	335.948	5	430.582	5	525.215	5	619.848	5	714.481	5	809.115	5	903.748
6	52.9947	6	147.628	6	242.261	6	336.895	6	431.528	6	526.161	6	620.794	6	715.428	6	810.061	6	904.694
7	53.9410	7	148.574	7	243.208	7	337.841	7	432.474	7	527.108	7	621.741	7	716.374	7	811.007	7	905.641
8	54.8873	8	149.521	8	244.154	8	338.787	8	433.421	8	528.054	8	622.687	8	717.320	8	811.954	8	906.587
9	55.8337	9	150.467	9	245.100	9	339.734	9	434.367	9	529.000	9	623.633	9	718.267	9	812.900	9	907.533
60	56.7800	160	151.413	260	246.047	360	340.680	460	435.313	560	529.947	660	624.580	760	719.213	860	813.846	960	908.480
1	57.7263	1	152.360	1	246.993	1	341.626	1	436.260	1	530.893	1	625.526	1	720.159	1	814.793	1	909.426
2	58.6727	2	153.306	2	247.939	2	342.573	2	437.206	2	531.839	2	626.472	2	721.106	2	815.739	2	910.372
3	59.6190	3	154.252	3	248.886	3	343.519	3	438.152	3	532.786	3	627.419	3	722.052	3	816.685	3	911.319
4	60.5653	4	155.199	4	249.832	4	344.465	4	439.099	4	533.732	4	628.365	4	722.998	4	817.632	4	912.265
5	61.5116	5	156.145	5	250.778	5	345.412	5	440.045	5	534.678	5	629.311	5	723.945	5	818.578	5	913.211
6	62.4580	6	157.091	6	251.725	6	346.358	6	440.991	6	535.625	6	630.258	6	724.891	6	819.524	6	914.158
7	63.4043	7	158.038	7	252.671	7	347.304	7	441.938	7	536.571	7	631.204	7	725.837	7	820.471	7	915.104
8	64.3506	8	158.984	8	253.617	8	348.251	8	442.884	8	537.517	8	632.150	8	726.784	8	821.417	8	916.050
9	65.2970	9	159.930	9	254.564	9	349.197	9	443.830	9	538.464	9	633.097	9	727.730	9	822.363	9	916.997
70	66.2433	170	160.877	270	255.510	370	350.143	470	444.777	570	539.410	670	634.043	770	728.676	870	823.310	970	917.943
1	67.1896	1	161.823	1	256.456	1	351.090	1	445.723	1	540.356	1	634.989	1	729.623	1	824.256	1	918.889
2	68.1360	2	162.769	2	257.403	2	352.036	2	446.669	2	541.303	2	635.936	2	730.569	2	825.202	2	919.836
3	69.0823	3	163.716	3	258.349	3	352.982	3	447.616	3	542.249	3	636.882	3	731.515	3	826.149	3	920.782
4	70.0286	4	164.662	4	259.295	4	353.929	4	448.562	4	543.195	4	637.828	4	732.462	4	827.095	4	921.728
5	70.9750	5	165.608	5	260.242	5	354.875	5	449.508	5	544.142	5	638.775	5	733.408	5	828.041	5	922.675
6	71.9213	6	166.555	6	261.188	6	355.821	6	450.455	6	545.088	6	639.721	6	734.354	6	828.988	6	923.621
7	72.8676	7	167.501	7	262.134	7	356.768	7	451.401	7	546.034	7	640.667	7	735.301	7	829.934	7	924.567
8	73.8140	8	168.447	8	263.081	8	357.714	8	452.347	8	546.981	8	641.614	8	736.247	8	830.880	8	925.514
9	74.7603	9	169.394	9	264.027	9	358.660	9	453.294	9	547.927	9	642.560	9	737.193	9	831.827	9	926.460
80	75.7066	180	170.340	280	264.973	380	359.607	480	454.240	580	548.873	680	643.506	780	738.140	880	832.773	980	927.406
1	76.6530	1	171.286	1	265.920	1	360.553	1	455.186	1	549.820	1	644.453	1	739.086	1	833.719	1	928.353
2	77.5993	2	172.233	2	266.866	2	361.499	2	456.133	2	550.766	2	645.399	2	740.032	2	834.666	2	929.299
3	78.5456	3	173.179	3	267.812	3	362.446	3	457.079	3	551.712	3	646.345	3	740.979	3	835.612	3	930.245
4	79.4920	4	174.125	4	268.759	4	363.392	4	458.025	4	552.659	4	647.292	4	741.925	4	836.558	4	931.192
5	80.4383	5	175.072	5	269.705	5	364.338	5	458.972	5	553.605	5	648.238	5	742.871	5	837.505	5	932.138
6	81.3846	6	176.018	6	270.651	6	365.285	6	459.918	6	554.551	6	649.184	6	743.818	6	838.451	6	933.084
7	82.3310	7	176.964	7	271.598	7	366.231	7	460.864	7	555.498	7	650.131	7	744.764	7	839.397	7	934.031
8	83.2773	8	177.911	8	272.544	8	367.177	8	461.811	8	556.444	8	651.077	8	745.710	8	840.344	8	934.977
9	84.2236	9	178.857	9	273.490	9	368.124	9	462.757	9	557.390	9	652.023	9	746.657	9	841.290	9	935.923
90	85.1700	190	179.803	290	274.437	390	369.070	490	463.703	590	558.337	690	652.970	790	747.603	890	842.236	990	936.870
1	86.1163	1	180.750	1	275.383	1	370.016	1	464.650	1	559.283	1	653.916	1	748.549	1	843.183	1	937.816
2	87.0626	2	181.696	2	276.329	2	370.963	2	465.596	2	560.229	2	654.862	2	749.496	2	844.129	2	938.762
3	88.0090	3	182.642	3	277.276	3	371.909	3	466.542	3	561.176	3	655.809	3	750.442	3	845.075	3	939.709
4	88.9553	4	183.589	4	278.222	4	372.855	4	467.489	4	562.122	4	656.755	4	751.388	4	846.022	4	940.655
5	89.9016	5	184.535	5	279.168	5	373.802	5	468.435	5	563.068	5	657.701	5	752.335	5	846.968	5	941.601
6	90.8480	6	185.481	6	280.115	6	374.748	6	469.381	6	564.015	6	658.648	6	753.281	6	847.914	6	942.548
7	91.7943	7	186.428	7	281.061	7	375.694	7	470.328	7	564.961	7	659.594	7	754.227	7	848.861	7	943.494
8	92.7406	8	187.374	8	282.007	8	376.641	8	471.274	8	565.907	8	660.540	8	755.174	8	849.807	8	944.440
9	93.6870	9	188.320	9	282.954	9	377.587	9	472.220	9	566.854	9	661.487	9	756.120	9	850.753	9	945.387

CAPACITY—LITERS TO LIQUID QUARTS

[Reduction factor: 1 liter=1.0567104 quarts]

Liters	Liquid quarts	Liters	Liquid quarts	Liters	Liquid quarts	Liters	Liquid quarts	Liters	Liquid quarts	Liters	Liquid quarts	Liters	Liquid quarts	Liters	Liquid quarts	Liters	Liquid quarts	Liters	Liquid quarts
0		100	105.671	200	211.342	300	317.013	400	422.684	500	528.355	600	634.026	700	739.697	800	845.368	900	951.039
1	1.0567	1	106.728	1	212.399	1	318.070	1	423.741	1	529.412	1	635.083	1	740.754	1	846.425	1	952.096
2	2.1134	2	107.784	2	213.456	2	319.127	2	424.798	2	530.469	2	636.140	2	741.811	2	847.482	2	953.153
3	3.1701	3	108.841	3	214.512	3	320.183	3	425.854	3	531.525	3	637.196	3	742.867	3	848.538	3	954.209
4	4.2268	4	109.898	4	215.569	4	321.240	4	426.911	4	532.582	4	638.253	4	743.924	4	849.595	4	955.266
5	5.2836	5	110.955	5	216.626	5	322.297	5	427.968	5	533.639	5	639.310	5	744.981	5	850.652	5	956.323
6	6.3403	6	112.011	6	217.682	6	323.353	6	429.024	6	534.695	6	640.367	6	746.038	6	851.709	6	957.380
7	7.3970	7	113.068	7	218.739	7	324.410	7	430.081	7	535.752	7	641.423	7	747.094	7	852.765	7	958.436
8	8.4537	8	114.125	8	219.796	8	325.467	8	431.138	8	536.809	8	642.480	8	748.151	8	853.822	8	959.493
9	9.5104	9	115.181	9	220.852	9	326.524	9	432.195	9	537.866	9	643.537	9	749.208	9	854.879	9	960.550
10	10.5671	110	116.238	210	221.909	310	327.580	410	433.251	510	538.922	610	644.593	710	750.264	810	855.935	910	961.606
1	11.6238	1	117.295	1	222.966	1	328.637	1	434.308	1	539.979	1	645.650	1	751.321	1	856.992	1	962.663
2	12.6805	2	118.352	2	224.023	2	329.694	2	435.365	2	541.036	2	646.707	2	752.378	2	858.049	2	963.720
3	13.7372	3	119.408	3	225.079	3	330.750	3	736.421	3	542.092	3	647.763	3	753.435	3	859.106	3	964.777
4	14.7939	4	120.465	4	226.136	4	331.807	4	437.478	4	543.149	4	648.820	4	754.491	4	860.162	4	965.833
5	15.8507	5	121.522	5	227.193	5	332 864	5	438.535	5	544.206	5	649.877	5	755.548	5	861.219	5	966.890
6	16.9074	6	122.578	6	228.249	6	333.920	6	439.592	6	545.263	6	650.934	6	756.605	6	862.276	6	967.947
7	17.9641	7	123.635	7	229.306	7	334.977	7	440.648	7	546.319	7	651.990	7	757.661	7	863.332	7	969.003
8	19.0208	8	124.692	8	230.363	8	336.034	8	441.705	8	547.376	8	653.047	8	758.718	8	864.389	8	970.060
9	20.0775	9	125.749	9	231.420	9	337.091	9	442.762	9	548.433	9	654.104	9	759.775	9	865.446	9	971.117
20	21.1342	120	126.805	220	232.476	320	338.147	420	443.818	520	549.489	620	655.160	720	760.831	820	866.503	920	972.174
1	22.1909	1	127.862	1	233.533	1	339.204	1	444.875	1	550.546	1	656.217	1	761.888	1	867 559	1	973.230
2	23.2476	2	128.919	2	234.590	2	340.261	2	445.932	2	551.603	2	657.274	2	762.945	2	868.616	2	974.287
3	24.3043	3	129.975	3	235.646	3	341.317	3	446.988	3	552.660	3	658.331	3	764.002	3	869.673	3	975.344
4	25.3610	4	131.032	4	236.703	4	342.374	4	448.045	4	553.716	4	659.387	4	765.058	4	870.729	4	976.400
5	26.4178	5	132.089	5	237.760	5	343.431	5	449.102	5	554.773	5	660.444	5	766.115	5	871.786	5	977.457
6	27.4745	6	133.146	6	238.817	6	344.488	6	450.159	6	555.830	6	661.501	6	767.172	6	872.843	6	978.514
7	28.5312	7	134.202	7	239.873	7	345.544	7	451.215	7	556.886	7	662.557	7	768.228	7	873.900	7	979.571
8	29.5879	8	135.259	8	240.930	8	346.601	8	452.272	8	557.943	8	663.614	8	769.285	8	874.956	8	980.627
9	30.6446	9	136.316	9	241.987	9	347.658	9	453.329	9	559.000	9	664.671	9	770.342	9	876.013	9	981.684
30	31.7013	130	137.372	230	243.043	330	348.714	430	454.385	530	560.057	630	665.728	730	771.399	830	877.070	930	982.741
1	32.7580	1	138.429	1	244.100	1	349.771	1	455.442	1	561.113	1	666.784	1	772.455	1	878.126	1	983.797
2	33.8147	2	139.486	2	245.157	2	350.828	2	456.499	2	562.170	2	667.841	2	773.512	2	879.183	2	984.854
3	34.8714	3	140.542	3	246.214	3	351.885	3	457.556	3	563.227	3	668.898	3	774.569	3	880.240	3	985.911
4	35.9282	4	141.599	4	247.270	4	352.941	4	458.612	4	564.283	4	669.954	4	775.625	4	881.296	4	986.968
5	36.9849	5	142.656	5	248.327	5	353.998	5	459.669	5	565.340	5	671.011	5	776.682	5	882.353	5	988.024
6	38.0416	6	143.713	6	249.384	6	355.055	6	460.726	6	566.397	6	672.068	6	777.739	6	883.410	6	989.081
7	39.0983	7	144.769	7	250.440	7	356.111	7	461.782	7	567.453	7	673.125	7	778.796	7	884.467	7	990.138
8	40.1550	8	145.826	8	251.497	8	357.168	8	462.839	8	568.510	8	674.181	8	779.852	8	885.523	8	991.194
9	41.2117	9	146.883	9	252.554	9	358.225	9	463.896	9	569.567	9	675.238	9	780.909	9	886.580	9	992.251
40	42.2684	140	147.939	240	253.610	340	359.282	440	464.953	540	570.624	640	676.295	740	781.966	840	887.637	940	993.308
1	43.3251	1	148.996	1	254.667	1	360.338	1	466.009	1	571.680	1	677.351	1	783.022	1	888.693	1	994.364
2	44.3818	2	150.053	2	255.724	2	361.395	2	467.066	2	572.737	2	678.408	2	784.079	2	889.750	2	995.421
3	45.4385	3	151.110	3	256.781	3	362.452	3	468.123	3	573.794	3	679.465	3	785.136	3	890.807	3	996.478
4	46.4953	4	152.166	4	257.837	4	363.508	4	469.179	4	574.850	4	680.521	4	786.193	4	891.864	4	997.535
5	47.5520	5	153.223	5	258.894	5	364.565	5	470.236	5	575.907	5	681.578	5	787.249	5	892.920	5	998.591
6	48.6087	6	154.280	6	259.951	6	365.622	6	471.293	6	576.964	6	682.635	6	788.306	6	893.977	6	999.648
7	49.6654	7	155.336	7	261.007	7	366.679	7	472.350	7	578.021	7	683.692	7	789.363	7	895.034	7	1,000.705
8	50.7221	8	156.393	8	262.064	8	367.735	8	473.406	8	579.077	8	684.748	8	790.419	8	896.090	8	1,001.761
9	51.7788	9	157.450	9	263.121	9	368.792	9	474.463	9	580.134	9	685.805	9	791.476	9	897.147	9	1,002.818

50	52.8355	**150**	158.507	**250**	264.178	**350**	369.849	**450**	475.520	**550**	581.191	**650**	686.862	**750**	792.533	**850**	898.204	**950**	1,003.875
1	53.8922	1	159.563	1	265.234	1	370.905	1	476.576	1	582.247	1	687.918	1	793.590	1	899.261	1	1,004.932
2	54.9489	2	160.620	2	266.291	2	371.962	2	477.633	2	583.304	2	688.975	2	794.646	2	900.317	2	1,005.988
3	56.0057	3	161.677	3	267.348	3	373.019	3	478.690	3	584.361	3	690.032	3	795.703	3	901.374	3	1,007.045
4	57.0624	4	162.733	4	268.404	4	374.075	4	479.747	4	585.418	4	691.089	4	796.760	4	902.431	4	1,008.102
5	58.1191	5	163.790	5	269.461	5	375.132	5	480.803	5	586.474	5	692.145	5	797.816	5	903.487	5	1,009.158
6	59.1758	6	164.847	6	270.518	6	376.189	6	481.860	6	587.531	6	693.202	6	798.873	6	904.544	6	1,010.215
7	60.2325	7	165.904	7	271.575	7	377.246	7	482.917	7	588.588	7	694.259	7	799.930	7	905.601	7	1,011.272
8	61.2892	8	166.960	8	272.631	8	378.302	8	483.973	8	589.644	8	695.315	8	800.986	8	906.658	8	1,012.329
9	62.3459	9	168.017	9	273.688	9	379.359	9	485.030	9	590.701	9	696.372	9	802.043	9	907.714	9	1,013.385
60	63.4026	**160**	169.074	**260**	274.745	**360**	380.416	**460**	486.087	**560**	591.758	**660**	697.429	**760**	803.100	**860**	908.771	**960**	1,014.442
1	64.4593	1	170.130	1	275.801	1	381.472	1	487.143	1	592.815	1	698.486	1	804.157	1	909.828	1	1,015.499
2	65.5160	2	171.187	2	276.858	2	382.529	2	488.200	2	593.871	2	699.542	2	805.213	2	910.884	2	1,016.555
3	66.5728	3	172.244	3	277.915	3	383.586	3	489.257	3	594.928	3	700.599	3	806.270	3	911.941	3	1,017.612
4	67.6295	4	173.301	4	278.972	4	384.643	4	490.314	4	595.985	4	701.656	4	807.327	4	912.998	4	1,018.669
5	68.6862	5	174.357	5	280.028	5	385.699	5	491.370	5	597.041	5	702.712	5	808.383	5	914.054	5	1,019.726
6	69.7429	6	175.414	6	281.085	6	386.756	6	492.427	6	598.098	6	703.769	6	809.440	6	915.111	6	1,020.782
7	70.7996	7	176.471	7	282.142	7	387.813	7	493.484	7	599.155	7	704.826	7	810.497	7	916.168	7	1,021.839
8	71.8563	8	177.527	8	283.198	8	388.869	8	494.540	8	600.212	8	705.883	8	811.554	8	917.225	8	1,022.896
9	72.9130	9	178.584	9	284.255	9	389.926	9	495.597	9	601.268	9	706.939	9	812.610	9	918.281	9	1,023.952
70	73.9697	**170**	179.641	**270**	285.312	**370**	390.983	**470**	496.654	**570**	602.325	**670**	707.996	**770**	813.667	**870**	919.338	**970**	1,025.009
1	75.0264	1	180.697	1	286.369	1	392.040	1	497.711	1	603.382	1	709.053	1	814.724	1	920.395	1	1,026.066
2	76.0831	2	181.754	2	287.425	2	393.096	2	498.767	2	604.438	2	710.109	2	815.780	2	921.451	2	1,027.123
3	77.1399	3	182.811	3	288.482	3	394.153	3	499.824	3	605.495	3	711.166	3	816.837	3	922.508	3	1,028.179
4	78.1966	4	183.868	4	289.539	4	395.210	4	500.881	4	606.552	4	712.223	4	817.894	4	923.565	4	1,029.236
5	79.2533	5	184.924	5	290.595	5	396.266	5	501.937	5	607.608	5	713.280	5	818.951	5	924.622	5	1,030.293
6	80.3100	6	185.981	6	291.652	6	397.323	6	502.994	6	608.665	6	714.336	6	820.007	6	925.678	6	1,031.349
7	81.3667	7	187.038	7	292.709	7	398.380	7	504.051	7	609.722	7	715.393	7	821.064	7	926.735	7	1,032.406
8	82.4234	8	188.094	8	293.765	8	399.437	8	505.108	8	610.779	8	716.450	8	822.121	8	927.792	8	1,033.463
9	83.4801	9	189.151	9	294.822	9	400.493	9	506.164	9	611.835	9	717.506	9	823.177	9	928.848	9	1,034.519
80	84.5368	**180**	190.208	**280**	295.879	**380**	401.550	**480**	507.221	**580**	612.892	**680**	718.563	**780**	824.234	**880**	929.905	**980**	1,035.576
1	85.5935	1	191.265	1	296.936	1	402.607	1	508.278	1	613.949	1	719.620	1	825.291	1	930.962	1	1,036.633
2	86.6503	2	192.321	2	297.992	2	403.663	2	509.334	2	615.005	2	720.676	2	826.348	2	932.019	2	1,037.690
3	87.7070	3	193.378	3	299.049	3	404.720	3	510.391	3	616.062	3	721.733	3	827.404	3	933.075	3	1,038.746
4	88.7637	4	194.435	4	300.106	4	405.777	4	511.448	4	617.119	4	722.790	4	828.461	4	934.132	4	1,039.803
5	89.8204	5	195.491	5	301.162	5	406.834	5	512.505	5	618.176	5	723.847	5	829.518	5	935.189	5	1,040.860
6	90.8771	6	196.548	6	302.219	6	407.890	6	513.561	6	619.232	6	724.903	6	830.574	6	936.245	6	1,041.916
7	91.9338	7	197.605	7	303.276	7	408.947	7	514.618	7	620.289	7	725.960	7	831.631	7	937.302	7	1,042.973
8	92.9905	8	198.662	8	304.333	8	410.004	8	515.675	8	621.346	8	727.017	8	832.688	8	938.359	8	1,044.030
9	94.0472	9	199.718	9	305.389	9	411.060	9	516.731	9	622.402	9	728.073	9	833.745	9	939.416	9	1,045.087
90	95.1039	**190**	200.775	**290**	306.446	**390**	412.117	**490**	517.788	**590**	623.459	**690**	729.130	**790**	834.801	**890**	940.472	**990**	1,046.143
1	96.1606	1	201.832	1	307.503	1	413.174	1	518.845	1	624.516	1	730.187	1	835.858	1	941.529	1	1,047.200
2	97.2174	2	202.888	2	308.559	2	414.230	2	519.902	2	625.573	2	731.244	2	836.915	2	942.586	2	1,048.257
3	98.2741	3	203.945	3	309.616	3	415.287	3	520.958	3	626.629	3	732.300	3	837.971	3	943.642	3	1,049.313
4	99.3308	4	205.002	4	310.673	4	416.344	4	522.015	4	627.686	4	733.357	4	839.028	4	944.699	4	1,050.370
5	100.3875	5	206.059	5	311.730	5	417.401	5	523.072	5	628.743	5	734.414	5	840.085	5	945.756	5	1,051.427
6	101.4442	6	207.115	6	312.786	6	418.457	6	524.128	6	629.799	6	735.470	6	841.141	6	946.813	6	1,052.484
7	102.5009	7	208.172	7	313.843	7	419.514	7	525.185	7	630.856	7	736.527	7	842.198	7	947.869	7	1,053.540
8	103.5576	8	209.229	8	314.900	8	420.571	8	526.242	8	631.913	8	737.584	8	843.255	8	948.926	8	1,054.597
9	104.6143	9	210.285	9	315.956	9	421.627	9	527.298	9	632.970	9	738.641	9	844.312	9	949.983	9	1,055.654

CAPACITY—GALLONS TO LITERS

[Reduction factor: 1 gallon=3.7853323 liters]

Gallons	Liters	Gallons	Liters	Gallons	Liters	Gallons	Liters	Gallons	Liters	Gallons	Liters	Gallons	Liters	Gallons	Liters	Gallons	Liters	Gallons	Liters
0		**100**	378.533	**200**	757.066	**300**	1,135.600	**400**	1,514.133	**500**	1,892.666	**600**	2,271.199	**700**	2,649.733	**800**	3,028.266	**900**	3,406.799
1	3.7853	1	382.319	1	760.852	1	1,139.385	1	1,517.918	1	1,896.451	1	2,274.985	1	2,653.518	1	3,032.051	1	3,410.584
2	7.5707	2	386.104	2	764.637	2	1,143.170	2	1,521.704	2	1,900.237	2	2,278.770	2	2,657.303	2	3,035.837	2	3,414.370
3	11.3560	3	389.889	3	768.422	3	1,146.956	3	1,525.489	3	1,904.022	3	2,282.555	3	2,661.089	3	3,039.622	3	3,418.155
4	15.1413	4	393.675	4	772.208	4	1,150.741	4	1,529.274	4	1,907.807	4	2,286.341	4	2,664.874	4	3,043.407	4	3,421.940
5	18.9267	5	397.460	5	775.993	5	1,154.526	5	1,533.060	5	1,911.593	5	2,290.126	5	2,668.659	5	3,047.193	5	3,425.726
6	22.7120	6	401.245	6	779.778	6	1,158.312	6	1,536.845	6	1,915.378	6	2,293.911	6	2,672.445	6	3,050.978	6	3,429.511
7	26.4973	7	405.031	7	783.564	7	1,162.097	7	1,540.630	7	1,919.163	7	2,297.697	7	2,676.230	7	3,054.763	7	3,433.296
8	30.2827	8	408.816	8	787.249	8	1,165.882	8	1,544.416	8	1,922.949	8	2,301.482	8	2,680.015	8	3,058.548	8	3,437.082
9	34.0680	9	412.601	9	791.134	9	1,169.668	9	1,548.201	9	1,926.734	9	2,305.267	9	2,683.801	9	3,062.334	9	3,440.867
10	37.8533	**110**	416.387	**210**	794.920	**310**	1,173.453	**410**	1,551.986	**510**	1,930.519	**610**	2,309.053	**710**	2,687.586	**810**	3,066.119	**910**	3,444.652
1	41.6387	1	420.172	1	798.705	1	1,177.238	1	1,555.772	1	1,934.305	1	2,312.838	1	2,691.371	1	3,069.904	1	3,448.438
2	45.4240	2	423.957	2	802.490	2	1,181.024	2	1,559.557	2	1,938.090	2	2,316.623	2	2,695.157	2	3,073.690	2	3,452.223
3	49.2093	3	427.743	3	806.276	3	1,184.809	3	1,563.342	3	1,941.875	3	2,320.409	3	2,698.942	3	3,077.475	3	3,456.008
4	52.9947	4	431.528	4	810.061	4	1,188.594	4	1,567.128	4	1,945.661	4	2,324.194	4	2,702.727	4	3,081.260	4	3,459.794
5	56.7800	5	435.313	5	813.846	5	1,192.380	5	1,570.913	5	1,949.446	5	2,327.979	5	2,706.513	5	3,085.046	5	3,463.579
6	60.5653	6	439.099	6	817.632	6	1,196.165	6	1,574.698	6	1,953.231	6	2,231.765	6	2,710.298	6	3,088.831	6	3,467.364
7	64.3506	7	442.884	7	821.417	7	1,199.950	7	1,578.484	7	1,957.017	7	2,335.550	7	2,714.083	7	3,092.616	7	3,471.150
8	68.1360	8	446.669	8	825.202	8	1,203.736	8	1,582.269	8	1,960.802	8	2,339.335	8	2,717.869	8	3,096.402	8	3,474.935
9	71.9213	9	450.455	9	828.988	9	1,207.521	9	1,586.054	9	1,964.587	9	2,343.121	9	2,721.654	9	3,100.187	9	3,478.720
20	75.7066	**120**	454.240	**220**	832.773	**320**	1,211.306	**420**	1,589.840	**520**	1,968.373	**620**	2,346.906	**720**	2,725.439	**820**	3,103.972	**920**	3,482.506
1	79.4920	1	458.025	1	836.558	1	1,215.092	1	1,593.625	1	1,972.158	1	2,350.691	1	2,729.225	1	3,107.758	1	3,486.291
2	83.2773	2	461.811	2	840.344	2	1,218.877	2	1,597.410	2	1,975.943	2	2,354.477	2	2,733.010	2	3,111.543	2	3,490.076
3	87.0626	3	465.596	3	844.129	3	1,222.662	3	1,601.196	3	1,979.729	3	2,358.262	3	2,736.795	3	3,115.328	3	3,493.862
4	90.8480	4	469.381	4	847.914	4	1,226.448	4	1,604.981	4	1,983.514	4	2,362.047	4	2,740.581	4	3,119.114	4	3,497.647
5	94.6333	5	473.167	5	851.700	5	1,230.233	5	1,608.766	5	1,987.299	5	2,365.833	5	2,744.366	5	3,122.899	5	3,501.432
6	98.4186	6	476.952	6	855.485	6	1,234.018	6	1,612.552	6	1,991.085	6	2,369.618	6	2,748.151	6	3,126.684	6	3,505.218
7	102.2040	7	480.737	7	859.270	7	1,237.804	7	1,616.337	7	1,994.870	7	2,373.403	7	2,751.937	7	3,130.470	7	3,509.003
8	105.9893	8	484.523	8	863.056	8	1,241.589	8	1,620.122	8	1,998.655	8	2,377.189	8	2,755.722	8	3,134.255	8	3,512.788
9	109.7746	9	488.308	9	866.841	9	1,245.374	9	1,623.908	9	2,002.441	9	2,380.974	9	2,759.507	9	3,138.040	9	3,516.574
30	113.5600	**130**	492.093	**230**	870.626	**330**	1,249.160	**430**	1,627.693	**530**	2,006.226	**630**	2,384.759	**730**	2,763.293	**830**	3,141.826	**930**	3,520.359
1	117.3453	1	495.879	1	874.412	1	1,252.945	1	1,631.478	1	2,010.011	1	2,388.545	1	2,767.078	1	3,145.611	1	3,524.144
2	121.1306	2	499.664	2	878.197	2	1,256.730	2	1,635.264	2	2,013.797	2	2,392.330	2	2,770.863	2	3,149.396	2	3,527.930
3	124.9160	3	503.449	3	881.982	3	1,260.516	3	1,639.049	3	2,017.582	3	2,396.115	3	2,774.649	3	3,153.182	3	3,531.715
4	128.7013	4	507.235	4	885.768	4	1,264.301	4	1,642.834	4	2,021.367	4	2,399.901	4	2,778.434	4	3,156.967	4	3,535.500
5	132.4866	5	511.020	5	889.553	5	1,268.086	5	1,646.620	5	2,025.153	5	2,403.686	5	2,782.219	5	3,160.752	5	3,539.286
6	136.2720	6	514.805	6	893.338	6	1,271.872	6	1,650.405	6	2,028.938	6	2,407.471	6	2,786.005	6	3,164.538	6	3,543.071
7	140.0573	7	518.591	7	897.124	7	1,275.657	7	1,654.190	7	2,032.723	7	2,411.257	7	2,789.790	7	3,168.323	7	3,546.856
8	143.8426	8	522.376	8	900.909	8	1,279.442	8	1,657.976	8	2,036.509	8	2,415.042	8	2,793.575	8	3,172.108	8	3,550.642
9	147.6280	9	526.161	9	904.694	9	1,283.228	9	1,661.761	9	2,040.294	9	2,418.827	9	2,797.361	9	3,175.894	9	3,554.427
40	151.4133	**140**	529.947	**240**	908.480	**340**	1,287.013	**440**	1,665.546	**540**	2,044.079	**640**	2,422.613	**740**	2,801.146	**840**	3,179.679	**940**	3,558.212
1	155.1986	1	533.732	1	912.265	1	1,290.798	1	1,669.332	1	2,047.865	1	2,426.398	1	2,804.931	1	3,183.464	1	3,561.998
2	158.9840	2	537.517	2	916.050	2	1,294.584	2	1,673.117	2	2,051.650	2	2,430.183	2	2,808.717	2	3,187.250	2	3,565.783
3	162.7693	3	541.303	3	919.836	3	1,298.369	3	1,676.902	3	2,055.435	3	2,433.969	3	2,812.502	3	3,191.035	3	3,569.568
4	166.5546	4	545.088	4	923.621	4	1,302.154	4	1,680.688	4	2,059.221	4	2,437.754	4	2,816.287	4	3,194.820	4	3,573.354
5	170.3400	5	548.873	5	927.406	5	1,305.940	5	1,684.473	5	2,063.006	5	2,441.539	5	2,820.073	5	3,198.606	5	3,577.139
6	174.1253	6	552.659	6	931.192	6	1,309.725	6	1,688.258	6	2,066.791	6	2,445.325	6	2,823.858	6	3,202.391	6	3,580.924
7	177.9106	7	556.444	7	934.977	7	1,313.510	7	1,692.044	7	2,070.577	7	2,449.110	7	2,827.643	7	3,206.176	7	3,584.710
8	181.6960	8	560.229	8	938.762	8	1,317.296	8	1,695.829	8	2,074.362	8	2,452.895	8	2,831.429	8	3,209.962	8	3,588.495
9	185.4813	9	564.015	9	942.548	9	1,321.081	9	1,699.614	9	2,078.147	9	2,456.681	9	2,835.214	9	3,213.747	9	3,592.280

50	189.2666	**150**	567.800	**250**	946.333	**350**	1,324.866	**450**	1,703.400	**550**	2,081.933	**650**	2,460.466	**750**	2,838.999	**850**	3,217.532	**950**	3,596.066
1	193.0519	1	571.585	1	950.118	1	1,328.652	1	1,707.185	1	2,085.718	1	2,464.251	1	2,842.785	1	3,221.318	1	3,599.851
2	196.8373	2	575.371	2	953.904	2	1,332.437	2	1,710.970	2	2,089.503	2	2,468.037	2	2,846.570	2	3,225.103	2	3,603.636
3	200.6226	3	579.156	3	957.689	3	1,336.222	3	1,714.756	3	2,093.289	3	2,471.822	3	2,850.355	3	3,228.888	3	3,607.422
4	204.4079	4	582.941	4	961.474	4	1,340.008	4	1,718.541	4	2,097.074	4	2,475.607	4	2,854.141	4	3,232.674	4	3,611.207
5	208.1933	5	586.727	5	965.260	5	1,343.793	5	1,722.326	5	2,100.859	5	2,479.393	5	2,857.926	5	3,236.459	5	3,614.992
6	211.9786	6	590.512	6	969.045	6	1,347.578	6	1,726.112	6	2,104.645	6	2,483.178	6	2,861.711	6	3,240.244	6	3,618.778
7	215.7639	7	594.297	7	972.830	7	1,351.364	7	1,729.897	7	2,108.430	7	2,486.963	7	2,865.497	7	3,244.030	7	3,622.563
8	219.5493	8	598.083	8	976.616	8	1,355.149	8	1,733.682	8	2,112.215	8	2,490.749	8	2,869.282	8	3,247.815	8	3,626.348
9	223.3346	9	601.868	9	980.401	9	1,358.934	9	1,737.468	9	2,116.001	9	2,494.534	9	2,873.067	9	3,251.600	9	3,630.134
60	227.1199	**160**	605.653	**260**	984.186	**360**	1,362.720	**460**	1,741.253	**560**	2,119.786	**660**	2,498.319	**760**	2,876.853	**860**	3,255.386	**960**	3,633.919
1	230.9053	1	609.438	1	987.972	1	1,366.525	1	1,745.038	1	2,123.571	1	2,502.105	1	2,880.638	1	3,259.171	1	3,637.704
2	234.6906	2	613.224	2	991.757	2	1,370.290	2	1,748.824	2	2,127.357	2	2,505.890	2	2,884.423	2	3,262.956	2	3,641.490
3	238.4759	3	617.009	3	995.542	3	1,374.076	3	1,752.609	3	2,131.142	3	2,509.675	3	2,888.209	3	3,266.742	3	3,645.275
4	242.2613	4	620.794	4	999.328	4	1,377.861	4	1,756.394	4	2,134.927	4	2,513.461	4	2,891.994	4	3,270.527	4	3,649.060
5	246.0466	5	624.580	5	1,003.113	5	1,381.646	5	1,760.180	5	2,138.713	5	2,517.246	5	2,895.779	5	3,274.312	5	3,652.846
6	249.8319	6	628.365	6	1,006.898	6	1,385.432	6	1,763.965	6	2,142.498	6	2,521.031	6	2,899.565	6	3,278.098	6	3,656.631
7	253.6173	7	632.150	7	1,010.684	7	1,389.217	7	1,767.750	7	2,146.283	7	2,524.817	7	2,903.350	7	3,281.883	7	3,660.416
8	257.4026	8	635.936	8	1,014.469	8	1,393.002	8	1,771.536	8	2,150.069	8	2,528.602	8	2,907.135	8	3,285.668	8	3,664.202
9	261.1879	9	639.721	9	1,018.254	9	1,396.788	9	1,775.321	9	2,153.854	9	2,532.387	9	2,910.921	9	3,289.454	9	3,667.987
70	264.9733	**170**	643.506	**270**	1,022.040	**370**	1,400.573	**470**	1,779.106	**570**	2,157.639	**670**	2,536.173	**770**	2,914.706	**870**	3,293.239	**970**	3,671.772
1	268.7586	1	647.292	1	1,025.825	1	1,404.358	1	1,782.892	1	2,161.425	1	2,539.958	1	2,918.491	1	3,297.024	1	3,675.558
2	272.5439	2	651.077	2	1,029.610	2	1,408.144	2	1,786.677	2	2,165.210	2	2,543.743	2	2,922.277	2	3,300.810	2	3,679.343
3	276.3293	3	654.862	3	1,033.396	3	1,411.929	3	1,790.462	3	2,168.995	3	2,547.529	3	2,926.062	3	3,304.595	3	3,683.128
4	280.1146	4	658.648	4	1,037.181	4	1,415.714	4	1,794.248	4	2,172.781	4	2,551.314	4	2,929.847	4	3,308.380	4	3,686.914
5	283.8999	5	662.433	5	1,040.966	5	1,419.500	5	1,798.033	5	2,176.566	5	2,555.099	5	2,933.633	5	3,312.166	5	3,690.699
6	287.6853	6	666.218	6	1,044.752	6	1,423.285	6	1,801.818	6	2,180.351	6	2,558.885	6	2,937.418	6	3,315.951	6	3,694.484
7	291.4706	7	670.004	7	1,048.537	7	1,427.070	7	1,805.604	7	2,184.137	7	2,562.670	7	2,941.203	7	3,319.736	7	3,698.270
8	295.2559	8	673.789	8	1,052.322	8	1,430.856	8	1,809.389	8	2,187.922	8	2,566.455	8	2,944.989	8	3,323.522	8	3,702.055
9	299.0413	9	677.574	9	1,056.108	9	1,434.641	9	1,813.174	9	2,191.707	9	2,570.241	9	2,948.774	9	3,327.307	9	3,705.840
80	302.8266	**180**	681.360	**280**	1,059.893	**380**	1,438.426	**480**	1,816.960	**580**	2,195.493	**680**	2,574.026	**780**	2,952.559	**880**	3,331.092	**980**	3,709.626
1	306.6119	1	685.145	1	1,063.678	1	1,442.212	1	1,820.745	1	2,199.278	1	2,577.811	1	2,956.345	1	3,334.878	1	3,713.411
2	310.3972	2	688.930	2	1,067.464	2	1,445.997	2	1,824.530	2	2,203.063	2	2,581.597	2	2,960.130	2	3,338.663	2	3,717.196
3	314.1826	3	692.716	3	1,071.249	3	1,449.782	3	1,828.316	3	2,206.849	3	2,585.382	3	2,963.915	3	3,342.448	3	3,720.982
4	317.9679	4	696.501	4	1,075.034	4	1,453.568	4	1,832.101	4	2,210.634	4	2,589.167	4	2,967.701	4	3,346.234	4	3,724.767
5	321.7532	5	700.286	5	1,078.820	5	1,457.353	5	1,835.886	5	2,214.419	5	2,592.953	5	2,971.486	5	3,350.019	5	3,728.552
6	325.5386	6	704.072	6	1,082.605	6	1,461.138	6	1,839.671	6	2,218.205	6	2,596.738	6	2,975.271	6	3,353.804	6	3,732.338
7	329.3239	7	707.857	7	1,086.390	7	1,464.924	7	1,843.457	7	2,221.990	7	2,600.523	7	2,979.057	7	3,357.590	7	3,736.123
8	333.1092	8	711.642	8	1,090.176	8	1,468.709	8	1,847.242	8	2,225.775	8	2,604.309	8	2,982.842	8	3,361.375	8	3,739.908
9	336.8946	9	715.428	9	1,093.961	9	1,472.494	9	1,851.027	9	2,229.561	9	2,608.094	9	2,986.627	9	3,365.160	9	3,743.694
90	340.6799	**190**	719.213	**290**	1,097.746	**390**	1,476.280	**490**	1,854.813	**590**	2,233.346	**690**	2,611.879	**790**	2,990.413	**890**	3,368.946	**990**	3,747.479
1	344.4652	1	722.998	1	1,101.532	1	1,480.065	1	1,858.598	1	2,237.131	1	2,615.665	1	2,994.198	1	3,372.731	1	3,751.264
2	348.2506	2	726.784	2	1,105.317	2	1,483.850	2	1,862.383	2	2,240.917	2	2,619.450	2	2,997.983	2	3,376.516	2	3,755.050
3	352.0359	3	730.569	3	1,109.102	3	1,487.636	3	1,866.169	3	2,244.702	3	2,623.235	3	3,001.769	3	3,380.302	3	3,758.835
4	355.8212	4	734.354	4	1,112.888	4	1,491.421	4	1,869.954	4	2,248.487	4	2,627.021	4	3,005.554	4	3,384.087	4	3,762.620
5	359.6066	5	738.140	5	1,116.673	5	1,495.206	5	1,873.739	5	2,252.273	5	2,630.806	5	3,009.339	5	3,387.872	5	3,766.406
6	363.3919	6	741.925	6	1,120.458	6	1,498.992	6	1,877.525	6	2,256.058	6	2,634.591	6	3,013.125	6	3,391.658	6	3,770.191
7	367.1772	7	745.710	7	1,124.244	7	1,502.777	7	1,881.310	7	2,259.843	7	2,638.377	7	3,016.910	7	3,395.443	7	3,773.976
8	370.9626	8	749.496	8	1,128.029	8	1,506.562	8	1,885.095	8	2,263.629	8	2,642.162	8	3,020.695	8	3,399.228	8	3,777.762
9	374.7479	9	753.281	9	1,131.814	9	1,510.348	9	1,888.881	9	2,267.414	9	2,645.947	9	3,024.481	9	3,403.014	9	3,781.547

CAPACITY—LITERS TO GALLONS

[Reduction factor: 1 liter=0.2641776o gallons]

Liters	Gallons	Liters	Gallons	Liters	Gallons	Liters	Gallons	Liters	Gallons	Liters	Gallons	Liters	Gallons	Liters	Gallons	Liters	Gallons	Liters	Gallons
0		**100**	26.4178	**200**	52.8355	**300**	79.2533	**400**	105.6710	**500**	132.0888	**600**	158.5066	**700**	184.9243	**800**	211.3421	**900**	237.7598
1	0.26418	1	26.6819	1	53.0997	1	79.5175	1	105.9352	1	132.3530	1	158.7707	1	185.1885	1	211.6063	1	238.0240
2	0.52836	2	26.9461	2	53.3639	2	79.7816	2	106.1994	2	132.6172	2	159.0349	2	185.4527	2	211.8704	2	238.2882
3	0.79253	3	27.2103	3	53.6281	3	80.0458	3	106.4636	3	132.8813	3	159.2991	3	185.7169	3	212.1346	3	238.5524
4	1.05671	4	27.4745	4	53.8922	4	80.3100	4	106.7278	4	133.1455	4	159.5633	4	185.9810	4	212.3988	4	238.8166
5	1.32089	5	27.7386	5	54.1564	5	80.5742	5	106.9919	5	133.4097	5	159.8274	5	186.2452	5	212.6630	5	239.0807
6	1.58507	6	28.0028	6	54.4206	6	80.8383	6	107.2561	6	133.6739	6	160.0916	6	186.5094	6	212.9271	6	239.3449
7	1.84924	7	28.2670	7	54.6848	7	81.1025	7	107.5203	7	133.9380	7	160.3558	7	186.7736	7	213.1913	7	239.6091
8	2.11342	8	28.5312	8	54.9489	8	81.3667	8	107.7845	8	134.2022	8	160.6200	8	187.0377	8	213.4555	8	239.8733
9	2.37760	9	28.7954	9	55.2131	9	81.6309	9	108.0486	9	134.4664	9	160.8842	9	187.3019	9	213.7197	9	240.1374
10	2.64178	**110**	29.0595	**210**	55.4773	**310**	81.8951	**410**	108.3128	**510**	134.7306	**610**	161.1483	**710**	187.5661	**810**	213.9839	**910**	240.4016
1	2.90595	1	29.3237	1	55.7415	1	82.1592	1	108.5770	1	134.9948	1	161.4125	1	187.8303	1	214.2480	1	240.6658
2	3.17013	2	29.5879	2	56.0057	2	82.4234	2	108.8412	2	135.2589	2	161.6767	2	188.0945	2	214.5122	2	240.9300
3	3.43431	3	29.8521	3	56.2698	3	82.6876	3	109.1053	3	135.5231	3	161.9409	3	188.3586	3	214.7764	3	241.1941
4	3.69849	4	30.1162	4	56.5340	4	82.9518	4	109.3695	4	135.7873	4	162.2050	4	188.6228	4	215.0406	4	241.4583
5	3.96266	5	30.3804	5	56.7982	5	83.2159	5	109.6337	5	136.0515	5	162.4692	5	188.8870	5	215.3047	5	241.7225
6	4.22684	6	30.6446	6	57.0624	6	83.4801	6	109.8979	6	136.3156	6	162.7334	6	189.1512	6	215.5689	6	241.9867
7	4.49102	7	30.9088	7	57.3265	7	83.7443	7	110.1621	7	136.5798	7	162.9976	7	189.4153	7	215.8331	7	242.2509
8	4.75520	8	31.1730	8	57.5907	8	84.0085	8	110.4262	8	136.8440	8	163.2618	8	189.6795	8	216.0973	8	242.5150
9	5.01937	9	31.4371	9	57.8549	9	84.2727	9	110.6904	9	137.1082	9	163.5259	9	189.9437	9	216.3615	9	242.7792
20	5.28355	**120**	31.7013	**220**	58.1191	**320**	84.5368	**420**	110.9546	**520**	137.3724	**620**	163.7901	**720**	190.2079	**820**	216.6256	**920**	243.0434
1	5.54773	1	31.9655	1	58.3832	1	84.8010	1	111.2188	1	137.6365	1	164.0543	1	190.4720	1	216.8898	1	243.3076
2	5.81191	2	32.2297	2	58.6474	2	85.0652	2	111.4829	2	137.9007	2	164.3185	2	190.7362	2	217.1540	2	243.5717
3	6.07608	3	32.4938	3	58.9116	3	85.3294	3	111.7471	3	138.1649	3	164.5826	3	191.0004	3	217.4182	3	243.8359
4	6.34026	4	32.7580	4	59.1758	4	85.5935	4	112.0113	4	138.4291	4	164.8468	4	191.2646	4	217.6823	4	244.1001
5	6.60444	5	33.0222	5	59.4400	5	85.8577	5	112.2755	5	138.6932	5	165.1110	5	191.5288	5	217.9465	5	244.3643
6	6.86862	6	33.2864	6	59.7041	6	86.1219	6	112.5397	6	138.9574	6	165.3752	6	191.7929	6	218.2107	6	244.6285
7	7.13280	7	33.5506	7	59.9683	7	86.3861	7	112.8038	7	139.2216	7	165.6394	7	192.0571	7	218.4749	7	244.8926
8	7.39697	8	33.8147	8	60.2325	8	86.6503	8	113.0680	8	139.4858	8	165.9035	8	192.3213	8	218.7391	8	245.1568
9	7.66115	9	34.0789	9	60.4967	9	86.9144	9	113.3322	9	139.7500	9	166.1677	9	192.5855	9	219.0032	9	245.4210
30	7.92533	**130**	34.3431	**230**	60.7608	**330**	87.1786	**430**	113.5964	**530**	140.0141	**630**	166.4319	**730**	192.8496	**830**	219.2674	**930**	245.6852
1	8.18951	1	34.6073	1	61.0250	1	87.4428	1	113.8605	1	140.2783	1	166.6961	1	193.1138	1	219.5316	1	245.9493
2	8.45368	2	34.8714	2	61.2892	2	87.7070	2	114.1247	2	140.5425	2	166.9602	2	193.3780	2	219.7958	2	246.2135
3	8.71786	3	35.1356	3	61.5534	3	87.9711	3	114.3889	3	140.8067	3	167.2244	3	193.6422	3	220.0599	3	246.4777
4	8.98204	4	35.3998	4	61.8176	4	88.2353	4	114.6531	4	141.0708	4	167.4886	4	193.9064	4	220.3241	4	246.7419
5	9.24622	5	35.6640	5	62.0817	5	88.4995	5	114.9173	5	141.3350	5	167.7528	5	194.1705	5	220.5883	5	247.0061
6	9.51039	6	35.9282	6	62.3459	6	88.7637	6	115.1814	6	141.5992	6	168.0170	6	194.4347	6	220.8525	6	247.2702
7	9.77457	7	36.1923	7	62.6101	7	89.0279	7	115.4456	7	141.8634	7	168.2811	7	194.6989	7	221.1167	7	247.5344
8	10.03875	8	36.4565	8	62.8743	8	89.2920	8	115.7098	8	142.1275	8	168.5453	8	194.9631	8	221.3808	8	247.7986
9	10.30293	9	36.7207	9	63.1384	9	89.5562	9	115.9740	9	142.3917	9	168.8095	9	195.2272	9	221.6450	9	248.0628
40	10.56710	**140**	36.9849	**240**	63.4026	**340**	89.8204	**440**	116.2381	**540**	142.6559	**640**	169.0737	**740**	195.4914	**840**	221.9092	**940**	248.3269
1	10.83128	1	37.2490	1	63.6668	1	90.0846	1	116.5023	1	142.9201	1	169.3378	1	195.7556	1	222.1734	1	248.5911
2	11.09546	2	37.5132	2	63.9310	2	90.3487	2	116.7665	2	143.1843	2	169.6020	2	196.0198	2	222.4375	2	248.8553
3	11.35964	3	37.7774	3	64.1952	3	90.6129	3	117.0307	3	143.4484	3	169.8662	3	196.2840	3	222.7017	3	249.1195
4	11.62381	4	38.0416	4	64.4593	4	90.8771	4	117.2949	4	143.7126	4	170.1304	4	196.5481	4	222.9659	4	249.3837
5	11.88799	5	38.3058	5	64.7235	5	91.1413	5	117.5590	5	143.9768	5	170.3946	5	196.8123	5	223.2301	5	249.6478
6	12.15217	6	38.5699	6	64.9877	6	91.4054	6	117.8232	6	144.2410	6	170.6587	6	197.0765	6	223.4942	6	249.9120
7	12.41635	7	38.8341	7	65.2519	7	91.6696	7	118.0874	7	144.5051	7	170.9229	7	197.3407	7	223.7584	7	250.1762
8	12.68052	8	39.0983	8	65.5160	8	91.9338	8	118.3516	8	144.7693	8	171.1871	8	197.6048	8	224.0226	8	250.4404
9	12.94470	9	39.3625	9	65.7802	9	92.1980	9	118.6157	9	145.0335	9	171.4513	9	197.8690	9	224.2868	9	250.7045

N	Value	N	Value	N	Value	N	Value	N	Value	N	Value	N	Value	N	Value	N	Value	N	Value
50	13.20888	**150**	39.6266	**250**	66.0444	**350**	92.4622	**450**	118.8799	**550**	145.2977	**650**	171.7154	**750**	198.1332	**850**	224.5510	**950**	250.9687
1	13.47306	1	39.8908	1	66.3086	1	92.7263	1	119.1441	1	145.5619	1	171.9796	1	198.3974	1	224.8151	1	251.2329
2	13.73724	2	40.1550	2	66.5728	2	92.9905	2	119.4083	2	145.8260	2	172.2438	2	198.6616	2	225.0793	2	251.4971
3	14.00141	3	40.4192	3	66.8369	3	93.2547	3	119.6725	3	146.0902	3	172.5080	3	198.9257	3	225.3435	3	251.7613
4	14.26559	4	40.6834	4	67.1011	4	93.5189	4	119.9366	4	146.3544	4	172.7722	4	199.1899	4	225.6077	4	252.0254
5	14.52977	5	40.9475	5	67.3653	5	93.7830	5	120.2008	5	146.6186	5	173.0363	5	199.4541	5	225.8718	5	252.2896
6	14.79395	6	41.2117	6	67.6295	6	94.0472	6	120.4650	6	146.8827	6	173.3005	6	199.7183	6	226.1360	6	252.5538
7	15.05812	7	41.4759	7	67.8936	7	94.3114	7	120.7292	7	147.1469	7	173.5647	7	199.9824	7	226.4002	7	252.8180
8	15.32230	8	41.7401	8	68.1578	8	94.5756	8	120.9933	8	147.4111	8	173.8289	8	200.2466	8	226.6644	8	253.0821
9	15.58648	9	42.0042	9	68.4220	9	94.8398	9	121.2575	9	147.6753	9	174.0930	9	200.5108	9	226.9286	9	253.3463
60	15.85066	**160**	42.2684	**260**	68.6862	**360**	95.1039	**460**	121.5217	**560**	147.9395	**660**	174.3572	**760**	200.7750	**860**	227.1927	**960**	253.6105
1	16.11483	1	42.5326	1	68.9504	1	95.3681	1	121.7859	1	148.2036	1	174.6214	1	201.0392	1	227.4569	1	253.8747
2	16.37901	2	42.7968	2	69.2145	2	95.6323	2	122.0501	2	148.4678	2	174.8856	2	201.3033	2	227.7211	2	254.1389
3	16.64319	3	43.0609	3	69.4787	3	95.8965	3	122.3142	3	148.7320	3	175.1497	3	201.5675	3	227.9853	3	254.4030
4	16.90737	4	43.3251	4	69.7429	4	96.1606	4	122.5784	4	148.9962	4	175.4139	4	201.8317	4	228.2494	4	254.6672
5	17.17154	5	43.5893	5	70.0071	5	96.4248	5	122.8426	5	149.2603	5	175.6781	5	202.0959	5	228.5136	5	254.9314
6	17.43572	6	43.8535	6	70.2712	6	96.6890	6	123.1068	6	149.5245	6	175.9423	6	202.3600	6	228.7778	6	255.1956
7	17.69990	7	44.1177	7	70.5354	7	96.9532	7	123.3709	7	149.7887	7	176.2065	7	202.6242	7	229.0420	7	255.4597
8	17.96408	8	44.3818	8	70.7996	8	97.2174	8	123.6351	8	150.0529	8	176.4706	8	202.8884	8	229.3062	8	255.7239
9	18.22825	9	44.6460	9	71.0638	9	97.4815	9	123.8993	9	150.3171	9	176.7348	9	203.1526	9	229.5703	9	255.9881
70	18.49243	**170**	44.9102	**270**	71.3280	**370**	97.7457	**470**	124.1635	**570**	150.5812	**670**	176.9990	**770**	203.4168	**870**	229.8345	**970**	256.2523
1	18.75661	1	45.1744	1	71.5921	1	98.0099	1	124.4276	1	150.8454	1	177.2632	1	203.6809	1	230.0987	1	256.5164
2	19.02079	2	45.4385	2	71.8563	2	98.2741	2	124.6918	2	151.1096	2	177.5273	2	203.9451	2	230.3629	2	256.7806
3	19.28496	3	45.7027	3	72.1205	3	98.5382	3	124.9560	3	151.3738	3	177.7915	3	204.2093	3	230.6270	3	257.0448
4	19.54914	4	45.9669	4	72.3847	4	98.8024	4	125.2202	4	151.6379	4	178.0557	4	204.4735	4	230.8912	4	257.3090
5	19.81332	5	46.2311	5	72.6488	5	99.0666	5	125.4844	5	151.9021	5	178.3199	5	204.7376	5	231.1554	5	257.5732
6	20.07750	6	46.4953	6	72.9130	6	99.3308	6	125.7485	6	152.1663	6	178.5841	6	205.0018	6	231.4196	6	257.8373
7	20.34168	7	46.7594	7	73.1772	7	99.5950	7	126.0127	7	152.4305	7	178.8482	7	205.2660	7	231.6838	7	258.1015
8	20.60585	8	47.0236	8	73.4414	8	99.8591	8	126.2769	8	152.6947	8	179.1124	8	205.5302	8	231.9479	8	258.3657
9	20.87003	9	47.2878	9	73.7056	9	100.1233	9	126.5411	9	152.9588	9	179.3766	9	205.7944	9	232.2121	9	258.6299
80	21.13421	**180**	47.5520	**280**	73.9697	**380**	100.3875	**480**	126.8052	**580**	153.2230	**680**	179.6408	**780**	206.0585	**880**	232.4763	**980**	258.8940
1	21.39839	1	47.8161	1	74.2339	1	100.6517	1	127.0694	1	153.4872	1	179.9049	1	206.3227	1	232.7405	1	259.1582
2	21.66256	2	48.0803	2	74.4981	2	100.9158	2	127.3336	2	153.7514	2	180.1691	2	206.5869	2	233.0046	2	259.4224
3	21.92674	3	48.3445	3	74.7623	3	101.1800	3	127.5978	3	154.0155	3	180.4333	3	206.8511	3	233.2688	3	259.6866
4	22.19092	4	48.6087	4	75.0264	4	101.4442	4	127.8620	4	154.2797	4	180.6975	4	207.1152	4	233.5330	4	259.9508
5	22.45510	5	48.8729	5	75.2906	5	101.7084	5	128.1261	5	154.5439	5	180.9617	5	207.3794	5	233.7972	5	260.2149
6	22.71927	6	49.1370	6	75.5548	6	101.9726	6	128.3903	6	154.8081	6	181.2258	6	207.6436	6	234.0614	6	260.4791
7	22.98345	7	49.4012	7	75.8190	7	102.2367	7	128.6545	7	155.0723	7	181.4900	7	207.9078	7	234.3255	7	260.7433
8	23.24763	8	49.6654	8	76.0831	8	102.5009	8	128.9187	8	155.3364	8	181.7542	8	208.1719	8	234.5897	8	261.0075
9	23.51181	9	49.9296	9	76.3473	9	102.7651	9	129.1828	9	155.6006	9	182.0184	9	208.4361	9	234.8539	9	261.2716
90	23.77598	**190**	50.1937	**290**	76.6115	**390**	103.0293	**490**	129.4470	**590**	155.8648	**690**	182.2825	**790**	208.7003	**890**	235.1181	**990**	261.5358
1	24.04016	1	50.4579	1	76.8757	1	103.2934	1	129.7112	1	156.1290	1	182.5467	1	208.9645	1	235.3822	1	261.8000
2	24.30434	2	50.7221	2	77.1399	2	103.5576	2	129.9754	2	156.3931	2	182.8109	2	209.2287	2	235.6464	2	262.0642
3	24.56852	3	50.9863	3	77.4040	3	103.8218	3	130.2396	3	156.6573	3	183.0751	3	209.4928	3	235.9106	3	262.3284
4	24.83269	4	51.2505	4	77.6682	4	104.0860	4	130.5037	4	156.9215	4	183.3393	4	209.7570	4	236.1748	4	262.5925
5	25.09687	5	51.5146	5	77.9324	5	104.3502	5	130.7679	5	157.1857	5	183.6034	5	210.0212	5	236.4390	5	262.8567
6	25.36105	6	51.7788	6	78.1966	6	104.6143	6	131.0321	6	157.4498	6	183.8676	6	210.2854	6	236.7031	6	263.1209
7	25.62523	7	52.0430	7	78.4607	7	104.8785	7	131.2963	7	157.7140	7	184.1318	7	210.5495	7	236.9673	7	263.3851
8	25.88940	8	52.3072	8	78.7249	8	105.1427	8	131.5604	8	157.9782	8	184.3960	8	210.8137	8	237.2315	8	263.6492
9	26.15358	9	52.5713	9	78.9891	9	105.4069	9	131.8246	9	158.2424	9	184.6601	9	211.0779	9	237.4957	9	263.9134

CAPACITY—BUSHELS TO HECTOLITERS

[Reduction factor: 1 bushel=0.35238330 hectoliters]

Bushels	Hectoliters	Bushels	Hectoliters	Bushels	Hectoliters	Bushels	Hectoliters	Bushels	Hectoliters	Bushels	Hectoliters	Bushels	Hectoliters	Bushels	Hectoliters	Bushels	Hectoliters	Bushels	Hectoliters
0		**100**	35.2383	**200**	70.4767	**300**	105.7150	**400**	140.9533	**500**	176.1917	**600**	211.4300	**700**	246.6683	**800**	281.9066	**900**	317.1450
1	0.35238	1	35.5907	1	70.8290	1	106.0674	1	141.3057	1	176.5440	1	211.7824	1	247.0207	1	282.2590	1	317.4974
2	0.70477	2	35.9431	2	71.1814	2	106.4198	2	141.6581	2	176.8964	2	212.1347	2	247.3731	2	282.6114	2	317.8497
3	1.05715	3	36.2955	3	71.5338	3	106.7721	3	142.0105	3	177.2488	3	212.4871	3	247.7255	3	282.9638	3	318.2021
4	1.40953	4	36.6479	4	71.8862	4	107.1245	4	142.3629	4	177.6012	4	212.8395	4	248.0778	4	283.3162	4	318.5545
5	1.76192	5	37.0002	5	72.2386	5	107.4769	5	142.7152	5	177.9536	5	213.1919	5	248.4302	5	283.6686	5	318.9069
6	2.11430	6	37.3526	6	72.5910	6	107.8293	6	143.0676	6	178.3060	6	213.5443	6	248.7826	6	284.0209	6	319.2593
7	2.46668	7	37.7050	7	72.9433	7	108.1817	7	143.4200	7	178.6583	7	213.8967	7	249.1350	7	284.3733	7	319.6117
8	2.81907	8	38.0574	8	73.2957	8	108.5341	8	143.7724	8	179.0107	8	214.2490	8	249.4874	8	284.7257	8	319.9640
9	3.17145	9	38.4098	9	73.6481	9	108.8864	9	144.1248	9	179.3631	9	214.6014	9	249.8398	9	285.0781	9	320.3164
10	3.52383	**110**	38.7622	**210**	74.0005	**310**	109.2388	**410**	144.4772	**510**	179.7155	**610**	214.9538	**710**	250.1921	**810**	285.4305	**910**	320.6688
1	3.87622	1	39.1145	1	74.3529	1	109.5912	1	144.8295	1	180.0679	1	215.3062	1	250.5445	1	285.7829	1	321.0212
2	4.22860	2	39.4669	2	74.7053	2	109.9436	2	145.1819	2	180.4203	2	215.6586	2	250.8969	2	286.1352	2	321.3736
3	4.58098	3	39.8193	3	75.0576	3	110.2960	3	145.5343	3	180.7726	3	216.0110	3	251.2493	3	286.4876	3	231.7260
4	4.93337	4	40.1717	4	75.4100	4	110.6484	4	145.8867	4	181.1250	4	216.3633	4	251.6017	4	286.8400	4	322.0783
5	5.28575	5	40.5241	5	75.7624	5	111.0007	5	146.2391	5	181.4774	5	216.7157	5	251.9541	5	287.1924	5	322.4307
6	5.63813	6	40.8765	6	76.1148	6	111.3531	6	146.5915	6	181.8298	6	217.0681	6	252.3064	6	287.5448	6	322.7831
7	5.99052	7	41.2288	7	76.4672	7	111.7055	7	146.9438	7	182.1822	7	217.4205	7	252.6588	7	287.8972	7	323.1355
8	6.34290	8	41.5812	8	76.8196	8	112.0579	8	147.2962	8	182.5346	8	217.7729	8	253.0112	8	288.2495	8	323.4879
9	6.69528	9	41.9336	9	77.1719	9	112.4103	9	147.6486	9	182.8869	9	218.1253	9	253.3636	9	288.6019	9	323.8403
20	7.04767	**120**	42.2860	**220**	77.5243	**320**	112.7627	**420**	148.0010	**520**	183.2393	**620**	218.4776	**720**	253.7160	**820**	288.9543	**920**	324.1926
1	7.40005	1	42.6384	1	77.8767	1	113.1150	1	148.3534	1	183.5917	1	218.8300	1	254.0684	1	289.3067	1	324.5450
2	7.75243	2	42.9908	2	78.2291	2	113.4674	2	148.7058	2	183.9441	2	219.1824	2	254.4207	2	289.6591	2	324.8974
3	8.10482	3	43.3431	3	78.5815	3	113.8198	3	149.0581	3	184.2965	3	219.5348	3	254.7731	3	290.0115	3	325.2498
4	8.45720	4	43.6955	4	78.9339	4	114.1722	4	149.4105	4	184.6489	4	219.8872	4	255.1255	4	290.3638	4	325.6022
5	8.80958	5	44.0479	5	79.2862	5	114.5246	5	149.7629	5	185.0012	5	220.2396	5	255.4779	5	290.7162	5	325.9546
6	9.16197	6	44.4003	6	79.6386	6	114.8770	6	150.1153	6	185.3536	6	220.5919	6	255.8303	6	291.0686	6	326.3069
7	9.51435	7	44.7527	7	79.9910	7	115.2293	7	150.4677	7	185.7060	7	220.9443	7	256.1827	7	291.4210	7	326.6593
8	9.86673	8	45.1051	8	80.3434	8	115.5817	8	150.8201	8	186.0584	8	221.2967	8	256.5350	8	291.7734	8	327.0117
9	10.21912	9	45.4574	9	80.6958	9	115.9341	9	151.1724	9	186.4108	9	221.6491	9	256.8874	9	292.1258	9	327.3641
30	10.57150	**130**	45.8098	**230**	81.0482	**330**	116.2865	**430**	151.5248	**530**	186.7631	**630**	222.0015	**730**	257.2398	**830**	292.4781	**930**	327.7165
1	10.92388	1	46.1622	1	81.4005	1	116.6389	1	151.8772	1	187.1155	1	222.3539	1	257.5922	1	292.8305	1	328.0689
2	11.27627	2	46.5146	2	81.7529	2	116.9913	2	152.2296	2	187.4679	2	222.7062	2	257.9446	2	293.1829	2	328.4212
3	11.62865	3	46.8670	3	82.1053	3	117.3436	3	152.5820	3	187.8203	3	223.0586	3	258.2970	3	293.5353	3	328.7736
4	11.98103	4	47.2194	4	82.4577	4	117.6960	4	152.9344	4	188.1727	4	223.4110	4	258.6493	4	293.8877	4	329.1260
5	12.33342	5	47.5717	5	82.8101	5	118.0484	5	153.2867	5	188.5251	5	223.7634	5	259 0017	5	294.2401	5	329.4784
6	12.68580	6	47.9241	6	83.1625	6	118.4008	6	153.6391	6	188.8774	6	224.1158	6	259.3541	6	294.5924	6	329.8308
7	13.03818	7	48.2765	7	83.5148	7	118.7532	7	153.9915	7	189.2298	7	224.4682	7	259.7065	7	294.9448	7	330.1832
8	13.39057	8	48.6289	8	83.8672	8	119.1056	8	154.3439	8	189.5822	8	224.8205	8	260.0589	8	295.2972	8	330.5355
9	13.74295	9	48.9813	9	84.2196	9	119.4579	9	154.6963	9	189.9346	9	225.1729	9	260.4113	9	295.6496	9	330.8879
40	14.09533	**140**	49.3337	**240**	84.5720	**340**	119.8103	**440**	155.0487	**540**	190.2870	**640**	225.5253	**740**	260.7636	**840**	296.0020	**940**	331.2403
1	14.44772	1	49.6860	1	84.9244	1	120.1627	1	155.4010	1	190.6394	1	225.8777	1	261.1160	1	296.3544	1	331.5927
2	14.80010	2	50.0384	2	85.2768	2	120.5151	2	155.7534	2	190.9917	2	226.2301	2	261.4684	2	296.7067	2	331.9451
3	15.15248	3	50.3908	3	85.6291	3	120.8675	3	156.1058	3	191.3441	3	226.5825	3	261.8208	3	297.0591	3	332.2975
4	15.50487	4	50.7432	4	85.9815	4	121.2199	4	156.4582	4	191.6965	4	226.9348	4	262.1732	4	297.4115	4	332.6498
5	15.85725	5	51.0956	5	86.3339	5	121.5722	5	156.8106	5	192.0489	5	227.2872	5	262.5256	5	297.7639	5	333.0022
6	16.20963	6	51.4480	6	86.6863	6	121.9246	6	157.1630	6	192.4013	6	227.6396	6	262.8779	6	298.1163	6	333.3546
7	16.56202	7	51.8003	7	87.0387	7	122.2770	7	157.5153	7	192.7537	7	227.9920	7	263.2303	7	298.4687	7	333.7070
8	16.91440	8	52.1527	8	87.3911	8	122.6294	8	157.8677	8	193.1060	8	228.3444	8	263.5827	8	298.8210	8	334.0594
9	17.26678	9	52.5051	9	87.7434	9	122.9818	9	158.2201	9	193.4584	9	228.6968	9	263.9351	9	299.1734	9	334.4118

50	17.61917	**150**	52.8575	**250**	88.0958	**350**	123.3342	**450**	158.5725	**550**	193.8108	**650**	229.0491	**750**	264.2875	**850**	299.5258	**950**	334.7641
1	17.97155	1	53.2099	1	88.4482	1	123.6865	1	158.9249	1	194.1632	1	229.4015	1	264.6399	1	299.8782	1	335.1165
2	18.32393	2	53.5623	2	88.8006	2	124.0389	2	159.2773	2	194.5156	2	229.7539	2	264.9922	2	300.2306	2	335.4689
3	18.67631	3	53.9146	3	89.1530	3	124.3913	3	159.6296	3	194.8680	3	230.1063	3	265.3446	3	300.5830	3	335.8213
4	19.02870	4	54.2670	4	89.5054	4	124.7437	4	159.9820	4	195.2203	4	230.4587	4	265.6970	4	300.9353	4	336.1737
5	19.38108	5	54.6194	5	89.8577	5	125.0961	5	160.3344	5	195.5727	5	230.8111	5	266.0494	5	301.2877	5	336.5261
6	19.73346	6	54.9718	6	90.2101	6	125.4485	6	160.6868	6	195.9251	6	231.1634	6	266.4018	6	301.6401	6	336.8784
7	20.08585	7	55.3242	7	90.5625	7	125.8008	7	161.0392	7	196.2775	7	231.5158	7	266.7542	7	301.9925	7	337.2308
8	20.43823	8	55.6766	8	90.9149	8	126.1532	8	161.3916	8	196.6299	8	231.8682	8	267.1065	8	302.3449	8	337.5832
9	20.79061	9	56.0289	9	91.2673	9	126.5056	9	161.7439	9	196.9823	9	232.2206	9	267.4589	9	302.6973	9	337.9356
60	21.14300	**160**	56.3813	**260**	91.6197	**360**	126.8580	**460**	162.0963	**560**	197.3346	**660**	232.5730	**760**	267.8113	**860**	303.0496	**960**	338.2880
1	21.49538	1	56.7337	1	91.9720	1	127.2104	1	162.4487	1	197.6870	1	232.9254	1	268.1637	1	303.4020	1	338.6404
2	21.84776	2	57.0861	2	92.3244	2	127.5628	2	162.8011	2	198.0394	2	233.2777	2	268.5161	2	303.7544	2	338.9927
3	22.20015	3	57.4385	3	92.6768	3	127.9151	3	163.1535	3	198.3918	3	233.6301	3	268.8685	3	304.1068	3	339.3451
4	22.55253	4	57.7909	4	93.0292	4	128.2675	4	163.5059	4	198.7442	4	233.9825	4	269.2208	4	304.4592	4	339.6975
5	22.90491	5	58.1432	5	93.3816	5	128.6199	5	163.8582	5	199.0966	5	234.3349	5	269.5732	5	304.8116	5	340.0499
6	23.25730	6	58.4956	6	93.7340	6	128.9723	6	164.2106	6	199.4489	6	234.6873	6	269.9256	6	305.1639	6	340.4023
7	23.60968	7	58.8480	7	94.0863	7	129.3247	7	164.5630	7	199.8013	7	235.0397	7	270.2780	7	305.5163	7	340.7547
8	23.96206	8	59.2004	8	94.4387	8	129.6771	8	164.9154	8	200.1537	8	235.3920	8	270.6304	8	305.8687	8	341.1070
9	24.31445	9	59.5528	9	94.7911	9	130.0294	9	165.2678	9	200.5061	9	235.7444	9	270.9828	9	306.2211	9	341.4594
70	24.66683	**170**	59.9052	**270**	95.1435	**370**	130.3818	**470**	165.6202	**570**	200.8585	**670**	236.0968	**770**	271.3351	**870**	306.5735	**970**	341.8118
1	25.01921	1	60.2575	1	95.4959	1	130.7342	1	165.9725	1	201.2109	1	236.4492	1	271.6875	1	306.9259	1	342.1642
2	25.37160	2	60.6099	2	95.8483	2	131.0866	2	166.3249	2	201.5632	2	236.8016	2	272.0399	2	307.2782	2	342.5166
3	25.72398	3	60.9623	3	96.2006	3	131.4390	3	166.6773	3	201.9156	3	237.1540	3	272.3923	3	307.6306	3	342.8690
4	26.07636	4	61.3147	4	96.5530	4	131.7914	4	167.0297	4	202.2680	4	237.5063	4	272.7447	4	307.9830	4	343.2213
5	26.42875	5	61.6671	5	96.9054	5	132.1437	5	167.3821	5	202.6204	5	237.8587	5	273.0971	5	308.3354	5	343.5737
6	26.78113	6	62.0195	6	97.2578	6	132.4961	6	167.7345	6	202.9728	6	238.2111	6	273.4494	6	308.6878	6	343.9261
7	27.13351	7	62.3718	7	97.6102	7	132.8485	7	168.0868	7	203.3252	7	238.5635	7	273.8018	7	309.0402	7	344.2785
8	27.48590	8	62.7242	8	97.9626	8	133.2009	8	168.4392	8	203.6775	8	238.9159	8	274.1542	8	309.3925	8	344.6309
9	27.83828	9	63.0766	9	98.3149	9	133.5533	9	168.7916	9	204.0299	9	239.2683	9	274.5066	9	309.7449	9	344.9833
80	28.19066	**180**	63.4290	**280**	98.6673	**380**	133.9057	**480**	169.1440	**580**	204.3823	**680**	239.6206	**780**	274.8590	**880**	310.0973	**980**	345.3356
1	28.54305	1	63.7814	1	99.0197	1	134.2580	1	169.4964	1	204.7347	1	239.9730	1	275.2114	1	310.4497	1	345.6880
2	28.89543	2	64.1338	2	99.3721	2	134.6104	2	169.8488	2	205.0871	2	240.3254	2	275.5637	2	310.8021	2	346.0404
3	29.24781	3	64.4861	3	99.7245	3	134.9628	3	170.2011	3	205.4395	3	240.6778	3	275.9161	3	311.1545	3	346.3928
4	29.60020	4	64.8385	4	100.0769	4	135.3152	4	170.5535	4	205.7918	4	241.0302	4	276.2685	4	311.5068	4	346.7452
5	29.95258	5	65.1909	5	100.4292	5	135.6676	5	170.9059	5	206.1442	5	241.3826	5	276.6209	5	311.8592	5	347.0976
6	30.30496	6	65.5433	6	100.7816	6	136.0200	6	171.2583	6	206.4966	6	241.7349	6	276.9733	6	312.2116	6	347.4499
7	30.65735	7	65.8957	7	101.1340	7	136.3723	7	171.6107	7	206.8490	7	242.0873	7	277.3257	7	312.5640	7	347.8023
8	31.00973	8	66.2481	8	101.4864	8	136.7247	8	171.9631	8	207.2014	8	242.4397	8	277.6780	8	312.9164	8	348.1547
9	31.36211	9	66.6004	9	101.8388	9	137.0771	9	172.3154	9	207.5538	9	242.7921	9	278.0304	9	313.2688	9	348.5071
90	31.71450	**190**	66.9528	**290**	102.1912	**390**	137.4295	**490**	172.6678	**590**	207.9061	**690**	243.1445	**790**	278.3828	**890**	313.6211	**990**	348.8595
1	32.06688	1	67.3052	1	102.5435	1	137.7819	1	173.0202	1	208.2585	1	243.4969	1	278.7352	1	313.9735	1	349.2119
2	32.41926	2	67.6576	2	102.8959	2	138.1343	2	173.3726	2	208.6109	2	243.8492	2	279.0876	2	314.3259	2	349.5642
3	32.77165	3	68.0100	3	103.2483	3	138.4866	3	173.7250	3	208.9633	3	244.2016	3	279.4400	3	314.6783	3	349.9166
4	33.12403	4	68.3624	4	103.6007	4	138.8390	4	174.0774	4	209.3157	4	244.5540	4	279.7923	4	315.0307	4	350.2690
5	33.47641	5	68.7147	5	103.9531	5	139.1914	5	174.4297	5	209.6681	5	244.9064	5	280.1447	5	315.3831	5	350.6214
6	33.82880	6	69.0671	6	104.3055	6	139.5438	6	174.7821	6	210.0204	6	245.2588	6	280.4971	6	315.7354	6	350.9738
7	34.18118	7	69.4195	7	104.6578	7	139.8962	7	175.1345	7	210.3728	7	245.6112	7	280.8495	7	316.0878	7	351.3262
8	34.53356	8	69.7719	8	105.0102	8	140.2486	8	175.4869	8	210.7252	8	245.9635	8	281.2019	8	316.4402	8	351.6785
9	34.88595	9	70.1243	9	105.3626	9	140.6009	9	175.8393	9	211.0776	9	246.3159	9	281.5543	9	316.7926	9	352.0309

CAPACITY—HECTOLITERS TO BUSHELS

[Reduction factor: 1 hectoliter=2.8378189 bushels]

Hecto-liters	Bush-els	Hecto-liters	Bushels	Hecto-liters	Bushels	Hecto-liters	Bushels	Hecto-liters	Bushels	Hecto-liters	Bushels	Hecto-liters	Bushels	Hecto-liters	Bushels	Hecto-liters	Bushels	Hecto-liters	Bushels
0		**100**	283.782	**200**	567.564	**300**	851.346	**400**	1,135.128	**500**	1,418.909	**600**	1,702.691	**700**	1,986.473	**800**	2,270.255	**900**	2,554.037
1	2.8378	1	286.620	1	570.402	1	854.183	1	1,137.965	1	1,421.747	1	1,705.529	1	1,989.311	1	2,273.093	1	2,556.875
2	5.6756	2	289.458	2	573.239	2	857.021	2	1,140.803	2	1,424.585	2	1,708.367	2	1,992.149	2	2,275.931	2	2,559.713
3	8.5135	3	292.295	3	576.077	3	859.859	3	1,143.641	3	1,427.423	3	1,711.205	3	1,994.987	3	2,278.769	3	2,562.550
4	11.3513	4	295.133	4	578.915	4	862.697	4	1,146.479	4	1,430.261	4	1,714.043	4	1,997.825	4	2,281.606	4	2,565.388
5	14.1891	5	297.971	5	581.753	5	865.535	5	1,149.317	5	1,433.099	5	1,716.880	5	2,000.662	5	2,284.444	5	2,568.226
6	17.0269	6	300.809	6	584.591	6	868.373	6	1,152.154	6	1,435.936	6	1,719.718	6	2,003.500	6	2,287.282	6	2,571.064
7	19.8647	7	303.647	7	587.429	7	871.210	7	1,154.992	7	1,438.774	7	1,722.556	7	2,006.338	7	2,290.120	7	2,573.902
8	22.7026	8	306.484	8	590.266	8	874.048	8	1,157.830	8	1,441.612	8	1,725.394	8	2,009.176	8	2,292.958	8	2,576.740
9	25.5404	9	309.322	9	593.104	9	876.886	9	1,160.668	9	1,444.450	9	1,728.232	9	2,012.014	9	2,295.796	9	2,579.577
10	28.3782	**110**	312.160	**210**	595.942	**310**	879.724	**410**	1,163.506	**510**	1,447.288	**610**	1,731.070	**710**	2,014.851	**810**	2,298.633	**910**	2,582.415
1	31.2160	1	314.998	1	598.780	1	882.562	1	1,166.344	1	1,450.125	1	1,733.907	1	2,017.689	1	2,301.471	1	2,585.253
2	34.0538	2	317.836	2	601.618	2	885.400	2	1,169.181	2	1,452.963	2	1,736.745	2	2,020.527	2	2,304.309	2	2,588.091
3	36.8916	3	320.674	3	604.455	3	888.237	3	1,172.019	3	1,455.801	3	1,739.583	3	2,023.365	3	2,307.147	3	2,590.929
4	39.7295	4	323.511	4	607.293	4	891.075	4	1,174.857	4	1,458.639	4	1,742.421	4	2,026.203	4	2,309.985	4	2,593.766
5	42.5673	5	326.349	5	610.131	5	893.913	5	1,177.695	5	1,461.477	5	1,745.259	5	2,029.041	5	2,312.822	5	2,596.604
6	45.4051	6	329.187	6	612.969	6	896.751	6	1,180.533	6	1,464.315	6	1,748.096	6	2,031.878	6	2,315.660	6	2,599.442
7	48.2429	7	332.025	7	615.807	7	899.589	7	1,183.370	7	1,467.152	7	1,750.934	7	2,034.716	7	2,318.498	7	2,602.280
8	51.0807	8	334.863	8	618.645	8	902.426	8	1,186.208	8	1,469.990	8	1,753.772	8	2,037.554	8	2,321.336	8	2,605.118
9	53.9186	9	337.700	9	621.482	9	905.264	9	1,189.046	9	1,472.828	9	1,756.610	9	2,040.392	9	2,324.174	9	2,607.956
20	56.7564	**120**	340.538	**220**	624.320	**320**	908.102	**420**	1,191.884	**520**	1,475.666	**620**	1,759.448	**720**	2,043.230	**820**	2,327.012	**920**	2,610.793
1	59.5942	1	343.376	1	627.158	1	910.940	1	1,194.722	1	1,478.504	1	1,762.286	1	2,046.067	1	2,329.849	1	2,613.631
2	62.4320	2	346.214	2	629.996	2	913.778	2	1,197.560	2	1,481.341	2	1,765.123	2	2,048.905	2	2,332.687	2	2,616.469
3	65.2698	3	349.052	3	632.834	3	916.616	3	1,200.397	3	1,484.179	3	1,767.961	3	2,051.743	3	2,335.525	3	2,619.307
4	68.1077	4	351.890	4	635.671	4	919.453	4	1,203.235	4	1,487.017	4	1,770.799	4	2,054.581	4	2,338.363	4	2,622.145
5	70.9455	5	354.727	5	638.509	5	922.291	5	1,206.073	5	1,489.855	5	1,773.637	5	2,057.419	5	2,341.201	5	2,624.982
6	73.7833	6	357.565	6	641.347	6	925.129	6	1,208.911	6	1,492.693	6	1,776.475	6	2,060.257	6	2,344.038	6	2,627.820
7	76.6211	7	360.403	7	644.185	7	927.967	7	1,211.749	7	1,495.531	7	1,779.312	7	2,063.094	7	2,346.876	7	2,630.658
8	79.4589	8	363.241	8	647.023	8	930.805	8	1,214.586	8	1,498.368	8	1,782.150	8	2,065.932	8	2,349.714	8	2,633.496
9	82.2967	9	366.079	9	649.861	9	933.642	9	1,217.424	9	1,501.206	9	1,784.988	9	2,068.770	9	2,352.552	9	2,636.334
30	85.1346	**130**	368.916	**230**	652.698	**330**	936.480	**430**	1,220.262	**530**	1,504.044	**630**	1,787.826	**730**	2,071.608	**830**	2,355.390	**930**	2,639.172
1	87.9724	1	371.754	1	655.536	1	939.318	1	1,223.100	1	1,506.882	1	1,790.664	1	2,074.446	1	2,358.228	1	2,642.009
2	90.8102	2	374.592	2	658.374	2	942.156	2	1,225.938	2	1,509.720	2	1,793.502	2	2,077.283	2	2,361.065	2	2,644.847
3	93.6480	3	377.430	3	661.212	3	944.994	3	1,228.776	3	1,512.557	3	1,796.339	3	2,080.121	3	2,363.903	3	2,647.685
4	96.4858	4	380.268	4	664.050	4	947.832	4	1,231.613	4	1,515.395	4	1,799.177	4	2,082.959	4	2,366.741	4	2,650.523
5	99.3237	5	383.106	5	666.887	5	950.669	5	1,234.451	5	1,518.233	5	1,802.015	5	2,085.797	5	2,369.579	5	2,653.361
6	102.1615	6	385.943	6	669.725	6	953.507	6	1,237.289	6	1,521.071	6	1,804.853	6	2,088.635	6	2,372.417	6	2,656.199
7	104.9993	7	388.781	7	672.563	7	956.345	7	1,240.127	7	1,523.909	7	1,807.691	7	2,091.473	7	2,375.254	7	2,659.036
8	107.8371	8	391.619	8	675.401	8	959.183	8	1,242.965	8	1,526.747	8	1,810.528	8	2,094.310	8	2,378.092	8	2,661.874
9	110.6749	9	394.457	9	678.239	9	962.021	9	1,245.803	9	1,529.584	9	1,813.366	9	2,097.148	9	2,380.930	9	2,664.712
40	113.5128	**140**	397.295	**240**	681.077	**340**	964.858	**440**	1,248.640	**540**	1,532.422	**640**	1,816.204	**740**	2,099.986	**840**	2,383.768	**940**	2,667.550
1	116.3506	1	400.132	1	683.914	1	967.696	1	1,251.478	1	1,535.260	1	1,819.042	1	2,102.824	1	2,386.606	1	2,670.388
2	119.1884	2	402.970	2	686.752	2	970.534	2	1,254.316	2	1,538.098	2	1,821.880	2	2,105.662	2	2,389.444	2	2,673.225
3	122.0262	3	405.808	3	689.590	3	973.372	3	1,257.154	3	1,540.936	3	1,824.718	3	2,108.499	3	2,392.281	3	2,676.063
4	124.8640	4	408.646	4	692.428	4	976.210	4	1,259.992	4	1,543.773	4	1,827.555	4	2,111.337	4	2,395.119	4	2,678.901
5	127.7019	5	411.484	5	695.266	5	979.048	5	1,262.829	5	1,546.611	5	1,830.393	5	2,114.175	5	2,397.957	5	2,681.739
6	130.5397	6	414.322	6	698.103	6	981.885	6	1,265.667	6	1,549.449	6	1,833.231	6	2,117.013	6	2,400.795	6	2,684.577
7	133.3775	7	417.159	7	700.941	7	984.723	7	1,268.505	7	1,552.287	7	1,836.069	7	2,119.851	7	2,403.633	7	2,687.415
8	136.2153	8	419.997	8	703.779	8	987.561	8	1,271.343	8	1,555.125	8	1,838.907	8	2,122.689	8	2,406.470	8	2,690.252
9	139.0531	9	422.835	9	706.617	9	990.399	9	1,274.181	9	1,557.963	9	1,841.744	9	2,125.526	9	2,409.308	9	2,693.090

50	141.8909	**150**	425.673	**250**	709.455	**350**	993.237	**450**	1,277.019	**550**	1,560.800	**650**	1,844.582	**750**	2,128.364	**850**	2,412.146	**950**	2,695.928
1	144.7288	1	428.511	1	712.293	1	996.074	1	1,279.856	1	1,563.638	1	1,847.420	1	2,131.202	1	2,414.984	1	2,698.766
2	147.5666	2	431.348	2	715.130	2	998.912	2	1,282.694	2	1,566.476	2	1,850.258	2	2,134.040	2	2,417.822	2	2,701.604
3	150.4044	3	434.186	3	717.968	3	1,001.750	3	1,285.532	3	1,569.314	3	1,853.096	3	2,136.878	3	2,420.660	3	2,704.441
4	153.2422	4	437.024	4	720.806	4	1,004.588	4	1,288.370	4	1,572.152	4	1,855.934	4	2,139.715	4	2,423.497	4	2,707.279
5	156.0800	5	439.862	5	723.644	5	1,007.426	5	1,291.208	5	1,574.989	5	1,858.771	5	2,142.553	5	2,426.335	5	2,710.117
6	158.9179	6	442.700	6	726.482	6	1,010.264	6	1,294.045	6	1,577.827	6	1,861.609	6	2,145.391	6	2,429.173	6	2,712.955
7	161.7557	7	445.538	7	729.319	7	1,013.101	7	1,296.883	7	1,580.665	7	1,864.447	7	2,148.229	7	2,432.011	7	2,715.793
8	164.5935	8	448.375	8	732.157	8	1,015.939	8	1,299.721	8	1,583.503	8	1,867.285	8	2,151.067	8	2,434.849	8	2,718.631
9	167.4313	9	451.213	9	734.995	9	1,018.777	9	1,302.559	9	1,586.341	9	1,870.123	9	2,153.905	9	2,437.686	9	2,721.468
60	170.2691	**160**	454.051	**260**	737.833	**360**	1,021.615	**460**	1,305.397	**560**	1,589.179	**660**	1,872.960	**760**	2,156.742	**860**	2,440.524	**960**	2,724.306
1	173.1070	1	456.889	1	740.671	1	1,024.453	1	1,308.235	1	1,592.016	1	1,875.798	1	2,159.580	1	2,443.362	1	2,727.144
2	175.9448	2	459.727	2	743.509	1	1,027.290	2	1,311.072	2	1,594.854	2	1,878.636	2	2,162.418	2	2,446.200	2	2,729.982
3	178.7826	3	462.564	3	746.346	3	1,030.128	3	1,313.910	3	1,597.692	3	1,881.474	3	2,165.256	3	2,449.038	3	2,732.820
4	181.6204	4	465.402	4	749.184	4	1,032.966	4	1,316.748	4	1,600.530	4	1,884.312	4	2,168.094	4	2,451.876	4	2,735.657
5	184.4582	5	468.240	5	752.022	5	1,035.804	5	1,319.586	5	1,603.368	5	1,887.150	5	2,170.931	5	2,454.713	5	2,738.495
6	187.2960	6	471.078	6	754.860	6	1,038.642	6	1,322.424	6	1,606.206	6	1,889.987	6	2,173.769	6	2,457.551	6	2,741.333
7	190.1339	7	473.916	7	757.698	7	1,041.480	7	1,325.261	7	1,609.043	7	1,892.825	7	2,176.607	7	2,460.389	7	2,744.171
8	192.9717	8	476.754	8	760.535	8	1,044.317	8	1,328.099	8	1,611.881	8	1,895.663	8	2,179.445	8	2,463.227	8	2,747.009
9	195.8095	9	479.591	9	763.373	9	1,047.155	9	1,330.937	9	1,614.719	9	1,898.501	9	2,182.283	9	2,466.065	9	2,749.847
70	198.6473	**170**	482.429	**270**	766.211	**370**	1,049.993	**470**	1,333.775	**570**	1,617.557	**670**	1,901.339	**770**	2,185.121	**870**	2,468.902	**970**	2,752.684
1	201.4851	1	485.267	1	769.049	1	1,052.831	1	1,336.613	1	1,620.395	1	1,904.176	1	2,187.958	1	2,471.740	1	2,755.522
2	204.3230	2	488.105	2	771.887	2	1,055.669	2	1,339.451	2	1,623.232	2	1,907.014	2	2,190.796	2	2,474.578	2	2,758.360
3	207.1608	3	490.943	3	774.725	3	1,058.506	3	1,342.288	3	1,626.070	3	1,909.852	3	2,193.634	3	2,477.416	3	2,761.198
4	209.9986	4	493.780	4	777.562	4	1,061.344	4	1,345.126	4	1,628.908	4	1,912.690	4	2,196.472	4	2,480.254	4	2,764.036
5	212.8364	5	496.618	5	780.400	5	1,064.182	5	1,347.964	5	1,631.746	5	1,915.528	5	2,199.310	5	2,483.092	5	2,766.873
6	215.6742	6	499.456	6	783.238	6	1,067.020	6	1,350.802	6	1,634.584	6	1,918.366	6	2,202.147	6	2,485.929	6	2,769.711
7	218.5121	7	502.294	7	786.076	7	1,069.858	7	1,353.640	7	1,637.422	7	1,921.203	7	2,204.985	7	2,488.767	7	2,772.549
8	221.3499	8	505.132	8	788.914	8	1,072.696	8	1,356.477	8	1,640.259	8	1,924.041	8	2,207.823	8	2,491.605	8	2,775.387
9	224.1877	9	507.970	9	791.751	9	1,075.533	9	1,359.315	9	1,643.097	9	1,926.879	9	2,210.661	9	2,494.443	9	2,778.225
80	227.0255	**180**	510.807	**280**	794.589	**380**	1,078.371	**480**	1,362.153	**580**	1,645.935	**680**	1,929.717	**780**	2,213.499	**880**	2,497.281	**980**	2,781.063
1	229.8633	1	513.645	1	797.427	1	1,081.209	1	1,364.991	1	1,648.773	1	1,932.555	1	2,216.337	1	2,500.118	1	2,783.900
2	232.7012	2	516.483	2	800.265	2	1,084.047	2	1,367.829	2	1,651.611	2	1,935.393	2	2,219.174	2	2,502.956	2	2,786.738
3	235.5390	3	519.321	3	803.103	3	1,086.885	3	1,370.667	3	1,654.448	3	1,938.230	3	2,222.012	3	2,505.794	3	2,789.576
4	238.3768	4	522.159	4	805.941	4	1,089.722	4	1,373.504	4	1,657.286	4	1,941.068	4	2,224.850	4	2,508.632	4	2,792.414
5	241.2146	5	524.996	5	808.778	5	1,092.560	5	1,376.342	5	1,660.124	5	1,943.906	5	2,227.688	5	2,511.470	5	2,795.252
6	244.0524	6	527.834	6	811.616	6	1,095.398	6	1,379.180	6	1,662.962	6	1,946.744	6	2,230.526	6	2,514.308	6	2,798.089
7	246.8902	7	530.672	7	814.454	7	1,098.236	7	1,382.018	7	1,665.800	7	1,949.582	7	2,233.363	7	2,517.145	7	2,800.927
8	249.7281	8	533.510	8	817.292	8	1,101.074	8	1,384.856	8	1,668.638	8	1,952.419	8	2,236.201	8	2,519.983	8	2,803.765
9	252.5659	9	536.348	9	820.130	9	1,103.912	9	1,387.693	9	1,671.475	9	1,955.257	9	2,239.039	9	2,522.821	9	2,806.603
90	255.4037	**190**	539.186	**290**	822.967	**390**	1,106.749	**490**	1,390.531	**590**	1,674.313	**690**	1,958.095	**790**	2,241.877	**890**	2,525.659	**990**	2,809.441
1	258.2415	1	542.023	1	825.805	1	1,109.587	1	1,393.369	1	1,677.151	1	1,960.933	1	2,244.715	1	2,528.497	1	2,812.279
2	261.0793	2	544.861	2	828.643	2	1,112.425	2	1,396.207	2	1,679.989	2	1,963.771	2	2,247.553	2	2,531.334	2	2,815.116
3	263.9172	3	547.699	3	831.481	3	1,115.263	3	1,399.045	3	1,682.827	3	1,966.609	3	2,250.390	3	2,534.172	3	2,817.954
4	266.7550	4	550.537	4	834.319	4	1,118.101	4	1,401.883	4	1,685.664	4	1,969.446	4	2,253.228	4	2,537.010	4	2,820.792
5	269.5928	5	553.375	5	837.157	5	1,120.938	5	1,404.720	5	1,688.502	5	1,972.284	5	2,256.066	5	2,539.848	5	2,823.630
6	272.4306	6	556.213	6	839.994	6	1,123.776	6	1,407.558	6	1,691.340	6	1,975.122	6	2,258.904	6	2,542.686	6	2,826.468
7	275.2684	7	559.050	7	842.832	7	1,126.614	7	1,410.396	7	1,694.178	7	1,977.960	7	2,261.742	7	2,545.524	7	2,829.305
8	278.1063	8	561.888	8	845.670	8	1,129.452	8	1,413.234	8	1,697.016	8	1,980.798	8	2,264.579	8	2,548.361	8	2,832.143
9	280.9441	9	564.726	9	848.508	9	1,132.290	9	1,416.072	9	1,699.854	9	1,983.635	9	2,267.417	9	2,551.199	9	2,834.981

MASS—AVOIRDUPOIS POUNDS TO KILOGRAMS

[Reduction factor: 1 avoirdupois pound=0.4535924277 kilogram]

Pounds	Kilos	Pounds	Kilos	Pounds	Kilos	Pounds	Kilos	Pounds	Kilos	Pounds	Kilos	Pounds	Kilos	Pounds	Kilos	Pounds	Kilos	Pounds	Kilos
0		**100**	45.35924	**200**	90.71849	**300**	136.07773	**400**	181.43697	**500**	226.79621	**600**	272.15546	**700**	317.51470	**800**	362.87394	**900**	408.23318
1	0.45359	1	45.81284	1	91.17208	1	136.53132	1	181.89056	1	227.24981	1	272.60905	1	317.96829	1	363.32753	1	408.68678
2	.90718	2	46.26643	2	91.62567	2	136.98491	2	182.34416	2	227.70340	2	273.06264	2	318.42188	2	363.78113	2	409.14037
3	1.36078	3	46.72002	3	92.07926	3	137.43851	3	182.79775	3	228.15699	3	273.51623	3	318.87548	3	364.23472	3	409.59396
4	1.81437	4	47.17361	4	92.53286	4	137.89210	4	183.25134	4	228.61058	4	273.96983	4	319.32907	4	364.68831	4	410.04755
5	2.26796	5	47.62720	5	92.98645	5	138.34569	5	183.70493	5	229.06418	5	274.42342	5	319.78266	5	365.14190	5	410.50115
6	2.72155	6	48.08080	6	93.44004	6	138.79928	6	184.15853	6	229.51777	6	274.87701	6	320.23625	6	365.59550	6	410.95474
7	3.17515	7	48.53439	7	93.89363	7	139.25288	7	184.61212	7	229.97136	7	275.33060	7	320.68985	7	366.04909	7	411.40833
8	3.62874	8	48.98798	8	94.34722	8	139.70647	8	185.06571	8	230.42495	8	275.78420	8	321.14344	8	366.50268	8	411.86192
9	4.08233	9	49.44157	9	94.80082	9	140.16006	9	185.51930	9	230.87855	9	276.23779	9	321.59703	9	366.95627	9	412.31552
10	4.53592	**110**	49.89517	**210**	95.25441	**310**	140.61365	**410**	185.97290	**510**	231.33214	**610**	276.69138	**710**	322.05062	**810**	367.40987	**910**	412.76911
1	4.98952	1	50.34876	1	95.70800	1	141.06725	1	186.42649	1	231.78573	1	277.14497	1	322.50422	1	367.86346	1	413.22270
2	5.44311	2	50.80235	2	96.16159	2	141.52084	2	186.88008	2	232.23932	2	277.59857	2	322.95781	2	368.31705	2	413.67629
3	5.89670	3	51.25594	3	96.61519	3	141.97443	3	187.33367	3	232.69292	3	278.05216	3	323.41140	3	368.77064	3	414.12989
4	6.35029	4	51.70954	4	97.06878	4	142.42802	4	187.78727	4	233.14651	4	278.50575	4	323.86499	4	369.22424	4	414.58348
5	6.80389	5	52.16313	5	97.52237	5	142.88161	5	188.24086	5	233.60010	5	278.95934	5	324.31859	5	369.67783	5	415.03707
6	7.25748	6	52.61672	6	97.97596	6	143.33521	6	188.69445	6	234.05369	6	279.41294	6	324.77218	6	370.13142	6	415.49066
7	7.71107	7	53.07031	7	98.42956	7	143.78880	7	189.14804	7	234.50729	7	279.86653	7	325.22577	7	370.58501	7	415.94426
8	8.16466	8	53.52391	8	98.88315	8	144.24239	8	189.60163	8	234.96088	8	280.32012	8	325.67936	8	371.03861	8	416.39785
9	8.61826	9	53.97750	9	99.33674	9	144.69598	9	190.05523	9	235.41447	9	280.77371	9	326.13296	9	371.49220	9	416.85144
20	9.07185	**120**	54.43109	**220**	99.79033	**320**	145.14958	**420**	190.50882	**520**	235.86806	**620**	281.22731	**720**	326.58655	**820**	371.94579	**920**	417.30503
1	9.52544	1	54.88468	1	100.24393	1	145.60317	1	190.96241	1	236.32165	1	281.68090	1	327.04014	1	372.39938	1	417.75863
2	9.97903	2	55.33828	2	100.69752	2	146.05676	2	191.41600	2	236.77525	2	282.13449	2	327.49373	2	372.85298	2	418.21222
3	10.43263	3	55.79187	3	101.15111	3	146.51035	3	191.86960	3	237.22884	3	282.58808	3	327.94733	3	373.30657	3	418.66581
4	10.88622	4	56.24546	4	101.60470	4	146.96395	4	192.32319	4	237.68243	4	283.04167	4	328.40092	4	373.76016	4	419.11940
5	11.33981	5	56.69905	5	102.05830	5	147.41754	5	192.77678	5	238.13602	5	283.49527	5	328.85451	5	374.21375	5	419.57300
6	11.79340	6	57.15265	6	102.51189	6	147.87113	6	193.23037	6	238.58962	6	283.94886	6	329.30810	6	374.66735	6	420.02659
7	12.24700	7	57.60624	7	102.96548	7	148.32472	7	193.68397	7	239.04321	7	284.40245	7	329.76169	7	375.12094	7	420.48018
8	12.70059	8	58.05983	8	103.41907	8	148.77832	8	194.13756	8	239.49680	8	284.85604	8	330.21529	8	375.57453	8	420.93377
9	13.15418	9	58.51342	9	103.87267	9	149.23191	9	194.59115	9	239.95039	9	285.30964	9	330.66888	9	376.02812	9	421.38737
30	13.60777	**130**	58.96702	**230**	104.32626	**330**	149.68550	**430**	195.04474	**530**	240.40399	**630**	285.76323	**730**	331.12247	**830**	376.48171	**930**	421.84096
1	14.06137	1	59.42061	1	104.77985	1	150.13909	1	195.49834	1	240.85758	1	286.21682	1	331.57606	1	376.93531	1	422.29455
2	14.51496	2	59.87420	2	105.23344	2	150.59269	2	195.95193	2	241.31117	2	286.67041	2	332.02966	2	377.38890	2	422.74814
3	14.96855	3	60.32779	3	105.68704	3	151.04628	3	196.40552	3	241.76476	3	287.12401	3	332.48325	3	377.84249	3	423.20174
4	15.42214	4	60.78139	4	106.14063	4	151.49987	4	196.85911	4	242.21836	4	287.57760	4	332.93684	4	378.29608	4	423.65533
5	15.87573	5	61.23498	5	106.59422	5	151.95346	5	197.31271	5	242.67195	5	288.03119	5	333.39043	5	378.74968	5	424.10892
6	16.32933	6	61.68857	6	107.04781	6	152.40706	6	197.76630	6	243.12554	6	288.48478	6	333.84403	6	379.20327	6	424.56251
7	16.78292	7	62.14216	7	107.50141	7	152.86065	7	198.21989	7	243.57913	7	288.93838	7	334.29762	7	379.65686	7	425.01610
8	17.23651	8	62.59576	8	107.95500	8	153.31424	8	198.67348	8	244.03273	8	289.39197	8	334.75121	8	380.11045	8	425.46970
9	17.69010	9	63.04935	9	108.40859	9	153.76783	9	199.12708	9	244.48632	9	289.84556	9	335.20480	9	380.56405	9	425.92329
40	18.14370	**140**	63.50294	**240**	108.86218	**340**	154.22143	**440**	199.58067	**540**	244.93991	**640**	290.29915	**740**	335.65840	**840**	381.01764	**940**	426.37688
1	18.59729	1	63.95653	1	109.31578	1	154.67502	1	200.03426	1	245.39350	1	290.75275	1	336.11199	1	381.47123	1	426.83047
2	19.05088	2	64.41012	2	109.76937	2	155.12861	2	200.48785	2	245.84710	2	291.20634	2	336.56558	2	381.92482	2	427.28407
3	19.50447	3	64.86372	3	110.22296	3	155.58220	3	200.94145	3	246.30069	3	291.65993	3	337.01917	3	382.37842	3	427.73766
4	19.95807	4	65.31731	4	110.67655	4	156.03580	4	201.39504	4	246.75428	4	292.11352	4	337.47277	4	382.83201	4	428.19125
5	20.41166	5	65.77090	5	111.13014	5	156.48939	5	201.84863	5	247.20787	5	292.56712	5	337.92636	5	383.28560	5	428.64484
6	20.86525	6	66.22449	6	111.58374	6	156.94298	6	202.30222	6	247.66147	6	293.02071	6	338.37995	6	383.73919	6	429.09844
7	21.31884	7	66.67809	7	112.03733	7	157.39657	7	202.75582	7	248.11506	7	293.47430	7	338.83354	7	384.19279	7	429.55203
8	21.77244	8	67.13168	8	112.49092	8	157.85016	8	203.20941	8	248.56865	8	293.92789	8	339.28714	8	384.64638	8	430.00562
9	22.22603	9	67.58527	9	112.94451	9	158.30376	9	203.66300	9	249.02224	9	294.38149	9	339.74073	9	385.09997	9	430.45921

50	22.67962	**150**	68.03886	**250**	113.39811	**350**	158.75735	**450**	204.11659	**550**	249.47584	**650**	294.83508	**750**	340.19432	**850**	385.55356	**950**	430.91281
1	23.13321	1	68.49246	1	113.85170	1	159.21094	1	204.57018	1	249.92943	1	295.28867	1	340.64791	1	386.00716	1	431.36640
2	23.58681	2	68.94605	2	114.30529	2	159.66453	2	205.02378	2	250.38302	2	295.74226	2	341.10151	2	386.46075	2	431.81999
3	24.04040	3	69.39964	3	114.75888	3	160.11813	3	205.47737	3	250.83661	3	296.19586	3	341.55510	3	386.91434	3	432.27358
4	24.49399	4	69.85323	4	115.21248	4	160.57172	4	205.93096	4	251.29020	4	296.64945	4	342.00869	4	387.36793	4	432.72718
5	24.94758	5	70.30683	5	115.66607	5	161.02531	5	206.38455	5	251.74380	5	297.10304	5	342.46228	5	387.82153	5	433.18077
6	25.40118	6	70.76042	6	116.11966	6	161.47890	6	206.83815	6	252.19739	6	297.55663	6	342.91588	6	388.27512	6	433.63436
7	25.85477	7	71.21401	7	116.57325	7	161.93250	7	207.29174	7	252.65098	7	298.01022	7	343.36947	7	388.72871	7	434.08795
8	26.30836	8	71.66760	8	117.02685	8	162.38609	8	207.74533	8	253.10457	8	298.46382	8	343.82306	8	389.18230	8	434.54155
9	26.76195	9	72.12120	9	117.48044	9	162.83968	9	208.19892	9	253.55817	9	298.91741	9	344.27665	9	389.63590	9	434.99514
60	27.21555	**160**	72.57479	**260**	117.93403	**360**	163.29327	**460**	208.65252	**560**	254.01176	**660**	299.37100	**760**	344.73025	**860**	390.08949	**960**	435.44873
1	27.66914	1	73.02838	1	118.38762	1	163.74687	1	209.10611	1	254.46535	1	299.82459	1	345.18384	1	390.54308	1	435.90232
2	28.12273	2	73.48197	2	118.84122	2	164.20046	2	209.55970	2	254.91894	2	300.27819	2	345.63743	2	390.99667	2	436.35592
3	28.57632	3	73.93557	3	119.29481	3	164.65405	3	210.01329	3	255.37254	3	300.73178	3	346.09102	3	391.45027	3	436.80951
4	29.02992	4	74.38916	4	119.74840	4	165.10764	4	210.46689	4	255.82613	4	301.18537	4	346.54461	4	391.90386	4	437.26310
5	29.48351	5	74.84275	5	120.20199	5	165.56124	5	210.92048	5	256.27972	5	301.63896	5	346.99821	5	392.35745	5	437.71669
6	29.93710	6	75.29634	6	120.65559	6	166.01483	6	211.37407	6	256.73331	6	302.09256	6	347.45180	6	392.81104	6	438.17029
7	30.39069	7	75.74994	7	121.10918	7	166.46842	7	211.82766	7	257.18691	7	302.54615	7	347.90539	7	393.26463	7	438.62388
8	30.84429	8	76.20353	8	121.56277	8	166.92201	8	212.28126	8	257.64050	8	302.99974	8	348.35898	8	393.71823	8	439.07747
9	31.29788	9	76.65712	9	122.01636	9	167.37561	9	212.73485	9	258.09409	9	303.45333	9	348.81258	9	394.17182	9	439.53106
70	31.75147	**170**	77.11071	**270**	122.46996	**370**	167.82920	**470**	213.18844	**570**	258.54768	**670**	303.90693	**770**	349.26617	**870**	394.62541	**970**	439.98465
1	32.20506	1	77.56431	1	122.92355	1	168.28279	1	213.64203	1	259.00128	1	304.36052	1	349.71976	1	395.07900	1	440.43825
2	32.65865	2	78.01790	2	123.37714	2	168.73638	2	214.09563	2	259.45487	2	304.81411	2	350.17335	2	395.53260	2	440.89184
3	33.11225	3	78.47149	3	123.83073	3	169.18998	3	214.54922	3	259.90846	3	305.26770	3	350.62695	3	395.98619	3	441.34543
4	33.56584	4	78.92509	4	124.28433	4	169.64357	4	215.00281	4	260.36205	4	305.72130	4	351.08054	4	396.43978	4	441.79902
5	34.01943	5	79.37867	5	124.73792	5	170.09716	5	215.45640	5	260.81565	5	306.17489	5	351.53413	5	396.89337	5	442.25262
6	34.47302	6	79.83227	6	125.19151	6	170.55075	6	215.91000	6	261.26924	6	306.62848	6	351.98772	6	397.34697	6	442.70621
7	34.92662	7	80.28586	7	125.64510	7	171.00435	7	216.36359	7	261.72283	7	307.08207	7	352.44132	7	397.80056	7	443.15980
8	35.38021	8	80.73945	8	126.09869	8	171.45794	8	216.81718	8	262.17642	8	307.53567	8	352.89491	8	398.25415	8	443.61339
9	35.83380	9	81.19304	9	126.55229	9	171.91153	9	217.27077	9	262.63002	9	307.98926	9	353.34850	9	398.70774	9	444.06699
80	36.28739	**180**	81.64664	**280**	127.00588	**380**	172.36512	**480**	217.72437	**580**	263.08361	**680**	308.44285	**780**	353.80209	**880**	399.16134	**980**	444.52058
1	36.74099	1	82.10023	1	127.45947	1	172.81871	1	218.17796	1	263.53720	1	308.89644	1	354.25569	1	399.61493	1	444.97417
2	37.19458	2	82.55382	2	127.91306	2	173.27231	2	218.63155	2	263.99079	2	309.35004	2	354.70928	2	400.06852	2	445.42776
3	37.64817	3	83.00741	3	128.36666	3	173.72590	3	219.08514	3	264.44439	3	309.80363	3	355.16287	3	400.52211	3	445.88136
4	38.10176	4	83.46101	4	128.82025	4	174.17949	4	219.53874	4	264.89798	4	310.25722	4	355.61646	4	400.97571	4	446.33495
5	38.55536	5	83.91460	5	129.27384	5	174.63308	5	219.99233	5	265.35157	5	310.71081	5	356.07006	5	401.42930	5	446.78854
6	39.00895	6	84.36819	6	129.72743	6	175.08668	6	220.44592	6	265.80516	6	311.16441	6	356.52365	6	401.88289	6	447.24213
7	39.46254	7	84.82178	7	130.18103	7	175.54027	7	220.89951	7	266.25876	7	311.61800	7	356.97724	7	402.33648	7	447.69573
8	39.91613	8	85.27538	8	130.63462	8	175.99386	8	221.35310	8	266.71235	8	312.07159	8	357.43083	8	402.79008	8	448.14932
9	40.36973	9	85.72897	9	131.08821	9	176.44745	9	221.80670	9	267.16594	9	312.52518	9	357.88443	9	403.24367	9	448.60291
90	40.82332	**190**	86.18256	**290**	131.54180	**390**	176.90105	**490**	222.26029	**590**	267.61953	**690**	312.97878	**790**	358.33802	**890**	403.69726	**990**	449.05650
1	41.27691	1	86.63615	1	131.99540	1	177.35464	1	222.71388	1	268.07312	1	313.43237	1	358.79161	1	404.15085	1	449.51010
2	41.73050	2	87.08975	2	132.44899	2	177.80823	2	223.16747	2	268.52672	2	313.88596	2	359.24520	2	404.60445	2	449.96369
3	42.18410	3	87.54334	3	132.90258	3	178.26182	3	223.62107	3	268.98031	3	314.33955	3	359.69880	3	405.05804	3	450.41728
4	42.63769	4	87.99693	4	133.35617	4	178.71542	4	224.07466	4	269.43390	4	314.79314	4	360.15239	4	405.51163	4	450.87087
5	43.09128	5	88.45052	5	133.80977	5	179.16901	5	224.52825	5	269.88749	5	315.24674	5	360.60598	5	405.96522	5	451.32447
6	43.54487	6	88.90412	6	134.26336	6	179.62260	6	224.98184	6	270.34109	6	315.70033	6	361.05957	6	406.41882	6	451.77806
7	43.99847	7	89.35771	7	134.71695	7	180.07619	7	225.43544	7	270.79468	7	316.15392	7	361.51316	7	406.87241	7	452.23165
8	44.45206	8	89.81130	8	135.17054	8	180.52979	8	225.88903	8	271.24827	8	316.60751	8	361.96676	8	407.32600	8	452.68524
9	44.90565	9	90.26489	9	135.62414	9	180.98338	9	226.34262	9	271.70186	9	317.06111	9	362.42035	9	407.77959	9	453.13884

MASS—KILOGRAMS TO AVOIRDUPOIS POUNDS

[Reduction factor: 1 kilogram=2.204622341 avoirdupois pounds]

Kilos	Pounds	Kilos	Pounds	Kilos	Pounds	Kilos	Pounds	Kilos	Pounds	Kilos	Pounds	Kilos	Pounds	Kilos	Pounds	Kilos	Pounds	Kilos	Pounds
0		100	220.4622	200	440.9245	300	661.3867	400	881.8489	500	1,102.3112	600	1,322.7734	700	1,543.2356	800	1,763.6979	900	1,984.1601
1	2.2046	1	222.6669	1	443.1291	1	663.5913	1	884.0536	1	1,104.5158	1	1,324.9780	1	1,545.4403	1	1,765.9025	1	1,986.3647
2	4.4092	2	224.8715	2	445.3337	2	665.7959	2	886.2582	2	1,106.7204	2	1,327.1826	2	1,547.6449	2	1,768.1071	2	1,988.5694
3	6.6139	3	227.0761	3	447.5383	3	668.0006	3	888.4628	3	1,108.9250	3	1,329.3873	3	1,549.8495	3	1,770.3117	3	1,990.7740
4	8.8185	4	229.2807	4	449.7430	4	670.2052	4	890.6674	4	1,111.1297	4	1,331.5919	4	1,552.0541	4	1,772.5164	4	1,992.9786
5	11.0231	5	231.4853	5	451.9476	5	672.4098	5	892.8720	5	1,113.3343	5	1,333.7965	5	1,554.2588	5	1,774.7210	5	1,995.1832
6	13.2277	6	233.6900	6	454.1522	6	674.6144	6	895.0767	6	1,115.5389	6	1,336.0011	6	1,556.4634	6	1,776.9256	6	1,997.3878
7	15.4324	7	235.8946	7	456.3568	7	676.8191	7	897.2813	7	1,117.7435	7	1,338.2058	7	1,558.6680	7	1,779.1302	7	1,999.5925
8	17.6370	8	238.0992	8	458.5614	8	679.0237	8	899.4859	8	1,119.9481	8	1,340.4104	8	1,560.8726	8	1,781.3349	8	2,001.7971
9	19.8416	9	240.3038	9	460.7661	9	681.2283	9	901.6905	9	1,122.1528	9	1,342.6150	9	1,563.0772	9	1,783.5395	9	2,004.0017
10	22.0462	110	242.5085	210	462.9707	310	683.4329	410	903.8952	510	1,124.3574	610	1,344.8196	710	1,565.2819	810	1,785.7441	910	2,006.2063
1	24.2508	1	244.7131	1	465.1753	1	685.6375	1	906.0998	1	1,126.5620	1	1,347.0243	1	1,567.4865	1	1,787.9487	1	2,008.4110
2	26.4555	2	246.9177	2	467.3799	2	687.8422	2	908.3044	2	1,128.7666	2	1,349.2289	2	1,569.6911	2	1,790.1533	2	2,010.6156
3	28.6601	3	249.1223	3	469.5846	3	690.0468	3	910.5090	3	1,130.9713	3	1,351.4335	3	1,571.8957	3	1,792.3580	3	2,012.8202
4	30.8647	4	251.3269	4	471.7892	4	692.2514	4	912.7136	4	1,133.1759	4	1,353.6381	4	1,574.1004	4	1,794.5626	4	2,015.0248
5	33.0693	5	253.5316	5	473.9938	5	694.4560	5	914.9183	5	1,135.3805	5	1,355.8427	5	1,576.3050	5	1,796.7672	5	2,017.2294
6	35.2740	6	255.7362	6	476.1984	6	696.6607	6	917.1229	6	1,137.5851	6	1,358.0474	6	1,578.5096	6	1,798.9718	6	2,019.4341
7	37.4786	7	257.9408	7	478.4030	7	698.8653	7	919.3275	7	1,139.7898	7	1,360.2520	7	1,580.7142	7	1,801.1765	7	2,021.6387
8	39.6832	8	260.1454	8	480.6077	8	701.0699	8	921.5321	8	1,141.9944	8	1,362.4566	8	1,582.9188	8	1,803.3811	8	2,023.8433
9	41.8878	9	262.3501	9	482.8123	9	703.2745	9	923.7368	9	1,144.1990	9	1,364.6612	9	1,585.1235	9	1,805.5857	9	2,026.0479
20	44.0924	120	264.5547	220	485.0169	320	705.4791	420	925.9414	520	1,146.4036	620	1,366.8659	720	1,587.3281	820	1,807.7903	920	2,028.2526
1	46.2971	1	266.7593	1	487.2215	1	707.6838	1	928.1460	1	1,148.6082	1	1,369.0705	1	1,589.5327	1	1,809.9949	1	2,030.4572
2	48.5017	2	268.9639	2	489.4262	2	709.8884	2	930.3506	2	1,150.8129	2	1,371.2751	2	1,591.7373	2	1,812.1996	2	2,032.6618
3	50.7063	3	271.1685	3	491.6308	3	712.0930	3	932.5553	3	1,153.0175	3	1,373.4797	3	1,593.9420	3	1,814.4042	3	2,034.8664
4	52.9109	4	273.3732	4	493.8354	4	714.2976	4	934.7599	4	1,155.2221	4	1,375.6843	4	1,596.1466	4	1,816.6088	4	2,037.0710
5	55.1156	5	275.5778	5	496.0400	5	716.5023	5	936.9645	5	1,157.4267	5	1,377.8890	5	1,598.3512	5	1,818.8134	5	2,039.2757
6	57.3202	6	277.7824	6	498.2446	6	718.7069	6	939.1691	6	1,159.6314	6	1,380.0936	6	1,600.5558	6	1,821.0181	6	2,041.4803
7	59.5248	7	279.9870	7	500.4493	7	720.9115	7	941.3737	7	1,161.8360	7	1,382.2982	7	1,602.7604	7	1,823.2227	7	2,043.6849
8	61.7294	8	282.1917	8	502.6539	8	723.1161	8	943.5784	8	1,164.0406	8	1,384.5028	8	1,604.9651	8	1,825.4273	8	2,045.8895
9	63.9340	9	284.3963	9	504.8585	9	725.3208	9	945.7830	9	1,166.2452	9	1,386.7075	9	1,607.1697	9	1,827.6319	9	2,048.0942
30	66.1387	130	286.6009	230	507.0631	330	727.5254	430	947.9876	530	1,168.4498	630	1,388.9121	730	1,609.3743	830	1,829.8365	930	2,050.2988
1	68.3433	1	288.8055	1	509.2678	1	729.7300	1	950.1922	1	1,170.6545	1	1,391.1167	1	1,611.5789	1	1,832.0412	1	2,052.5034
2	70.5479	2	291.0101	2	511.4724	2	731.9346	2	952.3969	2	1,172.8591	2	1,393.3213	2	1,613.7836	2	1,834.2458	2	2,054.7080
3	72.7525	3	293.2148	3	513.6770	3	734.1392	3	954.6015	3	1,175.0637	3	1,395.5259	3	1,615.9882	3	1,836.4504	3	2,056.9126
4	74.9572	4	295.4194	4	515.8816	4	736.3439	4	956.8061	4	1,177.2683	4	1,397.7306	4	1,618.1928	4	1,838.6550	4	2,059.1173
5	77.1618	5	297.6240	5	518.0863	5	738.5485	5	959.0107	5	1,179.4730	5	1,399.9352	5	1,620.3974	5	1,840.8597	5	2,061.3219
6	79.3664	6	299.8286	6	520.2909	6	740.7531	6	961.2153	6	1,181.6776	6	1,402.1398	6	1,622.6020	6	1,843.0643	6	2,063.5265
7	81.5710	7	302.0333	7	522.4955	7	742.9577	7	963.4200	7	1,183.8822	7	1,404.3444	7	1,624.8067	7	1,845.2689	7	2,065.7311
8	83.7756	8	304.2379	8	524.7001	8	745.1624	8	965.6246	8	1,186.0868	8	1,406.5491	8	1,627.0113	8	1,847.4735	8	2,067.9358
9	85.9803	9	306.4425	9	526.9047	9	747.3670	9	967.8292	9	1,188.2914	9	,408.7537	9	1,629.2159	9	1,849.6781	9	2,070.1404
40	88.1849	140	308.6471	240	529.1094	340	749.5716	440	970.0338	540	1,190.4961	640	1,410.9583	740	1,631.4205	840	1,851.8828	940	2,072.3450
1	90.3895	1	310.8518	1	531.3140	1	751.7762	1	972.2385	1	1,192.7007	1	1,413.1629	1	1,633.6252	1	1,854.0874	1	2,074.5496
2	92.5941	2	313.0564	2	533.5186	2	753.9808	2	974.4431	2	1,194.9053	2	1,415.3675	2	1,635.8298	2	1,856.2920	2	2,076.7542
3	94.7988	3	315.2610	3	535.7232	3	756.1855	3	976.6477	3	1,197.1099	3	1,417.5722	3	1,638.0344	3	1,858.4966	3	2,078.9589
4	97.0034	4	317.4656	4	537.9279	4	758.3901	4	978.8523	4	1,199.3146	4	1,419.7768	4	1,640.2390	4	1,860.7013	4	2,081.1635
5	99.2080	5	319.6702	5	540.1325	5	760.5947	5	981.0569	5	1,201.5192	5	1,421.9814	5	1,642.4436	5	1,862.9059	5	2,083.3681
6	101.4126	6	321.8749	6	542.3371	6	762.7993	6	983.2616	6	1,203.7238	6	1,424.1860	6	1,644.6483	6	1,865.1105	6	2,085.5727
7	103.6173	7	324.0795	7	544.5417	7	765.0040	7	985.4662	7	1,205.9284	7	1,426.3907	7	1,646.8529	7	1,867.3151	7	2,087.7774
8	105.8219	8	326.2841	8	546.7463	8	767.2086	8	987.6708	8	1,208.1330	8	1,428.5953	8	1,649.0575	8	1,869.5197	8	2,089.9820
9	108.0265	9	328.4887	9	548.9510	9	769.4132	9	989.8754	9	1,210.3377	9	1,430.7999	9	1,651.2621	9	1,871.7244	9	2,092.1866

50	110.2311	**150**	330.6934	**250**	551.1556	**350**	771.6178	**450**	992.0801	**550**	1,212.5423	**650**	1,433.0045	**750**	1,653.4668	**850**	1,873.9290	**950**	2,094.3912
1	112.4357	1	332.8980	1	553.3602	1	773.8224	1	994.2847	1	1,214.7469	1	1,435.2091	1	1,655.6714	1	1,876.1336	1	2,096.5958
2	114.6404	2	335.1026	2	555.5648	2	776.0271	2	996.4893	2	1,216.9515	2	1,437.4138	2	1,657.8760	2	1,878.3382	2	2,098.8005
3	116.8450	3	337.3072	3	557.7695	3	778.2317	3	998.6939	3	1,219.1562	3	1,439.6184	3	1,660.0806	3	1,880.5429	3	2,101.0051
4	119.0496	4	339.5118	4	559.9741	4	780.4363	4	1,000.8985	4	1,221.3608	4	1,441.8230	4	1,662.2852	4	1,882.7475	4	2,103.2097
5	121.2542	5	341.7165	5	562.1787	5	782.6409	5	1,003.1032	5	1,223.5654	5	1,444.0276	5	1,664.4899	5	1,884.9521	5	2,105.4143
6	123.4589	6	343.9211	6	564.3833	6	784.8456	6	1,005.3078	6	1,225.7700	6	1,446.2323	6	1,666.6945	6	1,887.1567	6	2,107.6190
7	125.6635	7	346.1257	7	566.5879	7	787.0502	7	1,007.5124	7	1,227.9746	7	1,448.4369	7	1,668.8991	7	1,889.3613	7	2,109.8236
8	127.8681	8	348.3303	8	568.7926	8	789.2548	8	1,009.7170	8	1,230.1793	8	1,450.6415	8	1,671.1037	8	1,891.5660	8	2,112.0282
9	130.0727	9	350.5350	9	570.9972	9	791.4594	9	1,011.9217	9	1,232.3839	9	1,452.8461	9	1,673.3084	9	1,893.7706	9	2,114.2328
60	132.2773	**160**	352.7396	**260**	573.2018	**360**	793.6640	**460**	1,014.1263	**560**	1,234.5885	**660**	1,455.0507	**760**	1,675.5130	**860**	1,895.9752	**960**	2,116.4374
1	134.4820	1	354.9442	1	575.4064	1	795.8687	1	1,016.3309	1	1,236.7931	1	1,457.2554	1	1,677.7176	1	1,898.1798	1	2,118.6421
2	136.6866	2	357.1488	2	577.6111	2	798.0733	2	1,018.5355	2	1,238.9978	2	1,459.4600	2	1,679.9222	2	1,900.3845	2	2,120.8467
3	138.8912	3	359.3534	3	579.8157	3	800.2779	3	1,020.7401	3	1,241.2024	3	1,461.6646	3	1,682.1268	3	1,902.5891	3	2,123.0513
4	141.0958	4	361.5581	4	582.0203	4	802.4825	4	1,022.9448	4	1,243.4070	4	1,463.8692	4	1,684.3315	4	1,904.7937	4	2,125.2559
5	143.3005	5	363.7627	5	584.2249	5	804.6872	5	1,025.1494	5	1,245.6116	5	1,466.0739	5	1,686.5361	5	1,906.9983	5	2,127.4606
6	145.5051	6	365.9673	6	586.4295	6	806.8918	6	1,027.3540	6	1,247.8162	6	1,468.2785	6	1,688.7407	6	1,909.2029	6	2,129.6652
7	147.7097	7	368.1719	7	588.6342	7	809.0964	7	1,029.5586	7	1,250.0209	7	1,470.4831	7	1,690.9453	7	1,911.4076	7	2,131.8698
8	149.9143	8	370.3766	8	590.8388	8	811.3010	8	1,031.7633	8	1,252.2255	8	1,472.6877	8	1,693.1500	8	1,913.6122	8	2,134.0744
9	152.1189	9	372.5812	9	593.0434	9	813.5056	9	1,033.9679	9	1,254.4301	9	1,474.8923	9	1,695.3546	9	1,915.8168	9	2,136.2790
70	154.3236	**170**	374.7858	**270**	595.2480	**370**	815.7103	**470**	1,036.1725	**570**	1,256.6347	**670**	1,477.0970	**770**	1,697.5592	**870**	1,918.0214	**970**	2,138.4837
1	156.5282	1	376.9904	1	597.4527	1	817.9149	1	1,038.3771	1	1,258.8394	1	1,479.3016	1	1,699.7638	1	1,920.2261	1	2,140.6883
2	158.7328	2	379.1950	2	599.6573	2	820.1195	2	1,040.5817	2	1,261.0440	2	1,481.5062	2	1,701.9684	2	1,922.4307	2	2,142.8929
3	160.9374	3	381.3997	3	601.8619	3	822.3241	3	1,042.7864	3	1,263.2486	3	1,483.7108	3	1,704.1731	3	1,924.6353	3	2,145.0975
4	163.1421	4	383.6043	4	604.0665	4	824.5288	4	1,044.9910	4	1,265.4532	4	1,485.9155	4	1,706.3777	4	1,926.8399	4	2,147.3022
5	165.3467	5	385.8089	5	606.2711	5	826.7334	5	1,047.1956	5	1,267.6578	5	1,488.1201	5	1,708.5823	5	1,929.0445	5	2,149.5068
6	167.5513	6	388.0135	6	608.4758	6	828.9380	6	1,049.4002	6	1,269.8625	6	1,490.3247	6	1,710.7869	6	1,931.2492	6	2,151.7114
7	169.7559	7	390.2182	7	610.6804	7	831.1426	7	1,051.6049	7	1,272.0671	7	1,492.5293	7	1,712.9916	7	1,933.4538	7	2,153.9160
8	171.9605	8	392.4228	8	612.8850	8	833.3472	8	1,053.8095	8	1,274.2717	8	1,494.7339	8	1,715.1962	8	1,935.6584	8	2,156.1206
9	174.1652	9	394.6274	9	615.0896	9	835.5519	9	1,056.0141	9	1,276.4763	9	1,496.9386	9	1,717.4008	9	1,937.8630	9	2,158.3253
80	176.3698	**180**	396.8320	**280**	617.2943	**380**	837.7565	**480**	1,058.2187	**580**	1,278.6810	**680**	1,499.1432	**780**	1,719.6054	**880**	1,940.0677	**980**	2,160.5299
1	178.5744	1	399.0366	1	619.4989	1	839.9611	1	1,060.4233	1	1,280.8856	1	1,501.3478	1	1,721.8100	1	1,942.2723	1	2,162.7345
2	180.7790	2	401.2413	2	621.7035	2	842.1657	2	1,062.6280	2	1,283.0902	2	1,503.5524	2	1,724.0147	2	1,944.4769	2	2,164.9391
3	182.9837	3	403.4459	3	623.9081	3	844.3704	3	1,064.8326	3	1,285.2948	3	1,505.7571	3	1,726.2193	3	1,946.6815	3	2,167.1438
4	185.1883	4	405.6505	4	626.1127	4	846.5750	4	1,067.0372	4	1,287.4994	4	1,507.9617	4	1,728.4239	4	1,948.8861	4	2,169.3484
5	187.3929	5	407.8551	5	628.3174	5	848.7796	5	1,069.2418	5	1,289.7041	5	1,510.1663	5	1,730.6285	5	1,951.0908	5	2,171.5530
6	189.5975	6	410.0598	6	630.5220	6	850.9842	6	1,071.4465	6	1,291.9087	6	1,512.3709	6	1,732.8332	6	1,953.2954	6	2,173.7576
7	191.8021	7	412.2644	7	632.7266	7	853.1888	7	1,073.6511	7	1,294.1133	7	1,514.5755	7	1,735.0378	7	1,955.5000	7	2,175.9623
8	194.0068	8	414.4690	8	634.9312	8	855.3935	8	1,075.8557	8	1,296.3179	8	1,516.7802	8	1,737.2424	8	1,957.7046	8	2,178.1669
9	196.2114	9	416.6736	9	637.1359	9	857.5981	9	1,078.0603	9	1,298.5226	9	1,518.9848	9	1,739.4470	9	1,959.9093	9	2,180.3715
90	198.4160	**190**	418.8782	**290**	639.3405	**390**	859.8027	**490**	1,080.2649	**590**	1,300.7272	**690**	1,521.1894	**790**	1,741.6516	**890**	1,962.1139	**990**	2,182.5761
1	200.6206	1	421.0829	1	641.5451	1	862.0073	1	1,082.4696	1	1,302.9318	1	1,523.3940	1	1,743.8563	1	1,964.3185	1	2,184.7807
2	202.8253	2	423.2875	2	643.7497	2	864.2120	2	1,084.6742	2	1,305.1364	2	1,525.5987	2	1,746.0609	2	1,966.5231	2	2,186.9854
3	205.0299	3	425.4921	3	645.9543	3	866.4166	3	1,086.8788	3	1,307.3410	3	1,527.8033	3	1,748.2655	3	1,968.7278	3	2,189.1900
4	207.2345	4	427.6967	4	648.1590	4	868.6212	4	1,089.0834	4	1,309.5457	4	1,530.0079	4	1,750.4701	4	1,970.9324	4	2,191.3946
5	209.4391	5	429.9014	5	650.3636	5	870.8258	5	1,091.2881	5	1,311.7503	5	1,532.2125	5	1,752.6748	5	1,973.1370	5	2,193.5992
6	211.6437	6	432.1060	6	652.5682	6	873.0304	6	1,093.4927	6	1,313.9549	6	1,534.4171	6	1,754.8794	6	1,975.3416	6	2,195.8039
7	213.8484	7	434.3106	7	654.7728	7	875.2351	7	1,095.6973	7	1,316.1595	7	1,536.6218	7	1,757.0840	7	1,977.5462	7	2,198.0085
8	216.0530	8	436.5152	8	656.9775	8	877.4397	8	1,097.9019	8	1,318.3642	8	1,538.8264	8	1,759.2886	8	1,979.7509	8	2,200.2131
9	218.2576	9	438.7198	9	659.1821	9	879.6443	9	1,100.1065	9	1,320.5688	9	1,541.0310	9	1,761.4933	9	1,981.9555	9	2,202.4177

METRIC AND ENGLISH DISTANCE EQUIVALENTS FOR ATHLETIC EVENTS (TRACK AND FIELD)

1. METRIC UNITS IN ATHLETICS

On November 22, 1932 the Amateur Athletic Union took official action adopting metric distances for track events to be run in athletic meets held under the jurisdiction of that body.

It may be assumed that other athletic bodies will follow the lead of the AAU in adopting metric distances for track events, and also that the use of metric units may be extended to include field events.

With this thought in mind, the following tables of equivalents have been prepared to aid in conversions between the two systems of units and to assist those not entirely familiar with the metric system to gain a mental picture of the various distances expressed in meters.

Explanation: In table 1 are given the equivalent metric and English distances for the principal track events of both indoor and outdoor meets.

In table 2 are given the metric equivalents, expressed in meters, for all distances expressed in feet, inches, and binary fractions of an inch, up to 1 000 feet. This table may be conveniently used to find the metric equivalent of any distance, given to the nearest ⅛ inch, over a range sufficiently wide to include all field events by the simple process of breaking the distance down into convenient parts, finding the equivalent of each part, and adding them together to obtain the total equivalent.

In order to avoid confusion and possible error, it is recommended that all metric distances in track and field events be expressed in meters and decimal fractions of the meter.

Example: Find the metric equivalent of 242 feet, 10⅛ inches.

200 feet	=	60.960 meters
40 feet	=	12.192 meters
2 feet	=	.610 meter
10 inches	=	.254 meter
⅛ inch	=	.003 meter
242 ft, 10⅛ inches	=	74.019 meters

2. DISTANCE EQUIVALENTS

Basis { 1 meter = 39.37 inches = 3.280 8 feet = 1.093 6 yards
1 kilometer = 1 000 meters = 0.621 370 mile }

TABLE 1.—*Track events*

Yards : Meters	Meters : Yards
40 = 36.58	50 = 54.68
50 = 45.72	60 = 65.62
60 = 54.86	65 = 71.08
70 = 64.01	80 = 87.49
75 = 68.58	100 = 109.36
100 = 91.44	110 = 120.30
110 = 100.58	200 = 218.72
120 = 109.73	300 = 328.08
220 = 201.17	400 = 437.44
300 = 274.32	500 = 546.81
440 = 402.34 = ¼ mi	600 = 656.16
600 = 548.64	800 = 874.89
880 = 804.67 = ½ mi	1 000 = 1 093.61
1 000 = 914.40	1 500 = 1 640.42
1 320 = 1 207.01 = ¾ mi	1 600 = 1 749.78

Miles : Meters	Meters : Miles	Yards and inches		Miles (approx.)
1 = 1 609.3	2 000 = 1	427	8	1.24
2 = 3 218.7	2 400 = 1	864	24	1.49
3 = 4 828.0	3 000 = 1	1 520	30	1.86
4 = 6 437.4	3 200 = 1	1 739	20	1.99
5 = 8 046.7				
	5 000 = 3	188	2	3.11
6 = 9 656.1	6 000 = 3	1 281	24	3.73
7 = 11 265.4	10 000 = 6	376	4	6.21
8 = 12 874.8	15 000 = 9	564	6	9.32
9 = 14 484.1				
	20 000 = 12	752	8	12.43
10 = 16 093.5	25 000 = 15	940	10	15.53
15 = 24 140.2	30 000 = 18	1 128	12	18.64
20 = 32 186.9	50 000 = 31	120	20	31.07
25 = 40 233.7				
26 mi and 385 yd = 42 195.1				

TABLE 2.—*Field events*

Feet	: Meters
1	= 0.305
2	= .610
3	= .914
4	= 1.219
5	= 1.524
6	= 1.829
7	= 2.134
8	= 2.438
9	= 2.743
10	= 3.048
20	= 6.096
30	= 9.144
40	= 12.192
50	= 15.240
60	= 18.288
70	= 21.336
80	= 24.384
90	= 27.432
100	= 30.480
200	= 60.960
300	= 91.440
400	= 121.920
500	= 152.400
600	= 182.880
700	= 213.360
800	= 243.840
900	= 274.321

Inches	: Meter
1	= 0.025
2	= .051
3	= .076
4	= .102
5	= .127
6	= .152
7	= .178
8	= .203
9	= .229
10	= .254
11	= .279
12	= .305

Fractions of an inch	: Meter
1/8	= 0.003
1/4	= .006
3/8	= .010
1/2	= .013
5/8	= .016
3/4	= .019
7/8	= .022
1	= .025

In deciding upon the number of decimal places to be retained in the accompanying tables consideration has been given to the precision of measurement of distance and of time as ordinarily carried out in connection with track and field events.

In field events it is customary to measure distances in feet and inches to the nearest 1/8 inch. This corresponds to about 3 millimeters, or 0.003 meter. In order not to sacrifice accuracy in converting these measured distances to meters the metric equivalents are given to the nearest 0.001 meter. This precision is ample, since the equivalents are more precise than the measurements themselves.

In the case of track events consideration has been given to the precision of measurement of both distance and time as ordinarily carried out in connection with these events. The time measurements are, in general, much the less precise of the two.

Time, when taken with stop watches, is ordinarily given to 1/5 second or to 1/10 second. When taken with electrical timing devices it may be given to 1/100 second.

In the dashes, where 1 second represents a distance of approximately 10 yards or 10 meters, 1/10 second represents roughly 1 yard or 1 meter, and 1/100 second a distance of 1/10 yard or 1/10 meter. Obviously, then, there is no need at present to give metric equivalents of distances more precisely than to the nearest 1/10 meter even when the most precise timing methods are used. For distances less than 1 mile they have, however, been given to the nearest 1/100 meter in order to allow for possible future improvement in timing equipment.

GUIDE TO THE USE OF THE INTERNATIONAL SYSTEM OF UNITS with SPECIAL TABLES

PART IV

This section has been included for the convenience of those engaged in the higher echelons of science and technology. It explains the practical application of the rudiments of the newly adopted International System of Units (SI). These tables are more extensive, contain more advanced formulations, and they are broader in scope than those in the earlier part of this book. They will be found to be of great utility by students, teachers, engineers, scientists, and technologists in checking work papers for errors in symbols, expressions of units, base terms of the (SI) system, supplementary units, and in avoiding deviations which may occur through human error.

CONTENTS

1. INTRODUCTION

A most significant advantage of the SI over other measuring systems is that it is coherent, i.e. all SI units are formed by simple multiplication and/or division within a set of seven so-called base units and two supplementary units, and no factors other than the number one are necessary.

For example, in the imperial system the pound-force is often used as a force unit. This means that the factor g (acceleration due to gravity) appears in all calculations involving force, mass and acceleration because one pound-force acting on a mass of one pound does not produce one unit of acceleration but g units. In a similar way, use of the horsepower (defined as 550 foot pounds-force per second) results in the factor 550 appearing in calculations involving force, time and power.

Because of the use of the kilogram-force, the old metric technical system suffered from similar disadvantages and even the centimetre-gram-second system is non-coherent where relationships between mechanical, electrical and heat units are concerned.

The property of coherence and the choice of base units (see Chapter 2) ensure that in the SI there is only one unit for each physical quantity (i.e. length, mass, energy, etc.). The measure of rationalization and simplification brought about by this is clearly illustrated if the joule, which is the SI unit of energy (irrespective of whether it is mechanical, electrical, chemical or any other form of energy) is compared with the present series of energy units in general use; e.g. The British Thermal Unit, foot pound-force, foot poundal, kilopond-metre, calorie, kilowatt-hour, ton (equivalent of TNT), electronvolt, erg, thermie, etc.

At present it is unfortunately not yet possible in practice to use only SI units. All the required units fall into the following three broad groups:

a) SI units

b) Multiples and submultiples of SI units

c) Units which do not constitute part of the SI but which may be used with the SI subject to certain restrictions.

The conversion factors appearing in this publication have been compiled with great care. The Metrication Advisory Board cannot however be held responsible for any errors which may occur.

2. SI-UNITS

2.1. The three classes of SI units

SI units can in turn be divided into three classes:

base units

derived units

supplementary units

The SI is based on a set of seven base units, the metre (m), the kilogram (kg), the second (s), the ampere (A), the kelvin (K), the candela (cd) and the mole (mol) (see 2.2).

The second class, derived units, contains units which are formed by combinations of the base units and supplementary units according to the algebraic relationships between the corresponding physical quantities (see 2.3). As already mentioned, no factor other than the number one is used. For example, speed is defined by the algebraic ratio distance divided by time and consequently the SI unit of speed is the metre per second. Some derived units have special names, and these can again be used to form further derived units.

The third class contains only two units, the radian and the steradian, units of plane angle and solid angle respectively. The CGPM has not yet classified these units as either base units or derived units but refers to them as supplementary units. When calculating with these units they may be regarded as either base units or derived units.

Units which fall into these three classes form a coherent system and only these units are known as SI units. Multiples and submultiples of the SI units which are formed by using the SI prefixes (see Chapter 3) are *not* SI units and should be referred to as *multiples and submultiples of SI units.*

Example:

The metre (m), kilogram (kg), newton (N), watt (W) are the SI units of length, mass, force and power respectively (not main units, principal units, basic units, or anything similar).

The millimetre (mm) and the kilometre (km) are respectively a submultiple and a multiple of the SI unit of length.

2.2. SI base units

TABLE 1

THE SEVEN SI BASE UNITS

Quantity	SI base unit		
	Name	International symbol	Definition (CGPM)
length	metre	m	The metre is the length equal to 1 650 763,73 wavelengths in vacuum of the radiation corresponding to the transition between the levels 2 p_{10} and 5 d_5 of the krypton-86 atom. [11th CGPM (1960), Resolution 6].
mass	kilogram	kg	The kilogram is the mass of the international prototype kilogram recognised by the CGPM and in the custody of the Bureau International des Poids et Mesures, Sèvres, France. [1st CGPM (1889)].
time	second	s	The second is the duration of 9 192 631 770 periods of the radiation corresponding to the transition between the two hyperfine levels of the ground state of the caesium-133 atom. [13th CGPM (1967), Resolution 1].
electric current	ampere	A	The ampere is that constant current which, if maintained in two straight parallel conductors of infinite length, of negligible circular cross-section, and placed one metre apart in vacuum would produce between these conductors a force equal to 2×10^{-7} newton per metre of length. [CIPM (1946), Resolution 2, approved by the 9th CGPM (1948)]
thermodynamic temperature	kelvin	K	The kelvin, unit of thermodynamic temperature, is the fraction 1/273,16 of the thermodynamic temperature of the triple point of water. [13th CGPM (1967), Resolution 4].
luminous intensity	candela	cd	The candela is the luminous intensity, in the perpendicular direction of a surface of 1/600 000 square metre of a blackbody at the temperature of freezing platinum under a pressure of 101 325 newtons per square metre. [13th CGPM (1967), Resolution 5].
amount of substance	mole	mol	The mole is the amount of substance of a system which contains as many elementary entities as there are atoms in 0,012 kg of carbon 12. [14th CGPM (1971), Resolution 3]

Notes:

(i) The unit kelvin and its symbol K are also used to indicate temperature intervals or temperature differences. Besides thermodynamic temperature (Symbol T), expressed in kelvins, Celsius temperature (Symbol t) is also used. Celsius temperature is defined by the equation: $t = T - T_0$
where T_0 = 273,15 K by definition. Celsius temperature is in general expressed in degrees Celsius (Symbol °C). The unit "degree Celsius" is therefore equal to the unit "kelvin" and an interval or difference in Celsius temperature is also expressed in degrees Celsius (°C).
Note that the Celsius temperature of the triple point of water is 0,01°C, which accounts for the factor 273,16 in the definition of the kelvin.

(ii) Whenever the mole is used, the elementary entities must be specified, and may be atoms, molecules, ions, electrons, other particles or specified groups of such particles.

2.3. SI derived units

SI derived units are formed according to the rule in 2.1 and may be devided into three classes. Examples of each are given in the following table.

2.3.1.

TABLE 2

EXAMPLES OF SI DERIVED UNITS WHICH ARE EXPRESSED IN TERMS OF SI BASE UNITS AND SI SUPPLEMENTARY UNITS

Quantity	SI unit	
	Name	Symbol
acceleration	metre per second squared	m/s^2
angular acceleration	radian per second squared	rad/s^2
angular momentum	kilogram metre squared per second	$kg{\cdot}m^2/s$
angular velocity	radian per second	rad/s
area	square metre	m^2
coefficient of linear expansion	1 per kelvin	K^{-1}
concentration (of amount of substance)	mole per cubic metre	mol/m^3
density	kilogram per cubic metre	kg/m^3
diffusion coefficient	metre squared per second	m^2/s
electric current density	ampere per square metre	A/m^2
exposure rate (ionising radiation)	ampere per kilogram	A/kg
kinematic viscosity	metre squared per second	m^2/s
luminance	candela per square metre	cd/m^2
magnetic field strength	ampere per metre	A/m
magnetic moment	ampere metre squared	$A{\cdot}m^2$
mass flow rate	kilogram per second	kg/s
mass per unit area	kilogram per square metre	kg/m^2
mass per unit length	kilogram per metre	kg/m
molality	mole per kilogram	mol/kg
molar mass	kilogram per mole	kg/mol
molar volume	cubic metre per mole	m^3/mol
moment of inertia	kilogram metre squared	$kg{\cdot}m^2$
moment of momentum	kilogram metre squared per second	$kg{\cdot}m^2/s$
momentum	kilogram metre per second	$kg{\cdot}m/s$
radioactivity (disintegration rate)	1 per second	s^{-1}
rotational frequency	1 per second	s^{-1}
specific volume	cubic metre per kilogram	m^3/kg
speed	metre per second	m/s
velocity	metre per second	m/s
volume	cubic metre	m^3
wave number	1 per metre	m^{-1}

2.3.2.

TABLE 3

SI DERIVED UNITS WITH SPECIAL NAMES

Quantity	SI unit			
	Name	Symbol	Expression in terms of other SI units and definition of unit	Expression in terms of SI base units
admittance	siemens	S	Ω^{-1}	$m^{-2}\cdot kg^{-1}\cdot s^{3}\cdot A^{2}$
capacitance	farad	F	C/V	$m^{-2}\cdot kg^{-1}\cdot s^{4}\cdot A^{2}$
conductance	siemens	S	Ω^{-1}	$m^{-2}\cdot kg^{-1}\cdot s^{3}\cdot A^{2}$
electrical resistance	ohm	Ω	V/A	$m^{2}\cdot kg\cdot s^{-3}\cdot A^{-2}$
electric charge	coulomb	C	A·s	s·A
electric flux	coulomb	C	A·s	s·A
electric potential	volt	V	W/A	$m^{2}\cdot kg\cdot s^{-3}\cdot A^{-1}$
electromotive force	volt	V	W/A	$m^{2}\cdot kg\cdot s^{-3}\cdot A^{-1}$
energy	joule	J	N·m	$m^{2}\cdot kg\cdot s^{-2}$
energy flux	watt	W	J/s	$m^{2}\cdot kg\cdot s^{-3}$
flux of displacement	coulomb	C	A·s	s·A
force	newton	N	$kg\cdot m/s^{2}$	$m\cdot kg\cdot s^{-2}$
frequency	hertz	Hz	s^{-1}	s^{-1}
illuminance	lux	lx	lm/m^{2}	$m^{-2}\cdot cd\cdot sr$
impedance	ohm	Ω	V/A	$m^{2}\cdot kg\cdot s^{-3}\cdot A^{-2}$
inductance	henry	H	Wb/A (V·s/A)	$m^{2}\cdot kg\cdot s^{-2}\cdot A^{-2}$
luminous flux	lumen	lm	cd·sr	cd·sr
magnetic flux	weber	Wb	V·s	$m^{2}\cdot kg\cdot s^{-2}\cdot A^{-1}$
magnetic flux density	tesla	T	Wb/m^{2}	$kg\cdot s^{-2}\cdot A^{-1}$
magnetic induction	tesla	T	Wb/m^{2}	$kg\cdot s^{-2}\cdot A^{-1}$
magnetic polarization	tesla	T	Wb/m^{2}	$kg\cdot s^{-2}\cdot A^{-1}$
permeance	henry	H	Wb/A (V·s/A)	$m^{2}\cdot kg\cdot s^{-2}\cdot A^{-2}$
potential difference	volt	V	W/A	$m^{2}\cdot kg\cdot s^{-3}\cdot A^{-1}$
power	watt	W	J/s	$m^{2}\cdot kg\cdot s^{-3}$
pressure	pascal	Pa	N/m^{2}	$m^{-1}\cdot kg\cdot s^{-2}$
quantity of electricity	coulomb	C	A·s	s·A
quantity of heat	joule	J	N·m	$m^{2}\cdot kg\cdot s^{-2}$
reactance	ohm	Ω	V/A	$m^{2}\cdot kg\cdot s^{-3}\cdot A^{-2}$
stress	pascal	Pa	N/m^{2}	$m^{-1}\cdot kg\cdot s^{-2}$
susceptance	siemens	S	Ω^{-1}	$m^{-2}\cdot kg^{-1}\cdot s^{3}\cdot A^{2}$
weight	newton	N	$kg\cdot m/s^{2}$	$m\cdot kg\cdot s^{-2}$
work	joule	J	N·m	$m^{2}\cdot kg\cdot s^{-2}$

Notes:

(i) The expressions in the fourth column represent the definitions of the respective units in symbolic form and were obtained in accordance with the rule in 2.1. For instance, the quantity force is defined as the product of mass and acceleration ($F = m\cdot a$) so the definition of the unit of force, the newton (N) is given by $1\ N = 1\ kg\cdot m/s^2$.

Where more than one algebraic relationship exists between applicable units, for example $V = I\cdot R$ (potential difference equals current multiplied by resistance) or $V = P/I$ (potential difference equals power divided by current) then the definition of the corresponding unit in the fourth column is the one which is officially accepted by the CGPM.

(ii) Mechanical energy must not be expressed in newton meters (N·m) but only in joules (J). The former unit is used only for torque or moment of force.

(iii) In the expressions for the lumen (lm) and lux (lx) in the fifth column, the steradian (sr) is treated as a base unit.

2.3.3. TABLE 4

EXAMPLES OF SI DERIVED UNITS WHICH ARE EXPRESSED IN TERMS OF SI DERIVED UNITS WITH SPECIAL NAMES AS WELL AS SI BASE UNITS AND SI SUPPLEMENTARY UNITS

Quantity	SI unit		
	Name	Symbol	Expression in terms of SI base units and SI supplementary units
absorbed dose	joule per kilogram	J/kg	$m^2 \cdot s^{-2}$
coefficient of heat transfer	watt per metre squared kelvin	$W/m^2 \cdot K$	$kg \cdot s^{-3} \cdot K^{-1}$
conductivity	siemens per metre	S/m	$m^{-3} \cdot kg^{-1} \cdot s^3 \cdot A^2$
dielectric polarization	coulomb per square metre	C/m^2	$m^{-2} \cdot s \cdot A$
displacement	coulomb per square metre	C/m^2	$m^{-2} \cdot s \cdot A$
dynamic viscosity	pascal second	Pa·s	$m^{-1} \cdot kg \cdot s^{-1}$
electric charge density	coulomb per cubic metre	C/m^3	$m^{-3} \cdot s \cdot A$
electric dipole moment	coulomb metre	C·m	m·s·A
electric field strength	volt per metre	V/m	$m \cdot kg \cdot s^{-3} \cdot A^{-1}$
energy density	joule per cubic metre	J/m^3	$m^{-1} \cdot kg \cdot s^{-2}$
entropy	joule per kelvin	J/K	$m^2 \cdot kg \cdot s^{-2} \cdot K^{-1}$
exposure (ionizing radiation)	coulomb per kilogram	C/kg	$kg^{-1} \cdot s \cdot A$
heat capacity	joule per kelvin	J/K	$m^2 \cdot kg \cdot s^{-2} \cdot K^{-1}$
heat flux density	watt per square metre	W/m^2	$kg \cdot s^{-3}$
magnetic dipole moment	weber metre	Wb·m	$m^3 \cdot kg \cdot s^{-2} \cdot A^{-1}$
molar energy	joule per mole	J/mol	$m^2 \cdot kg \cdot s^{-2} \cdot mol^{-1}$
molar entropy	joule per mole kelvin	J/mol·K	$m^2 \cdot kg \cdot s^{-2} \cdot K^{-1} \cdot mol^{-1}$
molar heat capacity	joule per mole kelvin	J/mol·K	$m^2 \cdot kg \cdot s^{-2} \cdot K^{-1} \cdot mol^{-1}$
moment of force	newton metre	N·m	$m^2 \cdot kg \cdot s^{-2}$
permeability	henry per metre	H/m	$m \cdot kg \cdot s^{-2} \cdot A^{-2}$
permittivity	farad per metre	F/m	$m^{-3} \cdot kg^{-1} \cdot s^4 \cdot A^2$
radiant intensity	watt per steradian	W/sr	$m^2 \cdot kg \cdot s^{-3} \cdot sr^{-1}$
reluctance	1 per henry	H^{-1}	$m^{-2} \cdot kg^{-1} \cdot s^2 \cdot A^2$
resistivity	ohm metre	Ω·m	$m^3 \cdot kg \cdot s^{-3} \cdot A^{-2}$
specific energy	joule per kilogram	J/kg	$m^2 \cdot s^{-2}$
specific entropy	joule per kilogram kelvin	J/kg·K	$m^2 \cdot s^{-2} \cdot K^{-1}$
specific heat capacity	joule per kilogram kelvin	J/kg·K	$m^2 \cdot s^{-2} \cdot K^{-1}$
specific latent heat	joule per kilogram	J/kg	$m^2 \cdot s^{-2}$
surface charge density	coulomb per square metre	C/m^2	$m^{-2} \cdot s \cdot A$
surface tension	newton per metre	N/m	$kg \cdot s^{-2}$
thermal conductivity	watt per metre kelvin	W/m·K	$m \cdot kg \cdot s^{-3} \cdot K^{-1}$
torque	newton metre	N·m	$m^2 \cdot kg \cdot s^{-2}$

Notes:

(i) In the interests of uniformity it is preferable to define, as far as possible, the SI derived units in accordance with the combinations given in the above tables. This does not, however, exclude the possibility of using other equivalent combinations in special cases. In education for example, it may be convenient to define electric field strength initially in terms of the force experienced by unit charge and to use the corresponding unit newton per coulomb (N/C) instead of volt per metre (V/m).
Note that: 1 V/m = 1 W/A·m = 1 J/s·A·m = 1 N·m/s·A·m = 1 N/C .

(ii) Torque or moment of force should not be expressed in joules (J) but only in newton metres (N·m) .

(iii) The values of certain so-called dimensionless quantities such as index of refraction, relative permeability and relative permittivity are expressed as pure numbers. Each of these quantities does have an SI unit but this consists of the ratio of two identical SI units and so may be expressed by the number 1.

2.4. SI supplementary units

TABLE 5

SI SUPPLEMENTARY UNITS

Quantity	SI unit		
	Name	Symbol	Definition
plane angle	radian	rad	The radian is the plane angle between two radii of a circle which cut off on the circumference an arc equal in length to the radius.
solid angle	steradian	sr	The steradian is the solid angle which, having its vertex in the centre of a sphere, cuts off an area of the surface of the sphere equal to that of a square with sides of length equal to the radius of the sphere.

2.5. Method of writing SI units and symbols

2.5.1. Symbols

2.5.1.1. Symbols are typed or printed in roman (upright) letters and are always lower case letters except if the name of the corresponding unit is derived from the name of a person, in which case the symbol, or the first letter thereof if it consists of more than one letter, is a capital letter. Examples: metre (m), second (s), but watt (W), hertz (Hz), etc. These rules apply always, even if the rest of the subject matter is printed in upper case or other letter types as in headings of tables, paragraphs, etc.

Examples:

LENGTH IN m or *Length in* m
FREQUENCY IN Hz or *Frequency in* Hz

2.5.1.2. Symbols have no plural, e.g. 1 m, 2 m, etc.

2.5.1.3. Note that it is incorrect to refer to the symbols as abbreviations. Symbols are therefore not followed by a fullstop unless they occur at the end of a sentence, in which case it is recommended that a space be left between the symbol and the fullstop to stress that the fullstop is not part of the symbol, e.g. The length of an object is 2,5 m .

2.5.1.4. The product of two or more symbols is preferably indicated by means of a point. This point is preferably raised above the line as in N·m but may be also printed on the line as in N.m if the former position is not easily reproduced, for example in typed work. If there is no risk of confusion with other symbols then the point may be replaced by a space (N m).

2.5.1.5. A solidus, horizontal line, or negative powers may be used to indicate division of symbols, e.g. m/s, $\frac{\mathrm{m}}{\mathrm{s}}$ or $\mathrm{m{\cdot}s^{-1}}$. Only one solidus may be used, e.g. J/mol·K or $\mathrm{m/s^2}$ but *not* J/mol/K or m/s/s. It is recommended that negative powers be used in more complicated cases which may lack clarity in other notation. Brackets may be also used.

Example:

$\mathrm{m{\cdot}kg{\cdot}s^{-3}{\cdot}A^{-1}}$ or $\mathrm{m{\cdot}kg/(s^3{\cdot}A)}$.

2.5.2. Units written in full

2.5.2.1. When written out in full the name of a unit is written in lower case letters irrespective of whether it is derived from the name of a person or not. Exceptions are if the unit appears at the beginning of a sentence in which case the first letter is a capital letter, and if the whole of the subject matter is printed in capital letters, e.g. LENGTH IN METRES.

2.5.2.2. The plural form is used in English but not in Afrikaans, e.g. 1 metre, 2 metres in English, but 1 meter, 2 meter in Afrikaans.

2.5.2.3. In English, multiplication of units is indicated by means of a space (newton metre), but in Afrikaans it is indicated by means of a hyphen (newton-meter).

2.5.2.4. Division is indicated by the word per in both English and Afrikaans, e.g. joule per mole kelvin (English) and joule per mol-kelvin (Afrikaans). Only one 'per' is used. Where this leads to complications such as with acceleration (metre per second 'per' second) expressions such as metre per second squared are recommended.

3. MULTIPLES AND SUBMULTIPLES OF SI UNITS

3.1. SI prefixes

The use of only SI units would sometimes lead to inconveniently large or small values. The distance from Johannesburg to Durban for example would be approximately 657 000 m or 6,57 x 10^5 m . Similarly, the wavelength of green light would be about 0,000 000 51 m or 5,1 x 10^{-7} m. To avoid this cumbersome notation a series of international prefixes is used to form decimal multiples and submultiples of SI units. Using these prefixes, the above examples become simply 657 km and 510 nm respectively.

3.1.1.

TABLE 6

PREFERRED SI PREFIXES

Factor		Factor in words	SI prefix	SI symbol
1 000 000 000 000	or 10^{12}	billion	tera-	T
1 000 000 000	or 10^{9}	milliard	giga-	G
1 000 000	or 10^{6}	million	mega-	M
1 000	or 10^{3}	thousand	kilo-	k
0,001	or 10^{-3}	thousandth	milli-	m
0,000 001	or 10^{-6}	millionth	micro-	μ
0,000 000 001	or 10^{-9}	milliardth	nano-	n
0,000 000 000 001	or 10^{-12}	billionth	pico-	p
0,000 000 000 000 001	or 10^{-15}	billiardth	femto-	f
0,000 000 000 000 000 001	or 10^{-18}	trillionth	atto-	a

3.1.2.

TABLE 7

OTHER SI PREFIXES

Factor			Factor in words	SI prefix	SI symbol
100	or	10^2	hundred	hecto-	h
10	or	10^1	ten	deca-	da
0,1	or	10^{-1}	tenth	deci-	d
0,01	or	10^{-2}	hundredth	centi-	c

3.2. Style and use of SI prefixes and their symbols

3.2.1. Base units, derived units with special names, and supplementary units

3.2.1.1. All of these SI units have single names and symbols i.e. they are not written as combinations of other units or their respective symbols. Multiples and submultiples of these SI units are formed by writing the prefixes and their symbols directly in front of the names of the relevant units or their symbols respectively, without a space, point or other mark of separation. In this manner new names and symbols are obtained, e.g. millimetre (mm), megagram (Mg), picofarad (pF), etc.

3.2.1.2. Combinations of prefixes must not be used, e.g.:

nanometre (nm) but *not* millimicrometre (mμm)
picofarad (pF) but *not* micromicrofarad ($\mu\mu$F)

3.2.1.3. As mentioned in 3.2.1.1, the combination of a unit and a prefix (or their symbols) is a new entity, consequently raising to a power has bearing on the combination as a whole and not merely on the unit which carries the exponent.

Examples:

1 $\mu s^{-1} = (10^{-6} s)^{-1} = 10^6 s^{-1}$ but not $10^{-6} s^{-1}$
1 $mm^2 = 1\ (mm)^2 = (10^{-3} m)^2 = 10^{-6} m^2$
and not $10^{-3} m^2$, i.e. mm^2 stands for square millimetre and *not* milli (square metre).

Note that multiples and submultiples of units which represent other units raised to powers (e.g. area: m^2, volume: m^3, etc.) can therefore not be formed by using rule 3.2.1.1. Multiples and submultiples of the SI units of area and volume are thus in effect powers of multiples and submultiples of the SI unit of length.

3.2.1.4. The kilogram (kg) represents a second exception to rule 3.2.1.1. This unit, although it is an SI unit, contains the prefix kilo- for historical reasons. It has thus far not been possible to remove this anomaly because agreement on a suitable name for the kilogram could as yet not be reached at an international level.

On account of rule 3.2.1.2, no prefixes may be attached to the word kilogram and multiples and submultiples of this unit are therefore formed by attaching the prefixes to the word gram (g), e.g. gram (g) and not a millikilogram (mkg) and milligram (mg) and not a microkilogram (μkg).

Note that the gram (g) is not an SI unit but that it is a submultiple of an SI unit.

3.2.2. Other derived units

3.2.2.1. If multiples and submultiples of units in this category (i.e. SI derived units which are written as combinations of other SI units) are formed in accordance with rule 3.2.1.1, they may be regarded as 'true' multiples and submultiples of SI units. The prefixes or their symbols are then written directly before the first unit or symbol of the combination respectively, e.g. millimetre per second (mm/s); micro-ohm metre (μΩ·m); kilowatt per square metre (kW/m^2).

In this form, the prefix may be regarded as applying to the combination (an SI unit) which follows it as a whole, i.e. kW/m^2 = k(W/m^2) and no further rules or exceptions other than those given in 3.2.1 are necessary.

3.2.2.2. In practice it is sometimes convenient to form combinations of multiples and submultiples of SI base units and SI supplementary units, or SI derived units with special names, e.g. micrometre per millisecond (μm/ms) or milliohm millimetre (mΩ·mm). Some experts are of the opinion that these forms are not true multiples and submultiples of SI units but only combinations of multiples and/or submultiples. It is recommended that they should be converted to forms such as in 3.2.2.1, e.g.

$1\ \mu m/ms = 10^{-6}\ m/10^{-3}\ s = 10^{-3}\ m/s = 1\ mm/s$
$1\ m\Omega \cdot mm = 10^{-3}\Omega \cdot 10^{-3}\ m = 10^{-6}\Omega \cdot m = 1\ \mu\Omega \cdot m$

This is recommended because the use of the combinations referred to can often lead to complications (see 3.2.3.4).

Note:

The use of prefixes in the denominator of combination units which consist of the quotients of SI units is particularly deprecated and must be avoided wherever possible [e.g. kilonewton per meter (kN/m) and not newton per millimetre (N/mm)], except of course in the case of the kilogram.

3.2.3. Preferred multiples and submultiples

3.2.3.1. One of the most important aims of the SI is the simplification and rationalization of units for measuring and calculations. (See for example the remarks on units of energy in the introduction). The number of multiples and submultiples is accordingly restricted by giving preference to the use of prefixes which represent steps of 1 000 (Table 6). Wherever possible, preference must be given to these prefixes and the other prefixes (Table 7) should be avoided. The latter are given merely for completeness and because they are still sometimes required in specialized fields. For example the centimetre (cm) is used in the clothing and textile industries and therefore also for the related dimensions of the human body. It should preferably not be introduced elsewhere.

3.2.3.2. Unfortunately it is not possible to apply the above preference consistently because SI units which are raised to the second or higher powers [e.g. area (m^2), volume (m^3), etc.] do not have special names and the preferred usage, because of 3.2.1.3, results in unpractically large steps, e.g.

Area:

$1\ km^2 = 10^6\ m^2$
$1\ m^2 = 10^6\ mm^2$, etc.

Volume:

$1\ km^3 = 10^9\ m^3$
$1\ m^3 = 10^9\ mm^3$, etc.

It is therefore sometimes necessary to use prefixes from Table 7, usually only centi- (c) and deci- (d). The other two are seldom required and should be avoided.

Note that in the case of volume the use of centi- (c) and deci- (d) again leads to steps of 1 000.

$1\ m^3 = 1\,000\ dm^3$
$1\ dm^3 = 1\,000\ cm^3$
$1\ cm^3 = 1\,000\ mm^3$

It is in particular, where multiples and submultiples of SI units are raised to powers of 2 or higher (area, volume, etc.) and are formed with non-preferred prefixes in combinations such as those in 3.2.2.2, that the complications which are mentioned there occur. If such combinations are used then additional factors of 10 and 100 are necessary to bring the data into a suitable form for calculations. (1 $g/dm^2 = 10^{-1} kg/m^2 = 100 \times 10^{-3} kg/m^2$; 1 $g{\cdot}cm^2 = 10^{-7} kg{\cdot}m^2$). Inevitably the chance of error is considerably greater than it would be if only factors of 1 000; 1 000 000; 1/1 000 etc. are used.

3.2.3.3. Although the above use does solve the problem of too big steps to a certain extent, it is never the less recommended that such data be given in terms of SI units where possible. The slight inconvenience brought about by the resulting small or large values is compensated for by the uniformity and simplicity of calculations (see 6.4).

Examples:

Write (a) 350 cm^2 as 0,035 m^2 or $3,5 \times 10^{-2}$ m^2
(b) 75 dm^3 as 0,075 m^3 or $7,5 \times 10^{-2}$ m^3

3.2.3.4. Because the consistent use of only SI units offers considerable advantages in calculations, it is preferable to express data in SI units or at least in terms of preferred multiples and submultiples of SI units. It is then only necessary to multiply by integral powers of 1 000 to bring the data into a suitable form for calculations.

4. UNITS WHICH DO NOT FORM PART OF THE SI BUT WHICH MAY BE USED WITH THE SI SUBJECT TO CERTAIN RESTRICTIONS

4.1. At this stage it is unfortunately not yet possible to use only SI units and multiples and submultiples of SI units. This chapter contains units which may be used with the SI as follows:

(a) Units which will probably be used with the SI for a long time yet;

(b) units which for the time being may be used with the SI in special fields;

(c) units which may be used together with the SI only temporarily.

Nevertheless, it is strongly recommended that where possible, and in particular where scientific and technological calculations are under consideration, such units be avoided in favour of the corresponding SI units, or where data are provided in such units that these data be first converted to the correct SI units before being used in calculations.

4.2. Units for general use together with the SI

The following units (Table 8) have attained such wide general usage that there is little hope of eliminating them within the foreseeable future. The CGPM recognises that they play such an important part that they must be retained for general use together with the SI. The CGPM does not list the gon in this table (see however note vii).

TABLE 8

UNITS WHICH ARE USED TOGETHER WITH THE SI

Quantity	Name and symbol	Value in terms of the SI	Note
volume fluid	litre (ℓ)	1 ℓ = 1 dm^3 = 0,001 m^3	(v)
mass	metric ton (t)	1 t = 1 Mg = 1 000 kg	(iv)
plane angle	degree (°)	1 ° = π/180 rad	(i), (ii)
plane angle	minute (′)	1 ′ = π/10 800 rad	(i)
plane angle	second (″)	1″ = π/648 000 rad	(i)
plane angle	gon (gon)	1 gon = π/200 rad	(iii), (vii)
time	minute (min)	1 min = 60 s	(i)
time	hour (h)	1 h = 3,6 ks = 3 600 s	(i)
time	day (d)	1 d = 86,4 ks = 86 400 s	(i)
time	week	1 week = 604,8 ks = 604 800 s	(i), (vi)
volume fluid	litre (ℓ)	1 ℓ = 1 dm^3 = 0,001 m^3	(v)

Notes:

(i) SI prefixes are not used with the minute, hour, day, week, degree, minute and second (plane angle).

(ii) When the degree (°) is used it is recommended that where possible decimal subdivisions be used in place of minutes (′) and seconds (″).

(iii) Note that there are 400 gon in a circle, i.e. 100 gon in a right angle.

(iv) The metric ton is a 'commercial' unit, i.e. it is intended for everyday use but not for use in scientific and technological calculations. Only the SI prefixes kilo-, mega-, giga-, and tera- are used with the metric ton.

(v) The litre is likewise a 'commercial' unit. Only SI prefixes from Table 6 and deca- and hecto- may be used with the litre. The litre is by definiton [Resolution 6, 12th CGPM, (1964)] exactly equal to 1 dm^3. The international symbol for the litre is a lower case l. This can however sometimes be confused with the number 1 in typed and printed matter and it is therefore recommended that a 'script' ℓ such as shown here be used for this purpose. Suppliers of typewriters supply this symbol and in addition also the degree- (^{0}), second power- (^{2}), and third power- (^{3}) symbols.

(vi) The month and year of the Gregorian calendar are also used with the SI although they are variable and therefore strictly speaking cannot be classified as units.

(vii) The gon enjoys equal status with the degree in the EEC directive. ISO refers to it under the following remark: The units degree and grade (or gon), with their decimal subdivisions, are recommended for use when the unit radian is not suitable. Note that ISO recognises both names (grade and gon) in English. In this document gon is used throughout for uniformity in English and Afrikaans.

4.3. Units for use with SI in specialised fields

The following units (Table 9) may be used with the SI in specialised fields of scientific research.

TABLE 9

UNITS WHICH MAY BE USED TOGETHER WITH THE SI IN SPECIALISED FIELDS

Quantity	Name	Symbol	*Approximate value in terms of the SI	Restriction
distance	astronomical unit	AU	1 AU = 149,6 Gm = 1,496 x 10^{11} m	Only in astronomy
distance	parsec	pc	1 pc = 30 857 Tm = 3,085 7 x 10^{16} m	Only in astronomy
energy	electronvolt	eV	1 eV = 0,160 219 aJ = 1,602 19 x 10^{-19} J	Only in the atomic and nuclear sciences
mass	atomic mass unit (unified)	u	1 u = 1,660 53 x 10^{-27} kg	Only in the natural sciences

*The values of these units in terms of the SI have to be determined experimentally and are therefore not known exactly.

Definitions of units in Table 9:

(i) 1 eV is the energy acquired by an electron in passing through a potential difference of 1 V in vacuum.

(ii) 1 u is equal to the fraction $\frac{1}{12}$ of the mass of an atom of the nuclide carbon-12.

(iii) 1 AU is the length of the radius of the unperturbed circular orbit of a body of negligible mass moving around the sun with a sidereal angular velocity of 17,202 098 950 mrad/d of 86 400 ephemeris seconds.

(iv) 1 pc is the distance at which 1 AU subtends an angle of 1″.

4.4. Units which are only recognised temporarily

Due to existing practice the following units (Table 10) are still used with the SI and some of them have been classified by the CGPM as units to be used with the SI for a limited time. In some respects South Africa is ahead of the rest of the world in its application of the SI so these units must be used only where absolutely necessary, and where they are already in use it is strongly recommended that steps be taken now to replace them by the correct SI units or multiples and submultiples thereof. Table 10 also contains units which are not recognised by the CGPM for use with the SI. They must of necessity be included at this stage because they are recognised by ISO, legislation in other countries, and in order to take account of South African conditions.

TABLE 10

UNITS WHICH MAY ONLY BE USED WITH SI FOR A LIMITED TIME

Quantity	Name	Symbol	Value in terms of the SI	Restriction
absorbed dose	rad	rad,rd	1 rad = 10 mJ/kg = 10^{-2} J/kg	Only in radiation dosimetry
acceleration	gal	Gal	1 Gal = 10 mm/s^2 = 0,01 m/s^2	Only in geodesy and geophysics
apparent power	volt ampere	VA		Only in electrical engineering
area	hectare*	ha	1 ha = 10 000 m^2	Only in surveying and mapping
area	barn	b	1 b = 100 fm^2 = 10^{-28} m^2	Only in nuclear and atomic physics
distance	nautical mile (international)		1 nautical mile = 1,852 km = 1 852 m	Only in nautical and aeronautical navigation
exposure	röntgen	R	1 R = 258 μC/kg = 2,58 x 10^{-4} C/kg	Only in radiation dosimetry
length	ångström	Å	1 Å = 0,1 nm = 10^{-10} m	
mass per unit length	tex	tex	1 tex = 1 mg/m = 10^{-6} kg/m	Only in the textile industry
pressure**	bar	bar	1 bar = 100 kPa = 10^5 Pa	
radioactivity	curie	Ci	1 Ci = 37 ns^{-1} = 3,7 x 10^{10} s^{-1}	Only in connection with radioactive radiation
reactive power	volt ampere reactive	var		Only in electrical engineering
refractive power	dioptre	δ	1 δ = 1 m^{-1}	Only in optics
rotational frequency	revolution per minute	r/min	1 r/min = 16,666 667 ks^{-1} = 16,666 667 x 10^{-3} s^{-1}	
speed, velocity	knot (international)		1 knot = 1 nautical mile per hour = 0,514 444 m/s	Only in nautical and aeronautical navigation

Notes:

* The are (a) has been omitted from this table because it is not recognised in South Africa in connection with surveying and mapping.

** The name pascal, symbol Pa, was given to the SI unit for pressure and stress (the newton per square metre) by the CGPM in October 1971 and the bar was classified amongst the units which may only be used with the SI for a limited time. The CGPM took this decision with the motivation that it should contribute to the elimination of the bar. It is therefore strongly recommended that from now on all new instruments for measuring pressure be ordered with pascal calibrations (or suitable multiples and submultiples thereof). Existing instruments which are already calibrated in bars can easily be changed by simply replacing the word bar on the dial with x 100 kPa without altering any of the calibrations.

5. UNITS WHICH MUST NOT BE USED

5.1. The following units which have already been referred to in the foreword and introduction must in no circumstances be introduced in South Africa and must be eliminated as quickly as possible if they are already in use. This is particularly important where new measuring instruments and equipment are concerned as such instruments and equipment will otherwise have to be converted again at a later stage. Some of these units are not recognised at all by the CGPM, the use of others is discouraged, and in many countries legislation has been adopted or is being prepared to provide for their systematic elimination.

TABLE 11

Quantity	Name and symbol	Correct SI unit	Correct SI symbol	Value in terms of the SI
dynamic viscosity	poise (P)	pascal second	Pa·s	1 P = 100 mPa·s = 0,1 Pa·s
[1] energy	kilogram-force metre (kgf·m)	joule	J	1 kgf·m = 9,806 65 J
energy	erg (erg)	joule	J	1 erg = 100 nJ = 10^{-7} J
energy	calorie (cal) (I.T.)	joule	J	1 cal = 4,186 8 J
energy	litre atmosphere (ℓ·atm)	joule	J	1 ℓ·atm = 101,328 J
[1] force	kilogram-force (kgf)	newton	N	1 kgf = 9,806 65 N
force	dyne (dyn)	newton	N	1 dyn = 10 μN = 10^{-5} N
frequency	cycle per second (c/s)	hertz	Hz	1 c/s = 1 Hz
illuminance	phot (ph)	lux	lx	1 ph = 10 klx = 10^4 lx
kinematic viscosity	stokes (St)	metre squared per second	m^2/s	1 St = 100 mm^2/s = 10^{-4} m^2/s
length	fermi	metre	m	1 fermi = 1 fm (femtometre) = 10^{-15} m
length	micron (μ)	metre	m	1 μ = μm = 10^{-6} m
length	X-unit	metre	m	1 X-unit = 100,2 fm = 1,002 x 10^{-13} m (approx.)
luminance	stilb (sb)	candela per square metre	cd/m^2	1 sb = 10 kcd/m^2 = 10^4 cd/m^2
[2] magnetic induction, magnetic flux	gauss (Gs, G)	tesla	T	1 Gs corresponds to 100 μT = 10^{-4} T
magnetic induction, magnetic flux	gamma (γ)	tesla	T	1 γ = 1 nT = 10^{-9} T
[2] magnetic field strength	oersted (Oe)	ampere per metre	A/m	1 Oe corresponds to 1 000/4π A/m
[2] magnetic flux	maxwell (Mx)	weber	Wb	1 Mx corresponds to 10 nWb = 10^{-8} Wb

Quantity	Name and symbol	Correct SI unit	Correct SI symbol	Value in terms of the SI
mass	gamma (γ)	kilogram	kg	$1\ \gamma = \mu g = 10^{-9}$ kg
[1] moment of force	metre kilogram-force (m·kgf)	newton metre	N·m	1 m·kgf = 9,806 65 N·m
power	metric horsepower (CH., PS)	watt	W	1 C.V. = 735,499 W
[1] pressure	kilogram-force per square centimetre (kgf/cm^2)	pascal	Pa	$1\ kgf/cm^2$ = 98,066 5 kPa = $9,806\ 65 \times 10^4$ Pa
pressure	[4] torr (Torr)	pascal	Pa	1 Torr = 133,322 37 Pa
pressure	[4] millimetre of mercury (mmHg)	pascal	Pa	1 mmHg = 133,322 39 Pa
pressure	[3] atmosphere (standard) (atm)	pascal	Pa	1 atm = 101,325 kPa = 101 325 Pa
pressure	pièze	pascal	Pa	1 pièze = 1 kPa = 10^3 Pa
[1] stress	kilogram-force per square centi-metre (kgf/cm^2)	pascal	Pa	$1\ kgf/cm^2$ = 98,066 5 kPa = $9,806\ 65 \times 10^4$ Pa
[1] torque	metre kilogram-force (m·kgf)	newton metre	N·m	1 m·kgf = 9,806 65 N·m
volume	stere (st)	cubic metre	m^3	1 st = 1 m^3
volume	lambda	cubic metre	m^3	$1\ \lambda = 1\ mm^3 = 10^{-9}\ m^3$
[1] weight	kilogram-force (kgf)	newton	N	1 kgf = 9,806 65 N
weight	dyne (dyn)	newton	N	1 dyn = 10 μN = 10^{-5} N

Notes:

[1] (i) The kilogram-force (kgf), kilogram-force per square centimetre (kgf/cm^2), metre kilogram-force (m·kgf), and kilogram-force metre (kgf·m) are often given incorrectly as kilogram (kg), kilogram per square centimetre (kg/cm^2), metre kilogram (m·kg), and kilogram metre (kg·m) respectively, which is one of the reasons for the confusion between mass and weight.

(ii) Where errors of the order of 2% are acceptable, measuring instruments and equipment calibrated in kgf, kgf/cm^2, m·kgf, or kgf·m are easily converted by replacing these symbols by x 10 N, x 100 kPa, x 10 N·m and 10 J respectively, without altering any calibrations.

(iii) The kilogram-force is also sometimes known as the kilopond, symbol kp. It therefore also occurs in the units kp/cm^2, m·kp, and kp·m .

[2] This unit is part of the so-called 'electromagnetic' 3 dimensional c.g.s. system and cannot strictly speaking be compared with the corresponding unit of the 4 dimensional SI .

[3] The CGPM classifies the atmosphere (standard) under the units which may be used together with the SI for a limited time and the EEC places it with units of which the status will be reviewed before 1978. These classifications are a recognition of the fact that this unit is still used to a large extent in metric countries. Since this applies to a much smaller degree to South Africa the introduction of this unit is strongly deprecated and it is recommended that it be eliminated as quickly as possible where already in use.

[4] See footnotes 5 and 7 of table 12.

6. ADDITIONAL RULES, NOTATIONS AND HINTS

6.1. Notation for numbers

6.1.1. South Africa adopted the comma as decimal indicator in April 1971 to conform with other metric countries. The comma is used for all numbers including amounts of money, e.g.

15,7 kg
R22,75

6.1.2. For numbers smaller than 1 a zero *must* precede the decimal indicator, e.g.

0,5 m (or 500 mm) but *not* ,5 m
R0,25 (or 25c) but *not* R,25

6.1.3. If there are more than three numerals on either side of the decimal indicator then these numerals are divided into groups of three by means of spaces (counting from the decimal sign) to promote readability, e.g.

1 725 352,684 901

6.1.4. This notation (using spaces) suggests a further advantage to be gained by using the preferred SI prefixes. Because these prefixes represent steps of 1 000 it follows that successive multiples and submultiples are obtained by simply moving the decimal indicator to the next applicable space, e.g.

1 725 352 mm = 1 725,352 m = 1,725 352 km .

By contrast, the use of other SI prefixes requires the regrouping of the numerals, e.g.

1 725 352 mm = 172 535,2 cm .

6.1.5. Whenever a number is written with the symbol of a unit, a space is left between the number and the symbol, e.g.

150 m, *not* 150m;
50 kg, *not* 50kg .

6.2. Style

6.2.1. Because all the SI symbols are now internationally recognized and are also the same in all languages, it is preferable to use the symbols rather than the fully written forms in the interests of simplification and to reduce the amount of writing, e.g.

500 m instead of five hundred metres.

6.2.2. Numbers and names of units or names of numbers and symbols should not be mixed, i.e. 50 kg but not 50 kilograms and fifty kilograms but not fifty kg .

6.2.3. It must again be emphasized that only SI symbols or recognized symbols of units which are used with the SI may be used. Here are some examples of incorrect practices which already occur frequently.

Incorrect	*Correct*
25 cc	25 cm^3
25 kgrm.	25 kg
3 000 r.p.m.	3 000 r/min
750 sq.m.	750 m^2
50 kilos	50 kg
25 L.	25 ℓ
100 kph or 100 kmh	100 km/h

6.2.4. Contractions (abbreviations) and fully written terms must never be used together with symbols. Contractions may be used with fully written terms however if the preferred description presents problems for any reason.

Examples:

Correct

5 m^2 or five square metres
8 m^3 or eight cubic metres
2 Mℓ or two megalitres
60 km/h or sixty kilometres per hour

Incorrect	*Acceptable*
5 sq. m	five sq. metres
8 cub. m	eight cub. metres
2 M litres	–
60 km p. hour or 60 km per hour	–

6.2.5. Vulgar fractions have little status in the decimal metric system and in particular they must not be used with units, i.e. 0,5 ℓ or 500 mℓ but not ½ ℓ, and 0,25 kg or 250 g but not ¼ kg .

However, this does obviously not apply to cases where units are not involved, i.e. it is not necessary to refer to half an apple as 0,5 of an apple or a quarter-line as a 0,25 line.

6.3. Direct translation

In the imperial system, pressure is usually expressed in pounds-force per square inch, or more often incorrectly as pounds per square inch. Direct translation yields kilogram-force per square centimetre (or kilogram per square centimetre), a unit which must *not* be used in South Africa. It is therefore always necessary to consult the list of correct SI units when quantities are being metricated.

6.4. The use of SI units in calculations

6.4.1. The simplification brought about by the consistent use of SI units can be illustrated by the following simple example:

If P = power of a single cylinder engine
p = mean effective piston pressure
l = length of piston stroke
A = piston area
n = number of working strokes per unit time,

then these quantities are usually expressed in the following units:

P in kW, p in kPa, l in mm, A in mm^2, and n in strokes per minute.

It is possible to compile a formula in which these values can be used directly without changing them, and it turns out to be:

$$P = \frac{p\,l\,A\,n}{6 \times 10^{10}} \text{ kW}$$

The factor below the line and the fact that the relevant multiple of each of the other factors has to be remembered make it more difficult to memorize the formula. Moreover, the factor below the line has to be changed if the formula is used with one or more of the data expressed in terms of a different multiple or submultiple of the respective unit.

A better approach is to compile a formula in which all the relevant quantities are in SI units, in this case:

$$P = p\,l\,A\,n\,,$$

which also ensures that the answer automatically comes out in the applicable SI unit. This also avoids the necessity of applying time-wasting dimensional checks. All that is now necessary is to remember a simple formula and to consistently apply the rule that all data have to be expressed in SI units.

6.4.2. Example:

	Given	expressed in SI units
P =	900 kPa	900×10^3 Pa
l =	85 mm	85×10^{-3} m
A =	7 000 mm^2	$7\,000 \times 10^{-6}$ m^2
n =	720 strokes per minute	12 s^{-1}

Then

$$P = 900 \times 10^3 \times 85 \times 10^{-3} \times 7\,000 \times 10^{-6} \times 12 = 6\,426 \text{ W}.$$

This answer is now easily expressed in terms of the preferred multiple (kW) by simply inserting a comma in the space, i.e.

$$P = 6{,}426 \text{ kW}.$$

7. METRICATING, CONVERTING, AND ROUNDING-OFF

7.1. Metricating

Metrication does not mean that for example an inch is now called 25,4 mm . This kind of direct conversion leads to complicated numbers and inhibits the utilization of the advantages presented by the decimal character of the SI; in particular it retains the ratios 3; 4; 8; 12; 16; 32; etc. which are an inherent property of the imperial system, at the cost of the much simpler 2; 5; 10 pattern of the decimal system.

The first requirement when metricating any product is therefore a critical examination of the practicality and/or economics of converting its dimensions, mass, volume or other applicable quantities to rounded "metric" values. Metric sizes of a large number of basic materials and other products have already been decided on and the relevant information may be obtained from the Metrication Department of the SABS. Taking full advantage of the opportunity presented by metrication to introduce standardization and to reduce unnecessary variety, together with the resulting simplification of calculations for which the SI is noted, result in advantages which exceed by far the short-term costs of the conversion.

7.2. Converting and rounding-off

Naturally it is not possible to achieve the above ideal in all cases, and some existing data have to be converted from sheer necessity.

However, it is imperative that the precision of the original value be maintained as closely as possible during conversion.

It is just as senseless and misleading to convert a value such as 3 lb 5 oz, which is accurate to the nearest ounce, to 1,402 524 73 kg, (which represents conversion to the nearest 10 μg, i.e. more than a million times finer than the original value), as it is to convert and round off to the nearest millimetre the dimensions of a machine part which has to be manufactured to an accuracy of say the nearest thousandth of an inch to fit. Meaningful rounding-off can be achieved as follows:

7.2.1. Definitions

7.2.1.1. Significant figures

Any numeral (including zero) which is necessary to describe a number fully is a significant figure.

Examples:

Value	Value known or ascertainable to the nearest	Significant figures	Number of significant figures
0,029 50	0,000 01	2; 9; 5 and 0	4
0,029 5	0,000 1	2; 9; 5	3
10,029 5	0,000 1	1; 0; 0; 2; 9 and 5	6
5 677,0	0,1	5; 6; 7; 7 and 0	5
567 700	1	5; 6; 7; 7; 0 and 0	6
	10	5; 6; 7; 7 and 0	5
	100	5; 6; 7 and 7	4
$5{,}677 \times 10^3$	1	5; 6; 7 and 7	4
5,677	0,001	5; 6; 7 and 7	4

Notes:

(i) Note that zeros which are not preceded by other numerals are *not* significant figures as these zeros merely determine the order of magnitude of the number, e.g. 0,029 5 can be written as $2{,}95 \times 10^{-2}$.

(ii) Zeros at the end of a number often give rise to lack of clarity if the significant portion is not specified, e.g. 567 700 can be known to the nearest 1; 10 or 100 , depending on the circumstances. To avoid this ambiguity it is recommended that such numbers be written in terms of powers of ten with the necessary number of decimal places to indicate the significant part. 567 000 becomes $5{,}677\,00 \times 10^5$; $5{,}677\,0 \times 10^5$ or $5{,}677 \times 10^5$ to indicate significant parts to the nearest 1; 10 or 100 respectively.

7.2.1.2. Precision

The precision of a value is equal to 0,5 of unity in the last significant figure.

Examples:

Value	Rewritten to indicate significant part	Precision
0,029 50	$2,950 \times 10^{-2}$	$0,000\,5 \times 10^{-2}$ or 0,000 005
0,029 5	$2,95 \times 10^{-2}$	$0,005 \times 10^{-2}$ or 0,000 05
1 250 000	say $1,250 \times 10^{6}$	$0,000\,5 \times 10^{6}$ or 500
$32 \pm 0,02$	$32,00 \pm 0,02$	0,005
$32^{+0,15}_{-0,2}$	$32,00^{+0,15}_{-0,20}$	0,005

7.2.1.3. Fineness of rounding

The fineness of rounding is the decimal multiple or submultiple to which the value has been rounded off.

Examples:

Fineness of rounding	Rounded-off values of	
	125,154 5 and	4 657,5
0,001	125,154	–
0,01	125,15	–
0,1	125,2	4 657,5
1	125	4 658
10	$1,3 \times 10^{2}$	$4,66 \times 10^{3}$
100	1×10^{2}	$4,7 \times 10^{3}$

7.2.1.4. Meaningful Conversion

The following method will ensure that the precision of a converted value is as close as possible to that of the original value.

Note:

Conversion of dimensions of components with tolerances in industry is a specialized field which is not treated here because the requirement of carrying over the precision has to be abandoned in order to ensure that components which will be manufactured according to the converted values will still fit. It is recommended that ISO/R370 'Conversion of toleranced dimensions from inches to millimetres and vice versa' be used for this purpose.

Procedure:

(i) Rewrite the value which has to be converted to indicate the significant part.

(ii) Convert the rewritten value with the aid of the applicable conversion factor. This factor may be rounded off to simplify the arithmetic but it must contain at least two more significant figures than the rewritten value which is being converted.

(iii) Convert also the fineness of rounding of the original value and write the result in the form $A \times 10^{b}$ with $1 \leqslant A < 10$, i.e. write it as a number between 1 and 10 with the appropriate power of 10.

(iv) Compare A with $\sqrt{10}$, i.e. 3,162 277 . . . .

The required fineness of rounding is

10^{b} if A is less than or equal to $\sqrt{10}$ and
10^{b+1} if A is greater than $\sqrt{10}$.

7.2.2. Examples

7.2.2.1. Value for which the precision is known exactly

Say the capacity of a drum is given as $50^{+2}_{-1,5}$ gallons (UK).

(i) Rewrite as $50,0^{+2,0}_{-1,5}$ (i.e. fineness of rounding is 0,1 and the precision is 0,05).

(ii) The conversion factor is 4,546 087 litres per gallon, which can be rounded off to 4,546 1 (five significant figures). Converted values are:
$50,0 \times 4,546\,1 = 227,305$
$2,0 \times 4,546\,1 = 9,092\,2$
$1,5 \times 4,546\,1 = 6,819\,15$

(iii) The converted fineness of rounding is:
$0,1 \times 4,546\,1 = 0,454\,61$
$= 4,546\,1 \times 10^{-1}$

(iv) 4,546 1 is greater than $\sqrt{10}$, therefore the fineness of rounding required is $10^{0} = 1$. Rounding is therefore to the nearest litre.

The converted value is thus:

227^{+9}_{-7} ℓ

7.2.2.2. Value for which the precision can be determined from the context or from experience

Say the tensile stress of a metal is given as 30 tons-force per square inch and the usual practice is to determine tensile stresses of this magnitude to the nearest 112 pounds-force (hundred weight-force) per square inch.

(i) Rewrite as 30,0 tons-force per square inch. (Precision is 112 pounds-force per square inch which is equal to 0,05 tons-force per square inch. Fineness of rounding is therefore $2 \times 0,05 = 0,1$).

(ii) The conversion factor is $1,544\,43 \times 10^7$ pascals per ton-force per square inch, which can be rounded off to $1,544\,4 \times 10^7$. The converted value is:
$30,0 \times 1,544\,4 \times 10^7 = 46,332 \times 10^7$ Pa

(iii) The converted fineness of rounding is:
$0,1 \times 1,544\,4 \times 10^7 = 1,544\,4 \times 10^6$

(iv) 1,544 4 is less than $\sqrt{10}$ so the required fineness of rounding is 10^6.

The converted value is thus
$46,3 \times 10^7$ Pa or 463 MPa.

Note:

Suppose that for some reason or another, the value of 30,0 tons-force per square inch represent an absolute minimum, i.e. the converted stress may not be less than this in any circumstances. Then the converted value of $46,332 \times 10^7$ Pa must be rounded to $46,4 \times 10^7$ Pa and *not* $46,3 \times 10^7$ as is obtained by applying the usual rules. In a similar manner, absolute maximum values must be rounded to the next smaller value applicable.

7.3. Summary of various cases

The following examples show how various cases may be approached.

7.3.1. Estimated values

A person who thinks in terms of the imperial system may estimate the height of a post to be say approximately 3 feet, but a person who is 'metrically' oriented will definitely not estimate the height of the post to be approximately 914,4 mm, and it is also unlikely that he would estimate it to be approximately 900 mm. 'Approximately 1 m' is more realistic. Values which are obviously rough estimates must not be converted and rounded-off by using the methods of 7.2.1.5 but can be metricated in a proper manner.

7.3.2. More precise values

7.3.2.1. If however the value of 3 feet was found by measuring to a known accuracy of $\pm \frac{1}{16}$ inch and it is required that the precision of the converted value be as close as possible to this then the method of 7.2.1.5 must be used and will result in the value of 914 mm.

7.3.2.2. If the value is required by a manufacturer for the purposes of metricating a product then consideration must first be given to the feasibility of converting to the logical metric length of 1 m, or if a standard metric series of sizes already exists for that product, to the nearest value in that series. Conversion and rounding-off must only be considered if none of the above procedures is possible.

7.3.3. It cannot be sufficiently emphasized that proper metrication and metric thinking are absolutely essential in order to be able to enjoy the full advantages of the SI, and that first consideration must always be given to proper metrication. This applies to the manufacturers of products, the people responsible for metricating laws, ordinances, and regulations, the teacher or lecturer who has to metricate problems, the journalist and writer, and in fact to every person in the Republic.

8. DEFINITIONS OF DERIVED SI UNITS HAVING SPECIAL NAMES

Quantity	SI unit and symbol	Definition
frequency	hertz, Hz	The hertz is the frequency of a periodic phenomenon of which the period is 1 s.
force	newton, N	The newton is that force which, when applied to a body having a mass of 1 kg, gives it an acceleration of 1 m/s^2.
pressure, stress	pascal, Pa	The pascal is the pressure or stress which results when a force of 1 N is applied evenly and perpendicularly to an area of 1 m^2.
energy, work	joule, J	The joule is the work done when the point of application of a force of 1 N is displaced through a distance of 1 m in the direction of the force.
power	watt, W	The watt is the power which results in the production of energy at the rate of 1 J/s.
electric charge	coulomb, C	The coulomb is the quantity of electric charge transported in 1 s by a current of 1 A.
electric potential difference (electromotive force)	volt, V	The volt is the potential difference between two points of a conducting wire carrying a constant current of 1 A, when the power dissipated between these points is equal to 1 W.
capacitance	farad, F	The farad is the capacitance of a capacitor between the plates of which there appears a potential difference of 1 V when it is charged with an electric charge equal to 1 C.
electric resistance	ohm, Ω	The ohm is the electric resistance between two points of a conductor when a constant potential difference of 1 V, applied between these two points, produces a current of 1 A, the conductor not being the source of any electromotive force.
conductance	siemens, S	The siemens is the conductance of a conductor of resistance 1 Ω and it is numerically equal to 1 Ω^{-1}.
magnetic flux	weber, Wb	The weber is the magnetic flux which, linking a circuit of one turn, produces in it an electromotive force of 1 V if it is reduced to zero at a uniform rate in 1 s.
magnetic flux density	tesla, T	The tesla is a magnetic flux density of 1 Wb/m^2.
inductance	henry, H	The henry is the inductance of a closed circuit in which an electromotive force of 1 V is produced if the electric current in the circuit varies uniformly at the rate of 1 A/s.
luminous flux	lumen, lm	The lumen is the luminous flux emitted within a solid angle of 1 sr by a point source having a uniform intensity of 1 cd.
illumination	lux, lx	The lux is an illumination of 1 lm/m^2.

9. CONVERSION FACTORS

9.1. Table 12 gives conversion factors in terms of the applicable SI units. After converting and rounding-off as described in Chapter 7, the relevant multiple or submultiple of the SI unit obtained may be chosen as discussed in Chapter 3.

Example:

Conversion of 31,00 tons-force (long tons-force) per square inch.

$31{,}00 \times 1{,}544\,43 \times 10^7 = 47{,}877\,33 \times 10^7$ Pa

which is rounded to $47{,}88 \times 10^7$ Pa .

This can be written as $478{,}8 \times 10^6$ Pa which is equal to 478,8 MPa .

9.2. Table 13 gives conversion factors in terms of some of the non-SI units which are still being used with the SI. As already mentioned, these units should be avoided if it is at all possible.

9.3. Conversion factors preceded by an asterisk are exact.

9.4. TABLE 12

ALPHABETICAL LIST OF CONVERSIONS TO SI UNITS (SI SYMBOLS ARE GIVEN IN BRACKETS)

To convert from	to	multiply by
abampere (biot)	ampere (A)	*1×10^1
abcoulomb	coulomb (C)	*1×10^1
abfarad	farad (F)	*1×10^9
abhenry	henry (H)	*1×10^{-9}
abmho	siemens (S)	*1×10^9
abohm	ohm (Ω)	*1×10^{-9}
abvolt	volt (V)	*1×10^{-8}
acre	square metre (m^2)	*$4{,}046\,86 \times 10^3$
acre foot	cubic metre (m^3)	$1{,}233\,482 \times 10^3$
ampere (international of 1948)	ampere (A)	$9{,}998\,35 \times 10^{-1}$
ampere hour	coulomb (C)	*$3{,}6 \times 10^3$
ångström	metre (m)	*1×10^{-10}
are	square metre (m^2)	*1×10^2
astronomical unit	metre (m)	$1{,}496 \times 10^{11}$
atmosphere (standard)	pascal (Pa)	*$1{,}013\,25 \times 10^5$
atmosphere (technical) (= 1 kgf/cm^2)	pascal (Pa)	$9{,}806\,65 \times 10^4$
atomic mass unit (unified)	kilogram (kg)	$1{,}660\,531 \times 10^{-27}$
bar	pascal (Pa)	*1×10^5
barn	square metre (m^2)	*1×10^{-28}
barrel (petroleum; 42 gallons (USA); liquid)	cubic metre (m^2)	$1{,}589\,873 \times 10^{-1}$
biot (abampere)	ampere (A)	*1×10^1
British thermal unit (mean)	joule (J)	$1{,}055\,87 \times 10^3$
British thermal unit (39°F)	joule (J)	$1{,}059\,67 \times 10^3$
British thermal unit (60°F)	joule (J)	$1{,}054\,68 \times 10^3$

To convert from	to	multiply by
British thermal unit (international table)	joule (J)	$1{,}055\,056 \times 10^{3}$
British thermal unit (thermochemical)	joule (J)	$1{,}054\,35 \times 10^{3}$
[1] British thermal unit (international table) inch per hour square foot degree Fahrenheit	watt per metre kelvin (W/m·K)	$1{,}442\,279 \times 10^{-1}$
[1] British thermal unit (thermochemical) inch per hour square foot degree Fahrenheit	watt per metre kelvin (W/m·K)	$1{,}441\,314 \times 10^{-1}$
British thermal unit (international table) per cubic foot	joule per cubic metre (J/m^3)	$3{,}725\,89 \times 10^{4}$
British thermal unit (thermochemical) per cubic foot	joule per cubic metre (J/m^3)	$3{,}723\,402 \times 10^{4}$
[1] British thermal unit (international table) per cubic foot degree Fahrenheit	joule per cubic metre kelvin ($J/m^3 \cdot K$)	$6{,}706\,607 \times 10^{4}$
[1] British thermal unit (thermochemical) per cubic foot degree Fahrenheit	joule per cubic metre kelvin ($J/m^3 \cdot K$)	$6{,}702\,118 \times 10^{4}$
British thermal unit (international table) per hour	watt (W)	$2{,}930\,711 \times 10^{-1}$
British thermal unit (thermochemical) per hour	watt (W)	$2{,}928\,751 \times 10^{-1}$
[1] British thermal unit (international table) per hour square foot degree Fahrenheit	watt per square metre kelvin ($W/m^2 \cdot K$)	$5{,}678\,263 \times 10^{0}$
[1] British thermal unit (thermochemical) per hour square foot degree Fahrenheit	watt per square metre kelvin ($W/m^2 \cdot K$)	$5{,}674\,466 \times 10^{0}$
British thermal unit (international table) per pound	joule per kilogram (J/kg)	$*2{,}326 \times 10^{3}$
British thermal unit (thermochemical) per pound	joule per kilogram (J/kg)	$2{,}324\,444 \times 10^{3}$
[1] British thermal unit (international table) per pound degree Fahrenheit	joule per kilogram kelvin (J/kg·K)	$*4{,}186\,8 \times 10^{3}$

[1] Because the degree Fahrenheit in the denominator represents a temperature interval it follows from 2.2 note (i) that precisely the same conversion factor applies if the kelvin (K) in the SI unit is replaced by the degree Celsius (°C).

To convert from	to	multiply by
[1] British thermal unit (thermochemical) per pound degree Fahrenheit	joule per kilogram kelvin (J/kg·K)	4,184 x 10^3
British thermal unit (international table) per square foot	joule per square metre (J/m^2)	1,135 653 x 10^4
British thermal unit (thermochemical) per square foot	joule per square metre (J/m^2)	1,134 893 x 10^4
British thermal unit (international table) per square foot hour	watt per square metre (W/m^2)	3,154 592 x 10^0
British thermal unit (thermochemical) per square foot hour	watt per square metre (W/m^2)	3,152 481 x 10^0
bushel (UK)	cubic metre (m^3)	3,636 872 x 10^{-2}
bushel (USA)	cubic metre (m^3)	3,523 907 x 10^{-2}
[2] calorie (15°C)	joule (J)	4,185 5 x 10^0
[2] calorie (20°C)	joule (J)	4,181 9 x 10^0
[2] calorie (mean)	joule (J)	4,190 02 x 10^0
[2] calorie (international table)	joule (J)	*4,186 8 x 10^0
[2] calorie (thermochemical)	joule (J)	*4,184 x 10^0
[2] calorie (international table) per hour	watt (W)	*1,163 x 10^{-3}
[2] calorie (thermochemical) per square centimetre minute	watt per square metre (W/m^2)	6,973 333 x 10^2
candela per square foot	candela per square metre (cd/m^2)	1,076 39 x 10^1
carat (metric)	kilogram (kg)	*2 x 10^{-4}
centimetre of mercury (0°C)	pascal (Pa)	1,333 223 9 x 10^3
centimetre of water (4°C)	pascal (Pa)	9,806 38 x 10^1
centipoise	pascal second (Pa·s)	*1 x 10^{-3}
centistokes	metre squared per second (m^2/s)	*1 x 10^{-6}
chain (Gunter's or surveyors')	metre (m)	*2,011 68 x 10^1
chain (Ramden's or engineers')	metre (m)	*3,048 x 10^1
cheval vapeur or metric horsepower	watt (W)	7,354 99 x 10^2
clo	kelvin square metre per watt ($K·m^2/W$)	2,003 712 x 10^{-}

[1] Because the degree Fahrenheit in the denominator represents a temperature interval it follows from 2.2 note (i) that precisely the same conversion factor applies if the kelvin (K) in the SI unit is replaced by the degree Celsius (°C).

[2] Note that reference is often made to the calorie when in actual fact the kilocalorie (also called Calorie, large calorie, or kilogram calorie) is meant. (1 kilocalorie = 1 000 calories).

To convert from	to	multiply by
clusec	pascal cubic metre per second (Pa·m³/s)	1,333 224 x 10^{-6}
coulomb (international of 1948)	coulomb (C)	9,998 35 x 10^{-1}
cubic foot	cubic metre (m³)	2,831 685 x 10^{-2}
cubic foot per minute	cubic metre per second (m³/s)	4,719 474 x 10^{-4}
cubic foot per second	cubic metre per second (m³/s)	2,831 685 x 10^{-2}
cubic inch	cubic metre (m³)	*1,638 706 4 x 10^{-5}
cubic inch per minute	cubic metre per second (m³/s)	2,731 177 x 10^{-7}
cubic inch per pound	cubic metre per kilogram (m³/kg)	3,612 729 x 10^{-5}
cubic yard	cubic metre (m³)	7,645 549 x 10^{-1}
cubic yard per minute	cubic metre per second (m³/s)	1,274 258 x 10^{-2}
cup (UK)	cubic metre (m³)	2,841 306 x 10^{-4}
cup (USA)	cubic metre (m³)	2,365 882 x 10^{-4}
curie	per second (s^{-1})	*3,7 x 10^{10}
cusec (see cubic foot per second)		
cusec hour	cubic metre (m³)	1,019 407 x 10^{2}
cycle per second	hertz (Hz)	*1 x 10^{0}
day (mean solar)	second (s)	8,64 x 10^{4}
day (sidereal)	second (s)	8,616 409 x 10^{4}
debye	coulomb metre (C·m)	3,335 64 x 10^{-30}
degree (angle)	radian (rad)	1,745 329 x 10^{-2} (=*π/180)
degree Celsius (particular temperature)	kelvin (K)	use $T = t_C + 273,15$
degree Celsius (temperature interval)	kelvin (K)	*1 x 10^{0}
degree Fahrenheit (particular temperature)	kelvin (K)	use $T = (t_F + 459,67)/1,8$
degree Fahrenheit (temperature interval)	kelvin (K)	0,555 556 (=*1/1,8)
degree Rankine (particular temperature and temperature interval)	kelvin (K)	0,555 556 (=*1/1,8)
dioptre	per metre (m^{-1})	*1 x 10^{0}
drachm (fluid) (UK)	cubic metre (m³)	3,551 633 x 10^{-6}
drachm (60 grains) (apothecaries)	kilogram (kg)	3,887 93 x 10^{-3}
dram ($\frac{1}{256}$ pound) (avoirdupois)	kilogram (kg)	1,771 85 x 10^{-3}

To convert from	to	multiply by
dyne	newton (N)	$*1 \times 10^{-5}$
dyne centimetre	newton metre (N·m)	$*1 \times 10^{-7}$
dyne per square centimetre	pascal (Pa)	$*1 \times 10^{-1}$
electromagnetic unit of capacitance	farad (F)	$*1 \times 10^{9}$
electromagnetic unit of charge	coulomb (C)	$*1 \times 10^{1}$
electromagnetic unit of current	ampere (A)	$*1 \times 10^{1}$
electromagnetic unit of inductance	henry (H)	$*1 \times 10^{-9}$
electromagnetic unit of potential	volt (V)	$*1 \times 10^{-8}$
electromagnetic unit of resistance	ohm (Ω)	$*1 \times 10^{-9}$
electronvolt	joule (J)	$1{,}602\ 191\ 7 \times 10^{-19}$
electrostatic unit of capacitance	farad (F)	$1{,}112\ 649 \times 10^{-12}$
electrostatic unit of charge (franklin)	coulomb (C)	$3{,}335\ 64 \times 10^{-10}$
electrostatic unit of current	ampere (A)	$3{,}335\ 64 \times 10^{-10}$
electrostatic unit of inductance	henry (H)	$8{,}987\ 554\ 31 \times 10^{11}$
electrostatic unit of potential	volt (V)	$2{,}997\ 925 \times 10^{2}$
electrostatic unit of resistance	ohm (Ω)	$8{,}987\ 554\ 31 \times 10^{11}$
erg	joule (J)	$*1 \times 10^{-7}$
farad (international of 1948)	farad (F)	$9{,}995\ 05 \times 10^{-1}$
faraday (based on carbon-12)	coulomb per mole (C/mol)	$9{,}648\ 67 \times 10^{4}$
faraday (chemical)	coulomb per mole (C/mol)	$9{,}649\ 57 \times 10^{4}$
faraday (physical)	coulomb per mole (C/mol)	$9{,}652\ 19 \times 10^{4}$
fathom	metre (m)	$*1{,}828\ 8 \times 10^{0}$
fermi	metre (m)	$*1 \times 10^{-15}$
fluid drachm (UK)	cubic metre (m^3)	$3{,}551\ 633 \times 10^{-6}$
fluid ounce (UK)	cubic metre (m^3)	$2{,}841\ 306 \times 10^{-5}$
fluid ounce (USA)	cubic metre (m^3)	$2{,}957\ 353 \times 10^{-5}$
foot	metre (m)	$*3{,}048 \times 10^{-1}$
foot (Cape)	metre (m)	$*3{,}148\ 581 \times 10^{-1}$
foot (geodetic Cape)	metre (m)	$*3{,}148\ 555\ 751\ 6 \times 10^{-1}$
[3] foot (South African geodetic)	metre (m)	$*3{,}047\ 972\ 654 \times 10^{-1}$

[3] Land surveyors also refer to this as the 'English foot'.

To convert from	to	multiply by
foot candle (lumen per square foot)	lux (lx)	1,076 391 x 10^{1}
foot lambert	candela per square metre (cd/m^2)	3,426 259 x 10^{0}
foot of water (39,2°F)	pascal (Pa)	2,988 98 x 10^{3}
foot per minute	metre per second (m/s)	*5,08 x 10^{-3}
foot per second squared	metre per second squared (m/s^2)	*3,048 x 10^{-1}
foot poundal (energy)	joule (J)	4,214 011 x 10^{-2}
foot poundal (torque)	newton metre (N·m)	4,214 011 x 10^{-2}
foot pound-force (energy)	joule (J)	1,355 818 x 10^{0}
foot pound-force (torque)	newton metre (N·m)	1,355 818 x 10^{0}
foot pound-force per second	watt (W)	1,355 818 x 10^{0}
foot to the fourth power (second moment of area)	metre to the fourth power (m^4)	8,630 975 x 10^{-3}
foot to the third power	(see cubic foot)	
franklin (electrostatic unit of charge)	coulomb (C)	3,335 64 x 10^{-10}
frigorie	watt (W)	1,162 639 x 10^{0}
furlong	metre (m)	*2,011 68 x 10^{2}
gal	metre per second squared (m/s^2)	*1 x 10^{-2}
gallon (Canada; liquid)	cubic metre (m^3)	4,546 122 x 10^{-3}
gallon (UK)	cubic metre (m^3)	*4,546 09 x 10^{-3}
gallon (USA; dry)	cubic metre (m^3)	4,404 884 x 10^{-3}
gallon (USA; liquid)	cubic metre (m^3)	3,785 412 x 10^{-3}
gallon (UK) per hour	cubic metre per second (m^3/s)	1,262 803 x 10^{-6}
gallon (UK) per pound	cubic metre per kilogram (m^3/kg)	1,002 24 x 10^{-2}
gamma (magnetic induction)	tesla (T)	*1 x 10^{-9}
gamma (mass)	kilogram (kg)	*1 x 10^{-9}
gauss	tesla (T)	*1 x 10^{-4}
gill (UK)	cubic metre (m^3)	1,420 653 x 10^{-4}
gill (USA)	cubic metre (m^3)	1,182 941 x 10^{-4}
[4] gon	radian (rad)	1,570 796 x 10^{-2} (=*π/200)
[4] grade (see gon)		
grain	kilogram (kg)	*6,479 891 x 10^{-5}
grain per gallon (UK)	kilogram per cubic metre (kg/m^3)	1,425 38 x 10^{-2}
grain per gallon (USA, liquid)	kilogram per cubic metre (kg/m^3)	1,711 806 x 10^{-2}

[4] See 4.2, note (vii).

To convert from	to	multiply by
hectare	square metre (m^2)	*1×10^4
henry (international of 1948)	henry (H)	$1{,}000\,495 \times 10^0$
horsepower (boiler)	watt (W)	$9{,}809\,5 \times 10^3$
horsepower (electrical)	watt (W)	*$7{,}46 \times 10^2$
horsepower (550 foot pounds-force per second)	watt (W)	$7{,}456\,999 \times 10^2$
horsepower (metric or cheval vapeur)	watt (W)	$7{,}345\,99 \times 10^2$
horsepower (UK)	watt (W)	$7{,}457 \times 10^2$
horsepower (water)	watt (W)	$7{,}460\,43 \times 10^2$
horsepower (550 foot pounds-force per second) hour	joule (J)	$2{,}684\,52 \times 10^6$
hour (mean; solar)	second (s)	$3{,}6 \times 10^3$
hour (sidereal)	second (s)	$3{,}590\,17 \times 10^3$
hundredweight (112 pounds)	kilogram (kg)	$5{,}080\,235 \times 10^1$
hundredweight (100 pounds)	kilogram (kg)	$4{,}535\,924 \times 10^1$
hundredweight (112 pounds) per acre	kilogram per square metre (kg/m^2)	$1{,}255\,35 \times 10^{-2}$
inch	metre (m)	*$2{,}54 \times 10^{-2}$
inch of mercury (32°F)	pascal (Pa)	$3{,}386\,389 \times 10^3$
inch of mercury (60°F)	pascal (Pa)	$3{,}376\,85 \times 10^3$
inch of water (39,2°F)	pascal (Pa)	$2{,}490\,82 \times 10^2$
inch of water (60°F)	pascal (Pa)	$2{,}488\,4 \times 10^2$
inch per minute	metre per second (m/s)	$4{,}233\,333 \times 10^{-4}$
inch to the fourth power	metre to the fourth power (m^4)	$4{,}162\,314 \times 10^{-7}$
inch to the third power	(see cubic inch)	
iron (shoes)	metre (m)	$5{,}3 \times 10^{-4}$
joule (international of 1948)	joule (J)	$1{,}000\,165 \times 10^0$
kayser	per metre (m^{-1})	*1×10^2
kilocalorie (international table)	joule (J)	*$4{,}186\,8 \times 10^3$
kilocalorie (mean)	joule (J)	$4{,}190\,02 \times 10^3$
kilocalorie (thermochemical)	joule (J)	*$4{,}184 \times 10^3$
kilogram-force	newton (N)	*$9{,}806\,65 \times 10^0$
kilogram-fórce metre (energy)	joule (J)	*$9{,}806\,65 \times 10^0$
kilogram-force metre (torque)	newton metre (N·m)	*$9{,}806\,65 \times 10^0$
kilogram-force per square centimetre	pascal (Pa)	*$9{,}806\,65 \times 10^4$
kilometre per hour	metre per second (m/s)	$2{,}777\,778 \times 10^{-1}$ (=*1/3,6)

To convert from	to	multiply by
kilopond (=kgf)	newton (N)	*9,806 65 x 10^0
kilopond metre (energy)	joule (J)	*9,806 65 x 10^0
kilopond metre (torque)	newton metre (N·m)	*9,806 65 x 10^0
kilopond per square centimetre	pascal (Pa)	*9,806 65 x 10^4
kilowatt hour	joule (J)	*3,6 x 10^6
kilowatt hour (international of 1948)	joule (J)	3,600 59 x 10^6
kip (1 000 pounds-force)	newton (N)	4,448 222 x 10^3
kip (1 000 pounds-force) per square inch	pascal (Pa)	6,894 757 x 10^6
knot (international)	metre per second (m/s)	5,144 444 x 10^{-1}
knot (UK)	metre per second (m/s)	5,147 733 x 10^{-1}
knot (USA)	metre per second (m/s)	5,144 444 x 10^{-1}
lambda	cubic metre (m^3)	1 x 10^{-9}
lambert	candela per square metre (cd/m^2)	3,183 099 x 10^3 (=*$10^4/\pi$)
langley	joule per square metre (J/m^2)	*4,184 x 10^4
leaguer	cubic metre (m^3)	5,773 534 x 10^{-1}
light year	metre (m)	9,460 55 x 10^{15}
ligne (buttons) ($\frac{1}{40}$ inch)	metre (m)	*6,35 x 10^{-4}
litre	cubic metre (m^3)	*1 x 10^{-3}
litre atmosphere	joule (J)	1,013 28 x 10^2
lumen per square foot (foot candle)	lux (lx)	1,076 391 x 10^1
lusec	pascal cubic metre per second ($Pa{\cdot}m^3/s$)	1,333 224 x 10^{-4}
maxwell	weber (Wb)	*1 x 10^{-8}
metre kilogram-force (energy)	joule (J)	*9,806 65 x 10^0
metre kilogram-force (torque)	newton metre (N·m)	*9,806 65 x 10^0
metre kilopond (energy)	joule (J)	*9,806 65 x 10^0
metre kilopond (torque)	newton metre (N·m)	*9,806 65 x 10^0
metre of water (4°C)	pascal (Pa)	9,806 38 x 10^3
metric horsepower or cheval vapeur	watt (W)	7,354 99 x 10^2
micron	metre (m)	*1 x 10^{-6}
mil (circular)	square metre (m^2)	5,067 075 x 10^{-10}
mil (thou)	metre	*2,54 x 10^{-5}
mile	metre (m)	*1,609 344 x 10^3
mile per hour	metre per second (m/s)	*4,470 4 x 10^{-1}
[5] millimetre of mercury (0°C)	pascal (Pa)	1,333 223 9 x 10^2

[5] Defined by: ρgh = 13 595,1 x 9,806 65 x 10^{-3}

To convert from	to	multiply by
minim	cubic metre (m^3)	$5{,}919\ 39 \times 10^{-8}$
minute (mean; solar)	second (s)	6×10^{1}
minute (sidereal)	second (s)	$5{,}983\ 617 \times 10^{1}$
minute (plane angle)	radian (rad)	$2{,}908\ 882 \times 10^{-4}$ (=*π/10 800)
month (mean; calendar)	second (s)	$2{,}628 \times 10^{6}$
morgen	square metre (m^2)	*$8{,}565\ 32 \times 10^{3}$
morgen foot	cubic metre (m^3)	$2{,}610\ 71 \times 10^{3}$
nautical mile (international)	metre (m)	*$1{,}852 \times 10^{3}$
nautical mile (telegraph)	metre (m)	$1{,}855\ 32 \times 10^{3}$
nautical mile (UK)	metre (m)	*$1{,}853\ 184 \times 10^{3}$
nautical mile (USA)	metre (m)	*$1{,}852 \times 10^{3}$
oersted	ampere per metre (A/m)	$7{,}957\ 747 \times 10^{1}$ (=*1 000/4 π)
ohm (international of 1948)	ohm (Ω)	$1{,}000\ 495 \times 10^{0}$
ohm centimetre	ohm metre (Ω·m)	*1×10^{-2}
ounce (avoirdupois)	kilogram (kg)	$2{,}834\ 952 \times 10^{-2}$
ounce (troy or apothecaries)	kilogram (kg)	$3{,}110\ 348 \times 10^{-2}$
ounce (fluid; UK)	cubic metre (m^3)	$2{,}841\ 306 \times 10^{-5}$
ounce (fluid; USA)	cubic metre (m^3)	$2{,}957\ 353 \times 10^{-5}$
ounce-force	newton (N)	$2{,}780\ 139 \times 10^{-1}$
ounce-force inch (torque)	newton metre (N·m)	$7{,}061\ 552 \times 10^{-3}$
ounce (avoirdupois) per cubic inch	kilogram per cubic metre (kg/m^3)	$1{,}729\ 994 \times 10^{3}$
ounce (avoirdupois) per gallon (UK)	kilogram per cubic metre (kg/m^3)	$6{,}236\ 023 \times 10^{0}$
ounce (avoirdupois) per gallon (USA; liquid)	kilogram per cubic metre (kg/m^3)	$7{,}489\ 152 \times 10^{0}$
ounce per inch	kilogram per metre (kg/m)	$1{,}116\ 12 \times 10^{0}$
ounce per square yard	kilogram per square metre (kg/m^2)	$3{,}390\ 575 \times 10^{-2}$
parsec	metre (m)	$3{,}085\ 7 \times 10^{16}$
peck (UK)	cubic metre (m^3)	*$9{,}092\ 18 \times 10^{-3}$
peck (USA)	cubic metre (m^3)	$8{,}809\ 768 \times 10^{-3}$
pennyweight	kilogram (kg)	$1{,}555\ 174 \times 10^{-3}$
perch (area)	square metre (m^2)	$2{,}529\ 29 \times 10^{1}$
perch (length)	metre (m)	*$5{,}029\ 2 \times 10^{0}$
perm (0°C)	kilogram per newton second (kg/N·s)	$5{,}721\ 35 \times 10^{-11}$
perm (23°C)	kilogram per newton second (kg/N·s)	$5{,}745\ 25 \times 10^{-11}$
perm inch (0°C)	kilogram metre per newton second (kg·m/N·s)	$1{,}453\ 22 \times 10^{-12}$
perm inch (23°C)	kilogram metre per newton second (kg·m/N·s)	$1{,}459\ 29 \times 10^{-12}$

To convert from	to	multiply by
phot	lux (lx)	*1×10^{4}
pica (printing)	metre (m)	$4{,}217\ 518 \times 10^{-3}$
pièze	pascal (Pa)	*1×10^{3}
pint (UK)	cubic metre (m^3)	$5{,}682\ 613 \times 10^{-4}$
pint (UK; reputed)	cubic metre (m^3)	$3{,}788\ 408 \times 10^{-4}$
pint (USA; dry)	cubic metre (m^3)	$5{,}506\ 105 \times 10^{-4}$
pint (USA; liquid)	cubic metre (m^3)	$4{,}731\ 765 \times 10^{-4}$
pint (USA; reputed)	cubic metre (m^3)	$3{,}154\ 51 \times 10^{-4}$
point (printing)	metre (m)	*$3{,}514\ 598 \times 10^{-4}$
poise	pascal second (Pa·s)	*1×10^{-1}
poiseuille	pascal second (Pa·s)	*1×10^{0}
pole (area)	square metre (m^2)	$2{,}529\ 29 \times 10^{1}$
pole (length)	metre (m)	*$5{,}029\ 2 \times 10^{0}$
pound	kilogram (kg)	*$4{,}535\ 923\ 7 \times 10^{-1}$
poundal	newton (N)	$1{,}382\ 55 \times 10^{-1}$
poundal per square foot	pascal (Pa)	$1{,}488\ 164 \times 10^{0}$
poundal per square inch	pascal (Pa)	$2{,}142\ 96 \times 10^{2}$
poundal second per square foot	pascal second (Pa·s)	$1{,}488\ 164 \times 10^{0}$
pound foot squared (moment of inertia)	kilogram metre squared ($kg \cdot m^2$)	$4{,}214\ 012 \times 10^{-2}$
pound-force	newton (N)	$4{,}448\ 222 \times 10^{0}$
pound-force foot (torque)	newton metre (N·m)	$1{,}355\ 818 \times 10^{0}$
pound-force inch (torque)	newton metre (N·m)	$1{,}129\ 848 \times 10^{-1}$
pound-force per foot	newton per metre (N/m)	$1{,}459\ 39 \times 10^{1}$
pound-force per square foot	pascal (Pa)	$4{,}788\ 026 \times 10^{1}$
pound-force per square inch	pascal (Pa)	$6{,}894\ 757 \times 10^{3}$
pound-force second per square foot	pascal second (Pa·s)	$4{,}788\ 026 \times 10^{1}$
pound inch squared (moment of inertia)	kilogram metre squared ($kg \cdot m^2$)	$2{,}926\ 397 \times 10^{-4}$
pound per cubic foot	kilogram per cubic metre (kg/m^3)	$1{,}601\ 846 \times 10^{1}$
pound per gallon (UK)	kilogram per cubic metre (kg/m^3)	$9{,}977\ 636 \times 10^{1}$
pound per gallon (USA; liquid)	kilogram per cubic metre (kg/m^3)	$1{,}198\ 264 \times 10^{2}$
pound per minute	kilogram per second (kg/s)	$7{,}559\ 873 \times 10^{-3}$
pound per square foot	kilogram per square metre (kg/m^2)	$4{,}882\ 428 \times 10^{0}$
pound per yard	kilogram per metre (kg/m)	$4{,}960\ 55 \times 10^{-1}$
quart (UK)	cubic metre (m^3)	$1{,}136\ 523 \times 10^{-3}$
quart (UK; reputed)	cubic metre (m^3)	$7{,}576\ 817 \times 10^{-4}$

To convert from	to	multiply by
quart (USA; dry)	cubic metre (m^3)	$1{,}101\ 221 \times 10^{-3}$
quart (USA; liquid)	cubic metre (m^3)	$9{,}463\ 529 \times 10^{-4}$
quart (USA; reputed)	cubic metre (m^3)	$6{,}309\ 02 \times 10^{-4}$
quarter (2 stone)	kilogram (kg)	$1{,}270\ 059 \times 10^{1}$
quintal	kilogram (kg)	$*1 \times 10^{2}$
rad (absorbed dose; ionising radiation)	joule per kilogram (J/kg)	$*1 \times 10^{-2}$
register ton	cubic metre (m^3)	$2{,}831\ 685 \times 10^{0}$
revolution per minute	per second (s^{-1})	$1{,}666\ 667 \times 10^{-2}$
rhe	per pascal second ($Pa^{-1} \cdot s^{-1}$)	$*1 \times 10^{1}$
rod (UK and USA)	metre (m)	$*5{,}029\ 2 \times 10^{0}$
röntgen	coulomb per kilogram (C/kg)	$*2{,}58 \times 10^{-4}$
rood (Cape)	metre (m)	$*3{,}778\ 297\ 2 \times 10^{0}$
rood (geodetic Cape)	metre (m)	$3{,}778\ 266\ 9 \times 10^{0}$
rood (UK)	square metre (m^2)	$*1{,}011\ 715 \times 10^{3}$
scruple	kilogram (kg)	$1{,}295\ 978 \times 10^{-3}$
second (plane angle)	radian (rad)	$4{,}848\ 137 \times 10^{-6}$ (= $*\pi/648\ 000$)
second (sidereal)	second (s)	$9{,}972\ 696 \times 10^{-1}$
shake	second (s)	$*1 \times 10^{-8}$
slug	kilogram (kg)	$1{,}459\ 39 \times 10^{1}$
slug per cubic foot	kilogram per cubic metre (kg/m^3)	$5{,}153\ 79 \times 10^{2}$
slug per foot second	pascal second (Pa·s)	$4{,}788\ 026 \times 10^{1}$
square foot	square metre (m^2)	$*9{,}290\ 304 \times 10^{-2}$
[1] square foot hour degree Fahrenheit per British thermal unit (international table) inch	metre kelvin per watt (m·K/W)	$6{,}933\ 471 \times 10^{0}$
[1] square foot hour degree Fahrenheit per British thermal unit (thermochemical) inch	metre kelvin per watt (m·K/W)	$6{,}938\ 113 \times 10^{0}$
square foot per hour	square metre per second (m^2/s)	$*2{,}580\ 64 \times 10^{-5}$
square inch	square metre (m^2)	$*6{,}451\ 6 \times 10^{-4}$
square mile	square metre (m^2)	$2{,}589\ 988 \times 10^{6}$
square mile per ton (2 240 pounds)	square metre per kilogram (m^2 kg)	$2{,}549\ 08 \times 10^{3}$
square yard	square metre (m^2)	$*8{,}361\ 273\ 6 \times 10^{-1}$
square yard per ton (2 240 pounds)	square metre per kilogram (m^2/kg)	$8{,}229\ 22 \times 10^{-4}$
statampere	ampere (A)	$3{,}335\ 64 \times 10^{-10}$

[1] Because the degree Fahrenheit in the denominator represents a temperature interval it follows from 2.2 note (i) that precisely the same conversion factor applies if the kelvin (K) in the SI unit is replaced by the degree Celsius (°C).

To convert from	to	multiply by
statcoulomb	coulomb (C)	$3{,}335\,64 \times 10^{-10}$
statfarad	farad (F)	$1{,}112\,649 \times 10^{-12}$
stathenry	henry (H)	$8{,}987\,554\,31 \times 10^{11}$
statmho	siemens (S)	$1{,}112\,649 \times 10^{-12}$
statohm	ohm (Ω)	$8{,}987\,554\,31 \times 10^{11}$
statvolt	volt (V)	$2{,}997\,925 \times 10^{2}$
stere	cubic metre (m^3)	*1×10^{0}
sthène	newton (N)	*1×10^{3}
stilb	candela per square metre (cd/m^2)	*1×10^{4}
stokes	metre squared per second (m^2/s)	*1×10^{-4}
stone	kilogram (kg)	$6{,}350\,293 \times 10^{0}$
tablespoon (UK)	cubic metre (m^3)	$1{,}420\,653 \times 10^{-5}$
tablespoon (USA)	cubic metre (m^3)	$1{,}478\,676 \times 10^{-5}$
teaspoon (UK)	cubic metre (m^3)	$4{,}735\,51 \times 10^{-6}$
teaspoon (USA)	cubic metre (m^3)	$4{,}928\,922 \times 10^{-6}$
tex	kilogram per metre (kg/m)	*1×10^{-6}
therm	joule (J)	$1{,}055\,06 \times 10^{8}$
thermie	joule (J)	$4{,}185\,5 \times 10^{6}$
therm per gallon (UK)	joule per cubic metre (J/m^3)	$2{,}320\,8 \times 10^{10}$
thou (mil)	metre (m)	*$2{,}54 \times 10^{-5}$
ton (2 240 pounds)	kilogram (kg)	$1{,}016\,047 \times 10^{3}$
ton (2 000 pounds)	kilogram (kg)	$9{,}071\,847 \times 10^{2}$
[6] ton (metric)	kilogram (kg)	*1×10^{3}
ton (nuclear equivalent of TNT)	joule (J)	$4{,}2 \times 10^{9}$
ton (refrigeration) (12 000 British thermal units per hour)	watt (W)	$3{,}516\,853 \times 10^{3}$
ton (refrigeration) (13 440 British thermal units per hour)	watt (W)	$3{,}938\,876 \times 10^{3}$
ton-force (2 240 pounds-force)	newton (N)	$9{,}964\,02 \times 10^{3}$
ton-force (2 000 pounds-force)	newton (N)	$8{,}896\,44 \times 10^{3}$
ton-force (metric)	newton (N)	*$9{,}806\,65 \times 10^{3}$
ton-force (2 240 pounds-force) per foot	newton per metre (N/m)	$3{,}269\,03 \times 10^{4}$
ton-force (2 240 pounds-force) per square foot	pascal (Pa)	$1{,}072\,52 \times 10^{5}$

[6] Also referred to overseas as 'tonne'.

To convert from	to	multiply by
ton-force (2 240 pounds-force) per square inch	pascal (Pa)	$1,544\ 43 \times 10^{7}$
ton-force (2 000 pounds-force) per square inch	pascal (Pa)	$1,378\ 95 \times 10^{7}$
ton (2 240 pound) mile	kilogram metre (kg·m)	$1,635\ 17 \times 10^{6}$
ton (2 240 pounds) per acre	kilogram per square metre (kg/m^2)	$2,510\ 71 \times 10^{-1}$
ton (2 000 pounds) per acre	kilogram per square metre (kg/m^2)	$2,241\ 7 \times 10^{-1}$
ton (2 240 pounds) per cubic yard	kilogram per cubic metre (kg/m^3)	$1,328\ 939 \times 10^{3}$
ton (2 000 pounds) per cubic yard	kilogram per cubic metre (kg/m^3)	$1,186\ 55 \times 10^{3}$
ton (2 240 pounds) per mile	kilogram per metre (kg/m)	$6,313\ 42 \times 10^{-1}$
ton (2 000 pounds) per mile	kilogram per metre (kg/m)	$5,636\ 98 \times 10^{-1}$
ton (2 000 pounds) per morgen	kilogram per square metre (kg/m^2)	$1,059\ 14 \times 10^{-1}$
ton (2 240 pounds) per square mile	kilogram per square metre (kg/m^2)	$3,922\ 98 \times 10^{-4}$
ton (2 240 pounds) per 1 000 yards	kilogram per metre (kg/m)	$1,111\ 16 \times 10^{0}$
ton (2 000 pounds) per 1 000 yards	kilogram per metre (kg/m)	$9,921\ 09 \times 10^{-1}$
torr	pascal (Pa)	[7] $1,333\ 223\ 7 \times 10^{2}$
unit pole	weber (Wb)	$1,256\ 637 \times 10^{-7}$
volt (international of 1948)	volt (V)	$1,000\ 33 \times 10^{0}$
watt (international of 1948)	watt (W)	$1,000\ 165 \times 10^{0}$
watt hour	joule (J)	*$3,6 \times 10^{3}$
yard	metre (m)	*$9,144 \times 10^{-1}$
yard per pound	metre per kilogram (m/kg)	$2,015\ 91 \times 10^{0}$
year (calendar)	second (s)	$3,153\ 6 \times 10^{7}$
year (sidereal)	second (s)	$3,155\ 815 \times 10^{7}$
year (tropical)	second (s)	$3,155\ 693 \times 10^{7}$

[7] The exact value is 101 325/760 Pa .

9.5. TABLE 13

ALPHABETICAL LIST OF CONVERSIONS TO NON-SI UNITS WHICH ARE STILL USED WITH THE SI

(INTERNATIONAL AND ONLY CORRECT SYMBOLS ARE GIVEN IN BRACKETS)

To convert from	to	multiply by
acre	hectare (ha)	*0,404 686
are	hectare (ha)	*0,01
barrel (petroleum; 42 gallons (USA); liquid)	litre (ℓ)	158,987 3
bushel (UK)	litre (ℓ)	36,368 72
bushel (USA)	litre (ℓ)	35,239 07
cup (UK)	millilitre (mℓ)	284,130 6
cup (USA)	millilitre (mℓ)	236,588 2
degree Fahrenheit (particular temperature)	degree Celsius (°C)	Use $t_C = (t_F - 32)/1,8$
degree Fahrenheit (temperature interval)	degree Celsius (°C)	0,555 556 (=*1/1,8)
degree Rankine (particular temperature	degree Celsius (°C)	Use $t_C = (t_R - 491,67)/1,8$
degree Rankine (temperature interval	degree Celsius (°C)	0,555 556 (=*1/1,8)
fluid drachm (UK)	millilitre (mℓ)	3,551 633
fluid ounce (UK)	millilitre (mℓ)	28,413 06
fluid ounce (USA)	millilitre (mℓ)	29,573 53
gallon (Canada; liquid)	litre (ℓ)	4,546 122
gallon (UK)	litre (ℓ)	*4,546 09
gallon (USA; dry)	litre (ℓ)	4,404 884
gallon (USA; liquid)	litre (ℓ)	3,785 412
gallon (UK) per hour	litre per hour (ℓ/h)	4,546 087
gallon (UK) per pound	litre per kilogram (ℓ/kg)	10,022 4
gill (UK)	millilitre (mℓ)	142,065 3
gill (USA)	millilitre (mℓ)	118,294 1
gon	degree (angle) (...°)	*0,9
grain per gallon (UK)	milligram per litre (mg/ℓ)	14,253 8
grain per gallon (USA; liquid)	milligram per litre (mg/ℓ)	17,118 06
hundredweight (112 pounds) per acre	kilogram per hectare (kg/ha)	125,535
knot (international)	kilometre per hour (km/h)	*1,852
knot (UK)	kilometre per hour (km/h)	*1,853 184
knot (USA)	kilometre per hour (km/h)	*1,852
lambda	microlitre (μℓ)	1
leaguer	litre (ℓ)	577,353 4
mile per gallon	litre per hundred kilometres (ℓ/100 km)	Divide 282,481 by the value in miles per gallon

To convert from	to	multiply by
mile per hour	kilometre per hour (km/h)	*1,609 344
minim	microlitre (μℓ)	59,193 9
morgen	hectare (ha)	*0,856 532
ounce (avoirdupois) per gallon (UK)	gram per litre (g/ℓ)	6,236 027
ounce (avoirdupois) per gallon (USA; liquid)	gram per litre (g/ℓ)	7,489 152
peck (UK)	litre (ℓ)	*9,092 18
peck (USA)	litre (ℓ)	8,809 768
pint (UK)	millilitre (mℓ)	568,261 3
pint (UK; reputed)	millilitre (mℓ)	378,840 8
pint (USA; dry)	millilitre (mℓ)	550,610 5
pint (USA; liquid)	millilitre (mℓ)	473,176 5
pint (USA; reputed)	millilitre (mℓ)	315,451
pound per gallon (UK)	gram per litre (g/ℓ)	99,776 36
pound per gallon (USA; liquid)	gram per litre (g/ℓ)	119,826 4
quart (UK)	litre (ℓ)	1,136 523
quart (UK; reputed)	millilitre (mℓ)	757,681 7
quart (USA; dry)	litre (ℓ)	1,101 221
quart (USA; liquid)	millilitre (mℓ)	946,352 9
quart (USA; reputed)	millilitre (mℓ)	630,902
square mile per ton (2 240 pounds)	square kilometre per metric ton (km^2/t)	2,549 08
square yard per ton (2 240 pounds)	square metre per metric ton (m^2/t)	0,822 922
tablespoon (UK)	millilitre (mℓ)	14,206 53
tablespoon (USA)	millilitre (mℓ)	14,786 76
teaspoon (UK)	millilitre (mℓ)	4,735 51
teaspoon (USA)	millilitre (mℓ)	4,928 922
ton (2 240 pounds)	metric ton (t)	1,016 047
ton (2 000 pounds)	metric ton (t)	0,907 184 7
ton (2 240 pounds) mile	metric ton kilometre (t·km)	1,635 17
ton (2 240 pounds) per acre	metric ton per hectare (t/ha)	2,510 71
ton (2 000 pounds) per acre	metric ton per hectare (t/ha)	2,241 7
ton (2 240 pounds) per cubic yard	metric ton per cubic metre (t/m^3)	1,328 939
ton (2 000 pounds) per cubic yard	metric ton per cubic metre (t/m^3)	1,186 55
ton (2 000 pounds) per morgen	metric ton per hectare (t/ha)	1,059 14
ton (2 240 pounds) per mile	metric ton per kilometre (t/km)	0,631 342
ton (2 000 pounds) per mile	metric ton per kilometre (t/km)	0,563 698
ton (2 240 pounds) per square mile	metric ton per square kilometre (t/km^2)	0,392 298

TEXT OF JOINT RESOLUTION

Proposed by Department of Commerce and Introduced in 92nd Congress as
Senate Joint Resolution 219 and House Joint Resolutions 1092, 1132, and 1169

To establish a national policy relating to conversion to the metric system in the United States.

Whereas the use of the metric system of weights and measures in the United States was authorized by the Act of July 28, 1866 (14 Stat. 339); and

Whereas the United States was one of the original signatories to the Convention of the Meter (20 Stat. 709), which established the General Conference of Weights and Measures, the International Committee of Weights and Measures, and the International Bureau of Weights and Measures; and

Whereas the metric measurement standards recognized and developed by the International Bureau of Weights and Measures have been adopted as the fundamental measurement standards of the United States and the customary units of weights and measures used in the United States have been since 1893 based upon such metric measurement standards; and

Whereas the Governments of Australia, Canada, Great Britain, India, Japan, New Zealand, and the Union of South Africa have determined to convert, are converting, or have converted to the use of the metric system in their respective jurisdictions; and

Whereas the United States is the only industrially developed nation which has not established a national policy committing itself to and facilitating conversion to the metric system; and

Whereas, as a result of the study to determine the advantages and disadvantages of increased use of the metric system in the United States authorized by Public Law 90-472 (82 Stat. 693), the Secretary of Commerce has found that increased use of the metric system in the United States is inevitable, and has concluded that a planned national program to achieve a metric changeover is desirable; that maximum efficiency will result and minimum costs to effect the conversion will be incurred if the conversion is carried out in general without Federal subsidies; that the changeover period be ten years, at the end of which the Nation would be predominantly, although not exclusively, metric; that a central coordinating body be established and assigned to coordinate the changeover in cooperation with all sectors of our society; and that immediate attention be given to public and formal education and to effective United States participation in international standardsmaking: Now, therefore, be it

Resolved by the Senate and House of Representatives of the United States of America in Congress assembled, That the policy of the United States shall be—

(1) to facilitate and encourage the substitution of metric measurement units for customary measurement units to education, trade, commerce, and all other sectors of the economy of the United States with a view to making metric units the predominant, although not exclusive, language of measurement with respect to transactions occurring after ten years from the date of the enactment of this resolution.

(2) to facilitate and encourage the development as rapidly as practicable of new or revised engineering standards based on metric measurement units in those specific fields or areas in the United States where such standards will result in rationalization or simplification of relationships, improvements of design, or increases in economy.

(3) to facilitate and encourage the retention in new metric language standards of those United States engineering designs, practices, and conventions that are internationally accepted or embody superior technology.

(4) to cooperate with foreign governments and public and private international organizations which are or become concerned with the encouragement and coordination of increased use of metric measurement units or engineering standards based on such units, or both, with a view to gaining international recognition for metric standards proposed by the United States and to encouraging retention of equivalent customary units in international recommendations during the United States changeover period.

(5) to assist the public through information and educational programs to become familiar with the meaning and applicability to metric terms and measures in daily life. Programs hereunder should include:

(a) Public information programs conducted by the Board through the use of newspapers, magazines, radio, television, other media, and through talks before appropriate citizens groups and public organizations.

(b) Counseling and consultation by the Secretary of Health, Education, and Welfare and the Director, National Science Foundation, with educational associations and groups so as to assure that the metric system of measurement is made a part of the curricula of the Nation's educational institutions and that teachers and other appropriate personnel are properly trained to teach the metric system of measurement.

(c) Consultation by the Secretary of Commerce with the National Conference of Weights and Measures so as to assure that State and local weights and measures officials are appropriately informed of the intended metric changeover and are thus assisted in their efforts to bring about timely amendments to weights and measures laws.

(d) Such other public information programs by any Federal agency in support of this resolution which relate to the mission of the agency.

Sec. 2. Definitions—For the purposes of this resolution—

(a) The term "metric system of measurement" means the international system of units as established by the General Conference of Weights and Measures in 1960 and interpreted or modified for the United States by the Secretary of Commerce.

(b) The term "engineering standard" means a standard which prescribes a concise set of conditions and requirements to be satisfied by a material, product, process, procedure, convention, test method, and the physical, functional, performance, and/or conformance characteristics thereof.

(c) The term "changeover period" means the length of time for the United States to become predominantly, although not exclusively, metric.

(d) The term "international recommendation" means a recommendation formulated and promulgated by an international organization and recommended for adoption by individual nations as a national standard of measurement.

Sec. 3. There is hereby established a National Metric Conversion Board (hereinafter referred to as the "Board") to implement the policy set out in this resolution.

Sec. 4. The composition of the Board shall be as follows:

(a) Not to exceed twenty-one persons appointed by the President who shall serve at his pleasure and for such terms as he shall specify and who shall be broadly representative of the American society. The President shall designate one of the members appointed by him to serve as Chairman and another to serve as the Vice Chairman of the Board;

(b) Two Members of the House of Representatives who shall not be members of the same political party shall be appointed by the Speaker of the House of Representatives; and

(c) Two Members of the Senate who shall not be members of the same political party shall be appointed by the President of the Senate.

Sec. 5. The Executive Director of the Board shall be appointed by the President and shall be responsible to the Board for carrying out its responsibilities according to the provisions of this resolution.

Sec. 6. (a) Within twelve months after funds have been appropriated to carry out the provisions of this resolution the Board shall, in furtherance and in support of the policy expressed in section 1 of this resolution, develop and submit to the Secretary of Commerce for his approval and transmittal to the President a comprehensive plan to accomplish a changeover to the metric system of measurement in the United States. If such plan is approved by the President, he shall transmit it to the Congress. Such plan may include recommendations for legislation deemed necessary and appropriate. In developing this plan the Board shall:

(1) Consult with and take into account the interests and views of United States commerce and industry, including small business; science; engineering; labor; education; consumers; government agencies at the Federal, State, and local level; nationally recognized standards developing and coordinating organizations; and such other individuals or groups as are considered appropriate by the Board to carry out the purposes of this section.

(2) Consult, to the extent deemed appropriate, with foreign governments, public international organizations, and, through appropriate member organizations, private international standards organizations. Contact with foreign governments and intergovernmental organizations shall be accomplished in consultation with the Department of State.

(b) Any amendment to an approved plan shall be submitted by the Board to the Secretary and the President under the provisions set out in subsection (a) of this section.

(c) Unless otherwise provided by the Congress, the Board shall have no compulsory powers.

Sec. 7. Upon approval of the plan by the President, the Board shall begin the implementation of the plan, except for those recommendations, if any, which require legislation.

Sec. 8. In carrying out its duties, the Board is authorized to:

(a) enter into contracts in accordance with the Federal Property and Administrative Services Act of 1949, as amended, with Federal or State agencies, private firms, institutions, and individuals for the conduct of research or surveys, the preparation of reports, and other activities necessary to the discharge of its duties;

(b) conduct hearings at such times and places as it deems appropriate;

(c) establish such committees and advisory panels as it deems necessary to work with the various sectors of the American economy and governmental agencies in the development and implementation of detailed changeover plans for those sectors; and

(d) perform such other acts as may be necessary to carry out the duties prescribed by this resolution.

Sec. 9. (a) Members of the Board who are not in the regular full-time employ of the United States shall, while attending meetings or conferences of the Board or otherwise engaged in the business of the Board, be entitled to receive compensation at a rate of $100 per day, including traveltime, and while so serving on the business of the Board away from their homes or regular places of business, they may be allowed travel expenses, including per diem in lieu of subsistence, as authorized by section 5703 of title 5, United States Code, for persons employed intermittently in the Government service. Payments under this section shall not render members of the Board employees or officials of the United States for any purpose.

(b) The Executive Director of the Board shall serve full time and receive basic pay at a rate not to exceed the rate provided for GS-18 in subchapter III of chapter 53 of title 5, United States Code.

Sec. 10. (a) The Board is authorized to appoint and fix the compensation of such staff personnel as may be necessary to carry out the provisions of this Act.

(b) The Board is authorized to employ experts and consultants or organizations thereof as authorized by section 3109 of title 5, United States Code, compensate individuals so employed at rates not in excess of the rate prescribed for grade 18 of the General Schedule under section 5332 of such title, including traveltime, and allow them, while away from their homes or regular places of business, travel expenses (including per diem in lieu of subsistence) as authorized by section 5703 of said title 5 for persons in the Government service employed: Provided, however, That contracts for such employment may be renewed annually.

Sec. 11. Financial and administrative services (including those related to budgeting, accounting, financial reporting, personnel, and procurement) and such other staff services as may be requested by the Board shall be provided the Board by the Secretary of Commerce, for which payment shall be made in advance, or by reimbursement, from funds of the Board in such amounts as may be agreed upon by the Chairman of the Board and the Secretary of Commerce.

In performing these functions for the Board, the Secretary is authorized to obtain such information and assistance from other Federal agencies as may be necessary.

Sec. 12. (a) The Board is hereby authorized to accept, hold, administer, and utilize gifts, donations, and bequests of property, both real and personal, and personal services, for the purpose of aiding or facilitating the work of the Board. Gifts and bequests of money and the proceeds from sales of other property received as gifts or bequests shall be deposited in the Treasury in a separate fund and shall be disbursed upon order of the Board.

(b) For the purpose of Federal income, estate, and gift taxes, property accepted under subsection (a) of this section shall be considered as a gift or bequest to or for the use of the United States.

(c) Upon the request of the Board, the Secretary of the Treasury may invest and reinvest in securities of the United States any moneys contained in the fund herein authorized. Income accruing from such securities, and from any other property accepted to the credit of the fund authorized herein, shall be disbursed upon the order of the Board.

Sec. 13. The Board shall cease to exist no later than ten years after approval by the President of the plan called for by section 6.

Sec. 14. The Board shall submit annual reports of its activities and progress under this resulution to the Secretary of Commerce for his approval and transmittal to the President and to the Congress.

Sec. 15. There are hereby authorized to be appropriated such sums as may be necessary to carry out the provisions of this resolution. Appropriations to carry out the provisions of this resolution may remain available for obligation and expenditure for such period or periods as may be specified in the Acts making such appropriations.

INDEX